Oxford & Cherwell Valley College
Bicester Campus
Telford House, Telford Road, Bicester, OX26 4LA
LIBRARY
Books must be returned by the last date shown below

...ewed by telephone or e-mail

..169 e-mail: library@ocvc.ac.uk

ERGONOMICS in the Automotive Design Process

Vivek D. Bhise

CRC Press
Taylor & Francis Group
Boca Raton London New York

CRC Press is an imprint of the
Taylor & Francis Group, an **informa** business

CRC Press
Taylor & Francis Group
6000 Broken Sound Parkway NW, Suite 300
Boca Raton, FL 33487-2742

© 2012 by Taylor & Francis Group, LLC
CRC Press is an imprint of Taylor & Francis Group, an Informa business

No claim to original U.S. Government works

International Standard Book Number: 978-1-4398-4210-2 (Hardback)

Library of Congress Cataloging-in-Publication Data

Bhise, Vivek D. (Vivek Dattatray), 1944-
   Ergonomics in the automotive design process / Vivek D. Bhise.
      p. cm.
   Includes bibliographical references.
   ISBN 978-1-4398-4210-2 (hardcover : alk. paper)
   1. Automobiles--Design and construction. 2. Human engineering. I. Title.

TL250.B48 2011
629.2'31--dc22   2011010491

Visit the Taylor & Francis Web site at
http://www.taylorandfrancis.com

and the CRC Press Web site at
http://www.crcpress.com

Contents

Preface...xiii
Website Materials xv
Acknowledgments..xvii
Author xix

PART I Ergonomics Concepts, Issues, and Methods in Vehicle Design

Chapter 1   Introduction to Automotive Ergonomics...3

    Ergonomics in Vehicle Design 3
        Objectives 3
        Ergonomics: What Is It? 3
        Ergonomics Approach 4
            Fitting the Equipment to the Users 4
            Designing for the Most 4
            Systems Approach 4
        Problem-Solving Methodologies 5
        Ergonomics Research Studies 7
        Ergonomics Engineer's Responsibilities in Vehicle Design.......................7
    History...8
        Origins of Ergonomics and Human Factors Engineering 8
            Prehistoric Times and Functional Changes in Products.........................8
            Air Force Research 8
            "Ergonomics" Coined 8
        History of Ergonomics in Automotive Product Design.............................8
    Importance of Ergonomics..9
        Characteristics of Ergonomically Designed Products, Systems, and Processes.....9
        Why Apply Ergonomics?...9
        Ergonomics Is Not Commonsense 9
    A Brief Overview of Human Characteristics and Capabilities.......................10
        Physical Capabilities..10
        Information-Processing Capabilities 10
    Implementing Ergonomics 10
    References 11

Chapter 2   Engineering Anthropometry and Biomechanics...13

    Introduction 13
    Use of Anthropometry in Designing Vehicles 13
        Computation of Percentile Values 18
    Applications of Biomechanics in Vehicle Design 19
        Basic Biomechanical Considerations 20
        Biomechanical Considerations in Seat Design...22

            Other Seat Design Considerations..22
            Seat Design Considerations Related to Driver Accommodation.........................25
        Recent Advances in Digital Manikins 28
        References 28

Chapter 3   Occupant Packaging...29

        What Is Vehicle Packaging?..29
            Occupant Package or Seating Package Layout...29
            Developing the Occupant Package: Design Considerations.............................29
        Sequence in Development of Vehicle Package...30
            Advanced Vehicle Design Stage...30
            Development of the "Accepted" Vehicle Concept 31
        Definition of Key Vehicle Dimensions and Reference Points..................................33
            Units, Dimensions, and Axes 33
            Package Dimensions, Reference Points, and Seat-Track-Related Dimensions......33
            Interior Dimensions..38
        Driver Package Development Procedures 42
            Other Issues and Dimensions 49
        References 49

Chapter 4   Driver Information Acquisition and Processing 51

        Introduction 51
        Importance of Time..51
        Understanding Driver Vision Considerations 52
            Structure of the Human Eye 52
            Visual Information Acquisition in Driving 54
            Accommodation 54
        Information Processing 55
            Some Information-Processing Issues and Considerations...............................55
            Human Information-Processing Models 56
            Human Memory 62
            Generic Model of Human Information Processing with the Three
            Memory Systems 63
        Human Errors...64
            Definition of an Error 64
            Types of Human Errors 64
            Understanding Human Errors with the SORE Model....................................65
        Psychophysics...65
        Visual Capabilities 66
            Visual Contrast Thresholds 66
            Visual Acuity..67
            Driver's Visual Fields 68
        Occlusion Studies 68
        Information Acquired through Other Sensory Modalities.......................................71
            Human Audition and Sound Measurements...71
            Other Sensory Information..72
        Applications of Information Processing for Vehicle Design.....................................72
            Some Design Guidelines Based on Driver Information-Processing
            Capabilities...72

Concluding Comments 74
References 74

Chapter 5   Controls, Displays, and Interior Layouts 77

Introduction 77
Controls and Displays Interface 77
    Characteristics of a Good Control 78
    Characteristics of a Good Visual Display 78
Types of Controls and Displays 79
    In-Vehicle Controls 79
    In-Vehicle Displays 83
Design Considerations, Issues, and Location Principles 85
    Some General Design Considerations 85
    Control Design Considerations 86
    Visual Display Design Considerations 87
    Control and Display Location Principles 88
Methods to Evaluate Controls and Displays 89
    Space Available to Locate Hand Controls and Displays 90
    Checklists for Evaluation of Controls and Displays 92
    Ergonomics Summary Chart 92
Some Examples of Control and Display Design Issues 99
Concluding Remarks 102
References 103

Chapter 6   Field of View from Automotive Vehicles 105

Introduction to Field of View 105
    Linking Vehicle Interior to Exterior 105
    What Is Field of View? 105
    Origins of Data to Support Required Fields of View 106
Types of Fields of View 106
    Systems Consideration of 360-Degree Visibility 107
    Monocular, Ambinocular, and Binocular Vision 108
Forward-Field-of-View Evaluations 111
    Up- and Down-Angle Evaluations 111
    Visibility of and over the Hood 111
    Command Sitting Position 112
    Short Driver Problems 112
    Tall Driver Problems 113
    Sun Visor Design Issues 113
    Wiper and Defroster Requirements 114
    Obstructions Caused by A-Pillars 115
Mirror Design Issues 116
    Requirements on Mirror Fields 116
    Mirror Locations 118
        Inside Mirror Locations 118
        Outside Mirror Locations 119
    Procedure for Determining Driver's Field of View through Mirrors 119
    Convex and Aspherical Mirrors 121
Methods to Measure Fields of View 121

Polar Plots.. 122
Other Visibility Issues 125
Light Transmissivity 125
Other Visibility-Degradation Causes 125
Shade Bands 125
Plane and Convex Combination Mirrors.. 125
Heavy-Truck Driver Issues 125
Cameras and Display Screens 126
Concluding Remarks 126
References 126

Chapter 7   Automotive Lighting 127

Introduction 127
Automotive Lighting Equipment... 127
Objectives 128
Headlamps and Signal Lamps: Purpose and Basic Ergonomic Issues 128
Headlamps.. 128
Signal Lamps 129
Headlighting Design Considerations.. 129
Target Visibility Considerations 130
Problems with Current Headlighting Systems 130
New Technological Advances in Headlighting 131
Signal Lighting Design Considerations.. 131
Signal Lighting Visibility Issues 132
Problems with Current Signal Lighting Systems 132
New Technology Advances and Related Issues in Signal Lighting 133
Photometric Measurements of Lamp Outputs 133
Light Measurement Units 133
Headlamp Photometry Test Points and Headlamp Beam Patterns 134
Low and High Beam Patterns 134
Pavement Luminance and Glare Illumination from Headlamps.................. 135
Photometric Requirements for Signal Lamps 139
Headlamp Evaluation Methods 139
Signal Lighting Evaluation Methods.. 144
CHMSL Fleet Study 147
Other Signal Lighting Studies 150
Concluding Remarks 151
References 152

Chapter 8   Entry and Exit from Automotive Vehicles 155

Introduction to Entry and Exit 155
Problems during Entry and Exit.. 155
Vehicle Features and Dimensions Related to Entry and Exit 156
Door Handles... 156
Lateral Sections at the SgRP and Foot Movement Areas 158
Body Opening Clearances from SgRP Locations.. 160
Door and Hinge Angles... 160
Seat Bolsters, Location, and Materials.. 161
Seat Hardware 161

Tires and Rocker Panels 161
Running Boards.. 161
Third Row and Rear Seat Entry from Two-Door Vehicles 162
Heavy-Truck Cab Entry and Exit 162
Methods to Evaluate Entry and Exit 163
Task Analysis 164
Effect of Vehicle Body Style on Vehicle Entry and Exit 166
Concluding Comments 167
References 167

Chapter 9 Automotive Exterior Interfaces: Service and Loading/Unloading Tasks 169

Introduction to Exterior Interfaces... 169
Exterior Interfacing Issues 169
Methods and Issues to Study 172
Standards, Design Guidelines, and Requirements 172
Checklists 173
Biomechanical Guidelines for Loading and Unloading Tasks......................... 173
Applications of Manual Lifting Models.. 174
Task Analysis... 174
Methods of Observation, Communication, and Experimentation..................... 174
Concluding Remarks 176
References 176

Chapter 10 Automotive Craftsmanship 177

Craftsmanship in Vehicle Design.. 177
Objectives 177
Craftsmanship: What Is It?.. 177
Importance of Craftsmanship.. 177
The Ring Model of Product Desirability... 178
Kano Model of Quality 179
Attributes of Craftsmanship.. 180
Visual Quality.. 181
Touch Feel Quality 181
Sound Quality.. 181
Harmony 181
Smell Quality... 182
Comfort and Convenience 182
Measurement Methods 182
Some Examples of Craftsmanship Evaluation Studies 182
Craftsmanship of Steering Wheels.. 182
Other Studies 184
Concluding Remarks 185
References 185

Chapter 11 Role of Ergonomics Engineers in the Automotive Design Process 187

Introduction 187
Systems Engineering Model Describing the Vehicle Development Process 187
Vehicle Evaluation.. 189

Goal of Ergonomics Engineers... 189
Evaluation Measures... 189
Tools, Methods, and Techniques 189
Ergonomics Engineer's Responsibilities 190
Steps in Ergonomics Support Process during Vehicle Development 190
Steps in the Early Design Process 191
Trade-offs in the Design Process 194
Problems and Challenges 194
Concluding Remarks 195
References 195

PART II   Advanced Topics, Measurements, Modeling, and Research

Chapter 12 Modeling Driver Vision 199

Use of Driver Vision Models in Vehicle Design 199
Systems Considerations Related to Visibility 199
Light Measurements... 200
Light Measurement Units 200
Photometry and Measurement Instruments 202
Visual Contrast Thresholds... 202
Blackwell Contrast Threshold Curves.. 203
Computation of Contrast Values.. 203
Computation of Threshold Contrast and Visibility Distance 204
Effect of Glare on Visual Contrast 205
Steps in Computing Visibility of a Target... 206
Discomfort Glare Prediction 209
Legibility 211
Factors Affecting Legibility 211
Modeling Legibility.. 212
Veiling Glare Caused by Reflection of the Instrument Panel into the Windshield.. 214
A Design Tool to Evaluate Veiling Glare Effects... 214
Veiling Glare Prediction Model 216
Model Applications Illustrating Effects of Driver's Age, Sun Illumination,
and Vehicle Design Parameters 218
Concluding Remarks.. 219
References 220

Chapter 13 Driver Performance Measurement 223

Introduction 223
Characteristics of Effective Performance Measures 223
Driving and Nondriving Tasks 224
Determining What to Measure.. 225
Driver Performance Measures 226
Types and Categories of Measures 226
Some Measures Used in the Literature 227
Range of Driving Performance Measures 228

Some Studies Illustrating Driver Behavioral and Performance Measurements........228
    Standard Deviation of Lateral Position 228
    Standard Deviation of Steering Wheel Angle...229
    Standard Deviation of Velocity 229
    Vehicle Speed 229
    Total Task Time, Glance Durations, and Number of Glances............................230
    Driver Errors...231
Some Driving Performance Measurement Applications...232
Concluding Remarks 233
References 233

Chapter 14 Driver Workload Measurement..235

Introduction 235
    Driver Tasks and Workload Assessment 235
    Present Situation in the Industry 236
Concepts Underlying Mental Workload..236
Methods to Measure Driver Workload..237
Some Studies Illustrating Driver Workload Measurements................................243
    Destination Entry in Navigation Systems...243
    Handheld versus Voice Interfaces for Cell Phones and MP3 Players 244
    Text Messaging during Simulated Driving 245
    Comparison of Driver Behavior and Performance in Two Driving Simulators ..245
    Applications of the ISO LCT 246
Concluding Comments 249
References 249

Chapter 15 Vehicle Evaluation Methods..253

Overview on Evaluation Issues 253
Ergonomic Evaluations during Vehicle Development..254
Evaluation Methods..254
Methods of Data Collection and Analysis...254
    Observational Methods..254
    Communication Methods 257
    Experimental Methods 258
Objective Measures and Data Analysis Methods...258
Subjective Methods and Data Analysis..258
    Rating on a Scale 259
    Paired-Comparison-Based Methods 259
    Thurstone's Method of Paired Comparisons 262
    Analytical Hierarchical Method..265
Some Applications of Evaluation Techniques in Automotive Designs 267
    Checklists 267
    Observational Studies..268
    Vehicle User Interviews..268
    Ratings on Interval Scales 268
    Studies Using Programmable Vehicle Bucks 269
    Driving Simulator Studies 269
    Field Studies and Drive Tests 269
References 269

Chapter 16 Special Driver and User Populations..271

    An Overview on Users and Their Needs 271
    Understanding Users: Issues and Considerations...271
        Vehicle Types and Body Styles 272
        Market Segments 272
        Female Drivers 273
        Older Drivers 274
        Effect of Geographic Locations of the Markets..275
        Drivers with Disabilities and Functional Limitations 275
    Issues in Designing Global Vehicles 276
    Futuring..276
    References 277

Chapter 17 Future Research and New Technology Issues...279

    Introduction 279
    Ergonomic Needs in Designing Vehicles...279
        Passenger Vehicles in the Near Future 280
    Future Research Needs and Challenges 280
        Enabling Technologies..280
        Currently Available New Technology Hardware and Applications 281
        A Possible Technology Implementation Plan 284
        Questions Related to Implementation of the Technologies 284
    Other Research Needs...287
    Concluding Remarks 289
    References 289

Appendix 1... 291

Appendix 2... 295

Appendix 3... 297

Appendix 4... 299

Index.. 301

Preface

The purpose of this book is to provide a thorough understanding of ergonomic issues and to provide background information, principles, design guidelines, and tools and methods used in designing and evaluating automotive products. This book has been written to satisfy the needs of both students and professionals who are genuinely interested in improving the usability of automotive products. Undergraduate and graduate students in engineering and industrial design will gain an understanding of the ergonomics engineer's work and the complex coordination and teamwork of many professionals in the automotive product development process. Students will learn the importance of timely information and recommendations provided by the ergonomics engineers and the methods and tools that are available to improve user acceptance. The professionals in the industry will realize that the days of considering ergonomics as a "commonsense" science and simply "winging-in" quick fixes to achieve user-friendliness are over. The auto industry is facing tough competition and severe economic constraints. Their products need to be designed "right the first time" with the right combinations of features that not only satisfy the customers but continually please and delight them by providing increased functionality, comfort, convenience, safety, and craftsmanship.

The book is based on my more than 40 years of experience as a human factors researcher, engineer, manager, and teacher who has performed numerous studies and analyses designed to provide answers to designers, engineers, and managers involved in designing car and truck products, primarily for the markets in the United States and Europe. The book is not like many ergonomics textbooks that compile a lot of information from a large number of references reported in the human factors and ergonomics literature. I have included only the topics and materials that I found to be useful in designing car and truck products, and I concentrated on the ergonomic issues generally discussed in the automotive design studios and product development teams. The book is really about what an ergonomics engineer should know and do after he or she becomes a member of an automotive product development team and is asked to create an ergonomically superior vehicle.

The book begins with the definitions and goals of ergonomics, historic background, and ergonomics approaches. It covers important human characteristics, capabilities, and limitations considered in vehicle design in key areas such as anthropometry, biomechanics, and human information processing. Next, the reader is led in understanding how the driver and the occupants are positioned in the vehicle space and how package drawings and/or computer-aided design models are created from key vehicle dimensions used in the automobile industry. Various design tools used in the industry for occupant packaging, driver vision, and applications of other psychophysical methods are described. The book covers important driver information processing concepts and models and driver error categories to understand key considerations and principles used in designing controls, displays, and their usages, including current issues related to driver workload and driver distractions.

A vehicle's interior dimensions are related to its exterior dimensions in terms of the required fields of view from the driver's eye points through various window openings and other indirect vision devices (e.g., mirrors, cameras). Various field-of-view measurements, analysis techniques, visibility requirements, and design areas such as windshield wiper zones, obscurations caused by car pillars, and the required indirect fields of views are described along with many trade-off considerations. To understand the basics of headlamp beam pattern design and signal lighting performance and their photometric requirements, human factor considerations and night visibility issues are presented. Other customer/user concerns and comfort issues related to entering and exiting the vehicle, seating, loading and unloading cargo, and other service-related issues (engine and trunk compartment, refueling the vehicle, etc.) are covered. They provide insights into user considerations in designing vehicle body and mechanical packaging in terms of important vehicle dimensions related to body/door openings, roof, rocker panels, and clearances for the user's hands, legs, feet, torso, head, and so on.

A chapter on craftsmanship covers a relatively new technical and increasingly important area for ergonomics engineers. The whole idea behind craftsmanship is that the vehicle should be designed and built such that the customers will perceive the vehicle to be built with a lot of attention to details by craftsmen who apply their skills to enhance the pleasing perceptual characteristics of the product related to its appearance, touch, feel, sounds, and ease during operations. Several examples of research studies on measurement of craftsmanship and relating product perception measures to physical characteristics of interior materials are presented.

In addition, for researchers, the second part of the book includes chapters on driver behavioral and performance measurement, vehicle evaluation methods, modeling of driver vision (which illustrates how the target detection distances and legibility of displays can be predicted to evaluate vehicle lighting and display systems), and driver workload to evaluate in-vehicle devices. Discussions on ergonomic issues for development of new technological features in areas such as telematics, night vision, and other driver-safety- and comfort-related devices are included. The second part of the book also presents data and discusses many issues associated with designing for different population segments, such as older drivers, women drivers, and drivers in different geographic parts of the world. Finally, the last chapter is focused on various issues related to future research needs in several specialized areas of ergonomics as well as vehicle systems and on implementation of available ergonomic design guidelines and tools at different stages of the automotive product design process.

The book can be used to form the basis of two courses in vehicle ergonomics. The first course would cover the basic ergonomic considerations needed in designing and evaluating vehicles that are included in Part I—the first eleven chapters of this book. The remaining chapters covered in Part II can be used for an advanced and more research-oriented course.

Website Materials

The following files are in the Download section of this book's web page on the CRC Press website (http://crcpress.com/product/isbn/9781439842102).

A. Computer programs and models

    1.   Computations of percentile value of normal distribution
    2.   Driver package parameters computations
    3.   Reaction time measurement program
    4.   Legibility model
    5.   Visibility prediction model
    6.   Discomfort glare and dimming request prediction model

B. Slides for lectures 1–17 (corresponding to Chapters 1–17)

Acknowledgments

This book is a culmination of my education, experience, and interactions with many individuals from the automotive industry, academia, and government agencies. While it is impossible for me to thank all the people who influenced my career and thinking, I must acknowledge the contributions of the following individuals.

My greatest thanks go to Professor Thomas H. Rockwell of the Ohio State University. Tom got me interested in human factors engineering and driving research. He was my advisor and mentor during my doctoral program. I learned many skills on how to conduct research studies, analyze data, and more important, he had me introduced to the technical committees of the Transportation Research Board and the Society of Automotive Engineers, Inc.

I would like to thank Lyman Forbes, Dave Turner, and Bob Himes from the Ford Motor Company. Lyman Forbes, manager of the Human Factors Engineering and Ergonomics Department at the Ford Motor Company in Dearborn, Michigan, spent hours with me discussing various approaches and methods of conducting research studies on various crash-avoidance research issues related to the development of motor vehicle safety standards. Dave Turner from the Advanced Design Studios helped anchoring ergonomics in the automotive design process and also created an environment to establish a human factors group in Europe. Bob Himes of the advanced vehicle engineering staff helped in incorporating ergonomics and vehicle packaging as a vehicle attribute in the vehicle development process.

The University of Michigan–Dearborn campus provided me with unique opportunities to develop and teach various courses. Our automotive systems engineering and engineering management programs allowed me to interact with hundreds of students who in turn implemented many of the techniques taught in our graduate programs in solving problems within many other automotive original equipment manufacturers and supplier companies. I would like to thank Professors Adnan Aswad, Munna Kachhal, and Armen Zakarian for giving me opportunities to develop and teach many courses in industrial and manufacturing systems engineering and Dean Subrata Sengupta for supporting the creation of the Vehicle Ergonomics Laboratory in the new Institute for Advanced Vehicle Systems Building. Roger Schulze, director of the Institute for Advanced Vehicle Systems, got me interested in working on a number of multidisciplinary programs in vehicle design. Together, we developed a number of vehicle concepts such as the low-mass vehicle, a new Model T concept for Ford's 100th anniversary, and a reconfigurable electric vehicle. We also created a number of design projects by forming teams of our engineering students with students from the College for Creative Studies in Detroit, Michigan. My special thanks also go to James Dowd from Collins and Aikman and team members of the Advanced Cockpit Enablers (ACE) for sponsoring a number of research projects on various automotive interior components and creation of a driving simulator to evaluate a number of advanced concepts in vehicle interiors.

Over the past 40-plus years, I was also fortunate to meet and discuss many automotive design issues with members of many committees of the Society of Automotive Engineers, Inc., the Motor Vehicle Manufacturers Association, the Transportation Research Board, and the Human Factors and Ergonomics Society.

I would like to also thank Cindy Carelli of CRC Press—a Taylor & Francis Company—for encouragement in preparing the proposal for this book and Ed Curtis and his production group for turning the manuscript into this book. My thanks also go to Louis Tijerina, Anjan Vincent, and Calvin Matle for reviewing the manuscript and providing valuable suggestions to improve this book.

Finally, I want thank my wife, Rekha, for her constant encouragement and her patience while I spent many hours writing the manuscript and creating the figures included in this book.

Author

Vivek D. Bhise is currently visiting professor and professor in postretirement of Industrial and Manufacturing Systems Engineering at the University of Michigan–Dearborn. He received his BTech in Mechanical Engineering (1965) from the Indian Institute of Technology, Bombay, India, MS in Industrial Engineering (1966) from the University of California, Berkeley, California, and PhD in Industrial and Systems Engineering (1971) from the Ohio State University, Columbus, Ohio.

During 1973 to 2001, he held a number of management and research positions at the Ford Motor Company in Dearborn, Michigan. He was the manager of Consumer Ergonomics Strategy and Technology within the Corporate Quality Office and the manager of the Human Factors Engineering and Ergonomics in the Corporate Design of the Ford Motor Company, where he was responsible for the ergonomics attribute in the design of car and truck products.

Dr. Bhise has taught graduate courses in vehicle ergonomics, vehicle package engineering, automotive systems engineering, human factors engineering, total quality management and Six Sigma, product design and evaluations, and safety engineering over the past 30 years (1980–2001 as an adjunct professor and 2001–2009 as a professor) at the University of Michigan-Dearborn. He also worked on a number of research projects on human factors with Professor Thomas Rockwell at the Driving Research Laboratory at the Ohio State University (1968–1973). His publications include more than 100 technical papers in the design and evaluation of automotive interiors, vehicle lighting systems, field of view from vehicles, and modeling of human performance in different driver/user tasks.

He received the Human Factors Society's A. R. Lauer Award for outstanding contributions to the understanding of driver behavior in 1987. He has served on a number of committees of the Society of Automotive Engineers Inc., Vehicle Manufacturers Association, Human Factors Society, and Transportation Research Board of the National Academies. He is a member of the Human Factors and Ergonomics Society, the Society of Automotive Engineers Inc., and Alpha Pi Mu.

Part I

Ergonomics Concepts, Issues, and Methods in Vehicle Design

1 Introduction to Automotive Ergonomics

ERGONOMICS IN VEHICLE DESIGN

Designing an automotive product such as a car or truck involves the integration of inputs from many disciplines (e.g., designers, body engineers, chassis engineers, powertrain engineers, manufacturing engineers, product planners, market researchers, ergonomics engineers, electronics engineers). The design activities are driven by the intricate coordination and simultaneous consideration of many requirements (e.g., customer requirements, engineering functional requirements, business requirements, government regulatory requirements, manufacturing requirements) and trade-offs between the requirements of different systems in the vehicle. The systems should not only function well, but they must also satisfy the customers who purchase and use the products. The field of ergonomics or human factors engineering in the automotive product development involves working with many different vehicle design teams (e.g., management teams, exterior design teams, interior design teams, package engineering teams, instrument panel teams, seat design teams) to assure that all important ergonomic requirements and issues are considered at the earliest time and resolved to accommodate the needs of the users (i.e., the drivers, passengers, personnel involved in assembly, maintenance, service) while using (or working on) the vehicle.

OBJECTIVES

The objective of this book is to provide the reader a thorough understanding of ergonomic issues, design guidelines, models, methods to measure user performance and preference, and the various analysis procedures used in designing and evaluating automotive products. To develop an understanding of ergonomic issues, this book is organized to familiarize the reader with important ergonomic principles, theories, past research studies, and reference data related to design of different areas and systems within the vehicle. The final and the key objective was to provide a background into how an ergonomics engineer should work with the specialists from other design disciplines (e.g., package engineering, body engineering, lighting engineering, climate control engineering, driver information and entertainment systems design) to assure that ergonomically superior vehicles are designed with a comprehensive understanding of their functional needs and trade-offs.

ERGONOMICS: WHAT IS IT?

Ergonomics is a multidisciplinary science involving fields that have information about people (e.g., psychology, anthropometry, biomechanics, anatomy, physiology, psychophysics). It involves studying human characteristics, capabilities, and limitations and applying this information to design and evaluate the equipment and systems that people use.

The basic goal of ergonomics is to design equipment that will achieve the best possible fit between the users (drivers) and the equipment (vehicle) such that the users' safety (freedom from harm, injury, and loss), comfort, convenience, performance, and efficiency (productivity or increasing output/input) are improved.

The field of ergonomics is also called "human engineering," "human factors engineering," "engineering psychology," "man–machine systems," or "human–machine interface design." After World

War II, the field of human factors emerged in the United States, mainly among psychologists, to study the equipment and process design problems primarily from the human information processing viewpoint. The field of ergonomics emerged in the European countries around 1949 to improve workplaces and jobs in the industries, with an emphasis on biomechanical applications. The word "ergonomics," the science of work laws (or the science of applying natural laws to design work), was coined by joining two Greek words: "ergon" (work) and "nomos" (laws). Over the past 25 years, the field that covers both the physical and information processing aspects is more commonly known as "human factors engineering" or "ergonomics," with about equal preference on the use of either name for the field. After the fuel economy crisis of the 1970s, the U.S. automobile industry began placing more emphasis on both the aerodynamics and ergonomics fields to satisfy customers' energy-saving and comfort/convenience needs. The use of both the somewhat similar-sounding terms, aerodynamics and ergonomics, was also perceived to be somewhat appealing in marketing the products.

ERGONOMICS APPROACH

Fitting the Equipment to the Users

Ergonomics involves "fitting the equipment to the people (or users)." This means that equipment should be designed such that people (population of users) can fit comfortably (naturally) within the equipment and they can use the equipment without any awkward body postures, movements, or errors.

It should be noted that ergonomics is not about fitting the people to equipment (i.e., equipment should not be designed first and then people are simply asked to somehow adapt or force-fit to use the equipment). In some cases, the equipment is designed such that only people with certain characteristics can fit or use them (which normally involve personnel selection strategy, that is, placing restrictions on the "type" of people who can use the equipment).

Designing for the Most

Ergonomics involves "designing for the most" (i.e., to assure that most users within the intended population of the users of the product can fit within the product). It should be noted that if we use other design strategies, like "designing for the average" or "designing for the extreme," only a few individuals within the user population will find the product to be "just right" (or fit very well) for them. Thus, designing for the most will involve making sure that the designer knows what the user population is and knows the distributions of characteristics, capabilities, and limitations of the individuals in that population.

Systems Approach

Another important consideration involves "humans as a systems component." This means that the designer must treat the human to be a component of the system that is being designed. The process for designing a vehicle should thus involve the considerations of the following major components: (a) the driver/user, (b) the vehicle, and (c) the environment (see Figure 1.1). The characteristics of all the components in the system must be considered in designing the vehicle. The vehicle design should involve not only designing all the physical components that fit and function well but also making sure that the user is considered to be a human component and the user's characteristics are measured and used in designing a car or a truck—to assure that the vehicle will meet the users' needs related to comfort, convenience, and safety. It should be noted that during the design process of a physical product, the engineer designs each part of the product by paying attention to all of its properties (e.g., dimensions, material, hardness, color, surface, how it fits/works with other components). Similarly, when the human is involved as an operator or the user of the product (e.g., a car or a truck), all relevant human characteristics must be studied and used in designing the product.

Thus, in designing a vehicle, a thorough understanding of the intended user population and the operating environment (which consists of the roadway, traffic, weather, and operating conditions such as dawn, day, dusk, and night) of the vehicle must be considered. Figure 1.1 shows that when a

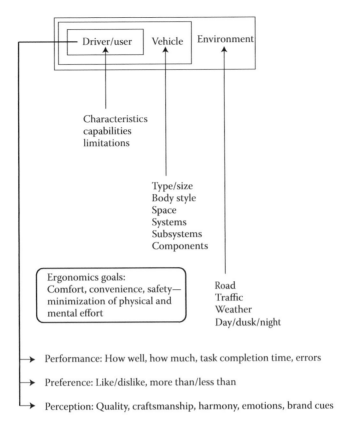

FIGURE 1.1 Ergonomics engineer's considerations related to characteristics of the driver, the vehicle, and the environment and their relationship to driver performance, preference, and perception.

driver/user operates a vehicle in a driving environment, the ergonomics engineer must consider the characteristics of all the components in the system and evaluate the following: (a) how the driver/user will perform various tasks; (b) the preferences of the driver/user in using the product; and (c) the pleasing perceptions created by experiencing the product, such as the quality, craftsmanship, emotions evoked by the product, and the resulting brand image. Bailey (1996) has proposed another but somewhat similar approach to conceptualize ergonomic problems. Bailey's approach takes into account the user, the activity (type of operation or usage), and the context (usage situation) in designing a system.

PROBLEM-SOLVING METHODOLOGIES

To solve different problems encountered during the development of a new automotive product, the ergonomics engineer relies on a number of different approaches. Figure 1.2 shows three pure or basic approaches in solving a problem. The least time-consuming approach shown is the middle branch of Figure 1.2. It uses the "guess" or the "guessed" answer made by the decision maker. The guessed answer is generally not regarded as an objective answer when compared with the other two approaches, namely, using a model or performing an experiment shown in the outer branches of Figure 1.2. The modeling approach shown in the left branch of Figure 1.2 assumes that a well-developed and validated model exists and that it can be exercised by inputting different combinations of the values of the input variables to arrive at a best solution. The experimental approach is used when an ergonomics engineer designs and conducts an experiment to determine the best combination of independent variables needed to obtain the required output.

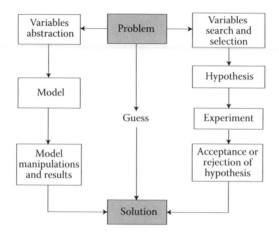

FIGURE 1.2   Problem-solving approaches.

It may seem that the guess approach is a dangerous approach. However, in many cases, when a design recommendation is needed immediately (which occurs quite frequently in the industry—where your boss wanted an answer yesterday), the ergonomics engineer needs to provide his or her best guessed answer. Of course, it is assumed that the ergonomics engineer does his or her homework, which generally involves steps such as (a) reviewing the problem along with available sketches, drawings, or hardware related to the problem; (b) reviewing past studies; (c) consulting other ergonomics experts; (d) applying available models; and (e) providing the best guessed answer with clearly specified assumptions. The "soundness" of the guessed answer will depend on the level of expertise possessed by the decision maker and the amount of information available while making the guess. If more time is available, then searching additional literature or data, benchmarking similar products developed by better competitors, conducting analyses using predeveloped models, or even running a quick-react experiment could help in proposing a solution.

Embedded within the thought processes involved in providing a solution is really the good old scientific method. This method is applied in all fields of sciences to develop solutions. The scientific method has been found to be a very effective method to solve ergonomic problems. The scientific method involves the following steps (remember the acronym DAMES suggested by Konz and Johnson, 2004):

D = Define the problem broadly.
A = Analyze all variables that can affect the performance of people.
M = Make search of solution space.
E = Evaluate alternatives—determine the best solution by applying the human factors knowledge and models or by experimental research.
S = Specify and sell solution.

Since each new problem faced by an ergonomics engineer is generally "new" for which a model to predict a solution is not available, the experimental approach may be the best. The experimental approach, however, is generally very time-consuming and costly. In reality, since no experiment or model is complete (because they involve the use of only a few key variables and have limited validity and applicability), a proficient ergonomics engineer will use a unique combination of all the three approaches shown in Figure 1.2 to solve each problem. Further, it should be realized that to resolve many design issues, a number of iterations using the combinations of the above three approaches and other variations such as structured walk-throughs with a team of experts may be needed.

ERGONOMICS RESEARCH STUDIES

Most research studies available in the ergonomics field can be categorized into the following three types or their combinations:

1. Descriptive research: This type of research generally provides data describing human characteristics of different populations (e.g., anthropometric measurements and their distributions).
2. Experimental research: This type of research generally involves experiments conducted to determine the effects of different combinations of independent variables on certain response variables under carefully manipulated and controlled experimental situations (e.g., determining the effects of different vehicle parameters on comfort ratings or performance measures during an entry into a vehicle).
3. Evaluative research: This type of research generally involves comparisons of user performance (and/or preference) in using different designs (vehicle systems or features; e.g., determining which of the four proposed radio designs would be most convenient to use and/or preferred by the drivers).

ERGONOMICS ENGINEER'S RESPONSIBILITIES IN VEHICLE DESIGN

The inclusion of ergonomics engineers as a part of the vehicle development team is now an accepted practice in the automobile industry. The ergonomics engineers work from the earliest stages of new vehicle concept creation to the periods when the customer uses the vehicle, disposes it, and is ready to purchase his or her next vehicle.

The ergonomics engineer's major tasks during the life cycle of the vehicle are as follows:

1. Provide the vehicle design teams with needed ergonomics design guidelines, information, data, analyses results, scorecards, and recommendations for product decisions at the right time (called the "gateways" or "milestones") in front of the right level of decision makers (involving design teams, program managers, chief engineers, senior management, etc.).
2. Apply available methods, models, and procedures (e.g., Society of Automotive Engineers Inc. [SAE, 2009] and company practices) to address issues raised in the vehicle development process.
3. Conduct quick-react studies (i.e., experiments) to answer questions raised during the vehicle development process.
4. Evaluate product/program assumptions, concepts, sketches, drawings, CAD models, physical models/mock-ups/bucks, mechanical prototypes (called "mules" in the auto industry), prototype vehicles, and production vehicles made by the manufacturer and its competitors.
5. Participate in the design and data collection phases of drive clinics and market research clinics involving concept vehicles and existing leading products as comparators (or controls).
6. Obtain, review, and act on the customer feedback data from complaints, warranty, customer satisfaction surveys, market research data (e.g., J. D. Power survey data [J. D. Power and Associates, 2010]), inspection surveys with owners, automotive magazines, press, etc.
7. Create ergonomics scorecards at selected program milestones during the vehicle development process.
8. Provide ergonomics consultations to members of the vehicle development teams.
9. Perform long-term tasks: conduct research, translate research results into design guidelines, and develop design tools.

The ergonomics engineer's roles and responsibilities are discussed in more detail in Chapter 11.

HISTORY

ORIGINS OF ERGONOMICS AND HUMAN FACTORS ENGINEERING

Prehistoric Times and Functional Changes in Products

The history of ergonomics goes back to the designing of tools used by the prehistoric man. These tools developed in the prehistoric era evolved over many years. Many generations of families mastered certain trades and became "craftsmen." Certain designs that were functional lived on. Useless tools were not remade. Changes were made for functional improvements. Thus, when considering future product improvements, we should not introduce any design change for the sake of change. Changes should be only made if functional improvements can be achieved.

Air Force Research

After World War II, the U.S. Air Force systematically studied many problems experienced by pilots. They found that pilot errors in using aircraft displays and operating controls were caused mainly by improperly designed equipment, and thus, the equipment could be better designed to reduce pilot errors (Fitts and Jones, 1947a, 1947b). (Note: This topic is discussed in more detail in Chapter 5.)

"Ergonomics" Coined

The word "ergonomics" was coined in England by K. F. H. Murrell in 1949 by joining two Greek words—"ergon," meaning work, and "nomos," meaning natural laws—to convey the philosophy of applying natural laws in designing for people (Murrell 1958). Thus, to design products for people, we should think about what comes naturally to people. In general, we have to think about what naturally conforms to people. The natural fit improves performance, that is, it reduces operation/task completion time and errors, reduces learning and training, and makes products more enjoyable, less difficult, and less boring.

HISTORY OF ERGONOMICS IN AUTOMOTIVE PRODUCT DESIGN

Some key historic events related to the applications of ergonomics to automotive design issues are presented below.

1918: SAE issued J585 standard on tail lamps and J587 standard on license plate illumination devices.
1927: SAE issued J588 standard on stop lamps.
1956: Ford Motor Company established the Human Factors Engineering Department.
1965: SAE published J941: motor vehicle driver's eye locations (eyellipse) recommended practice.
1966: Congress passed safety acts: National Traffic Safety and Motor Vehicle Safety Act and the Highway Safety Act.
1969: National Highway Traffic Safety Administration (NHTSA), Department of Transportation published notices of proposed rule making in crash avoidance area (e.g., vehicle lighting).
1976: SAE published J287: driver hand control reach recommended practice.
1978: NHTSA enacted Field of View Requirements on Motor Vehicles (Federal Motor Vehicle Safety Standard [FMVSS] 128 which rescinded later in 1979).
1984: Touch CRT introduced in cars (GM's Buick Riviera).
1986: Center high mounted stop lamp required on all vehicles.
1997: Toyota launched "Prius" (hybrid electric vehicle with state-of-power display).
1997: NHTSA published a report on investigation of the safety implications of wireless communications in vehicles (Goodman et al., 1997).
2000: Adjustable pedals introduced in the U.S. market on light truck products.
2000: NHTSA hosted the first Internet forum on driver distraction.
2001: Smart headlamps introduced in luxury vehicles.

2007: Ford introduced "sync" to connect cell phones and iPods and other USB-based systems.

2007: Rear seat entertainment systems with display screens available for rear passengers were introduced.

2010: Capacitive touch-screen technology introduced in vehicle displays.

2011: Plug-in electric vehicles expected to be sold in the U.S. market by major automotive manufacturers with power consumption gauges (eco-gauges).

Appendix 1 presents additional historic points in tracing the progress of human factors engineering.

IMPORTANCE OF ERGONOMICS

CHARACTERISTICS OF ERGONOMICALLY DESIGNED PRODUCTS, SYSTEMS, AND PROCESSES

If a product (or a system) is designed well (i.e., meets ergonomics requirements), the following effects or outcomes are expected:

1. An ergonomically designed product should fit people well (like a well-fitting suit). (Note: One would use a well-fitting suit much more often than a poor-fitting one). Thus, when it is time to replace an old product, a customer will most likely purchase a newer version of the same product that fits him or her well. This suggests that ergonomically designed products more likely will be repurchased.
2. An ergonomically designed product can be used with minimal mental and/or physical work. Thus, as product usage increases, the customer will realize the ease, comfort, and convenience features and the absence of problems while using the product.
3. An ergonomically designed product is easy to learn. (Note: Owner's manuals of easy-to-learn products are seldom used. Easy-to-learn products work in an "expected" manner.)
4. A product with usability problems (i.e., the absence of ergonomics) can be quickly noticed—usually after use. Thus, ergonomic characteristics of many products are not generally noticed in the showrooms where the customer does not have an opportunity to use them.
5. Ergonomically designed products are generally more efficient (productive) and safer (less injurious).

WHY APPLY ERGONOMICS?

1. It creates functionally superior products, processes, or systems.
2. Costly and time-consuming redesigns can be avoided (with early incorporation of ergonomics inputs in the design process, superior products or systems can be developed without additional design iterations).
3. There are thousands of ways to design a product, but only a few designs are truly outstanding. (You want to find those "outstanding" designs quickly.)

ERGONOMICS IS NOT COMMONSENSE

1. Commonsense ideas/solutions are often wrong. For example, a designer wanted to create an instrument panel illuminated with "deep red" lighting for a new hot sports car. The ergonomics engineer reminded him that about 8% of males have a color deficiency in perceiving the color red. The designer said, "But the air force uses red-colored instrument panels in airplanes." The ergonomics engineer reminded the designer, "Color-deficient persons cannot get a pilot's license, but a car is a consumer product, and you don't want to annoy these color-deficient males in using your vehicle. If you want red, then we should add some yellow in it and make it orangish-red so that the red-color-deficient people can still read the instruments."

2. Knowledge-based decisions are superior as they minimize usability problems. (The ergonomist brings his specialized knowledge and data about users at an early point in the design process.)

A BRIEF OVERVIEW OF HUMAN CHARACTERISTICS AND CAPABILITIES

The human characteristics and capabilities used in equipment design can be classified as follows:

Physical Capabilities

These can be measured by use of physical instruments (e.g., measuring tapes, rulers, calipers, weighing scales, strength/force measuring gauges).

1. Anthropometric characteristics (which involve measurements of human body dimensions). The measurements made when a human subject is stationary (not moving) are called "static" dimensions, which generally are taken when a subject is standing erect or sitting in an anthropometric measurement chair (with vertical torso and lower legs and horizontal upper legs). The human body dimensions measured when a subject is in a work posture (e.g., sitting in a car seat and performing a task) are called "functional" anthropometric dimensions. Other measurements of human body (and body segments) such as surface areas, volumes, center of gravity, and weights are also considered to be part of anthropometry (science of human body dimensions; see Chapter 2 for more details).
2. Biomechanical characteristics (e.g., ability to produce forces/strength and body movements; see Chapter 2 for more details).

Information-Processing Capabilities

3. These are mental (cognitive) capabilities involving the acquisition of information through various sensors (eye, ear, joint, vestibular tissues, etc.), transmitting this sensed information to the brain, recalling information stored in the memory, processing the information to make decisions (detecting, recognizing, comparing, selecting, etc.), and making responses (e.g., motor action—generating a body movement, activating a control, or making a verbal response; see Chapter 4 for more details).

In general, many human abilities degrade as people get older. The degradation in most human abilities is about 5%–10% per decade after about 25 years old. With practice, humans can perform complex tasks with very little or no conscious effort. However, humans are not consistent and precise in performing tasks like machines. Thus, human performance in performing most tasks varies considerably. The variability in the same subject performing the same task is called the "within-subject variability," whereas the "between-subject variability" is the difference in performance when a different subject performs the same task. When designing a vehicle, the ergonomics engineers must make sure that most users in the population can perform the tasks associated with the vehicle.

IMPLEMENTING ERGONOMICS

The following chapters of this book will cover ergonomic concepts, issues, and methods used in designing different automotive products and their features. Part I of this book deals with how the anthropometric, biomechanical, and information-processing considerations are used in designing aspects of occupant package, seats, controls, displays, instrument panels, window openings, entry/exit, vehicle-service-related issues, craftsmanship, and descriptions on how the ergonomics engineers work with the teams involved in the vehicle design process. Part II of this book covers more advanced and research-oriented issues, such as modeling of human vision to predict visibility and legibility, performance measurement, evaluation of driver workload, dealing with special user populations, and research needs to enable implementation of new technologies in future vehicles

REFERENCES

Bailey, R. W. 1996. *Human Performance Engineering*. Upper Saddle River, NJ: Prentice Hall PTR.

Fitts, P. M., and R. E. Jones. 1947a. Analysis of factors contributing to 460 "pilot errors" experiences in operating aircraft controls. Memorandum Report TSEAA-694-12. Aero Medical Laboratory, Air Materiel Command, Wright-Patterson Air Force Base, Dayton, OH. In *Selected Papers on Human Factors in the Design and Use of Control Systems*, ed. H. Wallace Sanaiko, New York: Dover Publications Inc.

Fitts, P. M., and R. E. Jones. 1947b. Psychological aspects of instrument display. I: Analysis of factors contributing to 270 "pilot errors" experiences in reading and interpreting aircraft instruments. Memorandum Report TSEAA-694-12A. Aero Medical Laboratory, Air Materiel Command, Wright-Patterson Air Force Base, Dayton, OH. In *Selected Papers on Human Factors in the Design and Use of Control Systems*, ed. H. Wallace Sanaiko, New York: Dover Publications Inc.

Goodman. M. J., F. Bents, L. Tijerina, W. W. Wierwille, N. A. Lerner, and D. Benel. 1997. An investigation of the safety implications of Wireless Communications in vehicles. Report DOT HS 808 635. Washington, DC: U.S. Department of Transportation, the National Highway Traffic Safety Administration.

J. D. Power and Associates. 2010. *Customer Surveys on Initial Quality, In-Service and Product Appeal*, accessed June 23, 2011, http://www.jdpower.com/autos/car-ratings/.

Konz, S., and S. Johnson. 2004. *Work Design: Occupational Ergonomics*. 6th ed. Scottsdale, AZ: Holcomb Hathaway Publishers Inc.

Murrell, K. F. H. 1958. The Term "Ergonomics." *American Psychologist*, 13, Issue 10, p. 602.

Society of Automotive Engineers Inc. 2010. *SAE Handbook*. Warrendale, PA: SAE.

2 Engineering Anthropometry and Biomechanics

INTRODUCTION

The process of vehicle design begins with a discussion on the size and type of the vehicle and the number of occupants that the vehicle should accommodate. To assure that the required number of occupants can be accommodated, the designers must consider the dimensions of drivers and the passengers and their posture in the vehicle space. Therefore, in this chapter, we will review basic concepts, principles, and data related to human anthropometric and biomechanical characteristics, along with the considerations used in vehicle design, with an emphasis on occupant package and seating design.

Anthropometry and biomechanics are related fields in the sense that both depend on the dimensions of humans and the ability of humans to assume different postures while working or using vehicles. The two fields can be defined as follows.

Engineering anthropometry is the science of measurement of human body dimensions of different populations. It deals with skeletal dimensions (which are measured from certain reference points on the bones that are less flexible as compared with skin tissues), shape, contours, area, volumes, centers of gravity, weights, and so forth, of the entire human body and body segments. Engineering anthropometry involves applications of the anthropometric measurement data to design and evaluate products to accommodate people.

Biomechanics deals primarily with dimensions, composition, and mass properties of body segments, joints linking the body segments, muscles that produce body movements, mobility of joints, mechanical reactions of the body to force fields (e.g., static and dynamic force applications, vibrations, impacts), and voluntary body movements in applying forces (torques, energy/power) to external objects (e.g., controls, tools, handles). It is used to evaluate if the human body and body parts will be comfortable (e.g., internal forces well below strength and tolerance limits) and safe (avoidance of injuries) while operating or using machines and equipment (or vehicles).

USE OF ANTHROPOMETRY IN DESIGNING VEHICLES

The very first step in designing a vehicle is to determine the user population(s) and their anthropometric and biomechanical characteristics. The anthropometric data of the user population will help in determining many basic dimensions of the vehicle. The biomechanical data will help design the vehicle such that the users will not be required to exert or be subjected to forces that are above their tolerance or comfort levels.

Figure 2.1 shows some basic dimensions of people in standing and seated postures. A number of populations from different countries and different types of vehicles are measured (available in many human factors books, e.g., Pheasant and Haslegrave, 2006). Such data are also obtained by automotive companies by measuring dimensions of participants invited to attend market research clinics to evaluate new vehicle concepts or early prototypes.

Table 2.1 provides anthropometric data on U.S. adults for dimensions that are useful in accommodating occupants and in evaluating interior spaces and clearances. The table presents 5th, 50th, and 95th percentile values and standard deviations of various anthropometric dimensions of females and males compiled from different sources. It should be noted that these values do not take into

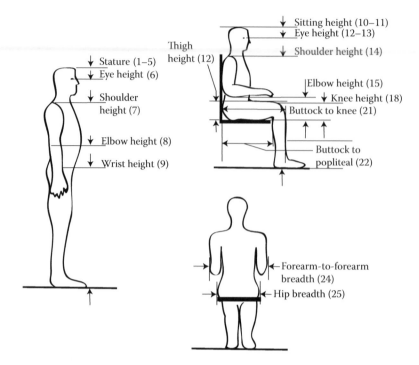

FIGURE 2.1 Static anthropometric measurements in standing and seated postures. (Note: The numbers in the parentheses refer to the dimension presented in Table 2.1.)

account the effects of attire (i.e., clothes, shoes, caps). The data from Kroemer et al. (1994), Pheasant and Haslegrave (2006), Jurgens et al. (1990), and McDowell et al. (2008) are for U.S. civilian adult population. The hand finger data from Garrett (1971) are from the U.S. Air Force flight personnel, and the data by Sanders (1983) are from measurement surveys of the U.S. heavy-truck drivers. The parameters of the normal distributions provided by different sources will also differ somewhat due to differences in the samples (due to differences in age, race, ethnicity, year when the samples were measured, etc.) used to measure the dimensions and the differences in the measurement process used by different sources. Further, since the anthropometric data needed for designing vehicles for different market segments differ from the U.S. civilian population, most automotive manufacturers maintain anthropometric databases on customers in different market segments (e.g., economy passenger cars, luxury passenger cars, pickup trucks). Additional anthropometric data on populations from different countries are provided in Appendix 2 and Chapter 16.

The majority of human anthropometric characteristics have been found to be normally distributed. Therefore, the normal distribution is generally used to compute the percentile values of populations that can be accommodated by a given vehicle dimension.

The normal distribution of a random variable x is defined as follows:

$$f(x) = \frac{1}{s\sqrt{2p}} e^{-(x-m)^2/2s^2}$$

where,

x = anthropometric dimension (e.g., standing height)

$f(x)$ = probability density function of x

μ = mean of the normal distribution of x

σ = standard deviation of the distribution of x

TABLE 2.1
Static Body Dimensions of United States Adults (Values Are in Millimeters)

No.	Measurement	Gender	5th	50th	95th	Std. Dev.	Reference
1	Stature: vertical distance from the floor to the vertex	Male	1647	1756	1855	67	Kroemer et al., 1994
		Female	1415	1516	1621	63	
2	Stature: vertical distance from the floor to the vertex	Male	1640	1755	1870	71	Pheasant and Haslegrave, 2006
		Female	1520	1625	1730	64	
3	Stature: vertical distance from the floor to the vertex	Male	1670	1790	1900	70	Jurgens, Aune, and Pieper, 1990
		Female	1540	1650	1760	67	
4	Stature: vertical distance from the floor to the vertex	Male	1636	1763	1887	76	McDowell et al., 2008
		Female	1507	1622	1731	68	
5	Stature: vertical distance from the floor to the vertex	Male	1665	1756	1880	62	Sanders, 1983
		Female	1572	1643	1708	64	
6	Eye height: vertical distance from the floor to the eyes	Male	1528	1634	1743	66	Kroemer et al., 1994
		Female	1415	1516	1621	62	
7	Shoulder height: hertical distance from the floor to the acromion	Male	1342	1442	1546	62	Kroemer et al., 1994
		Female	1241	1334	1432	58	
8	Elbow height: vertical distance from the floor to the radiale	Male	995	1072	1153	48	Kroemer et al., 1994
		Female	926	998	1074	45	
9	Wrist height: vertical distance from the floor to the wrist	Male	778	846	915	41	Kroemer et al., 1994
		Female	728	790	855	39	
10	Sitting height: vertical distance from the sitting surface to the vertex	Male	854	914	972	36	Kroemer et al., 1994
		Female	795	852	910	25	
11	Sitting height: vertical distance from the sitting surface to the vertex	Male	855	915	975	36	Pheasant and Haslegrave, 2006
		Female	800	860	920	36	
12	Sitting eye height: vertical distance from the sitting surface to the eyes	Male	735	792	848	34	Kroemer et al., 1994
		Female	685	739	794	33	
13	Sitting eye height: vertical distance from the sitting surface to the eyes	Male	749	811	863	38	Sanders, 1983
		Female	736	761	849	41	
14	Sitting shoulder height: vertical distance from the sitting surface to the acromion	Male	548	598	646	30	Kroemer et al., 1994
		Female	509	555	604	29	

continued

TABLE 2.1 (Continued)
Static Body Dimensions of United States Adults (Values Are in Millimeters)

No.	Measurement	Gender	5th	50th	95th	Std. Dev.	Reference
15	Sitting elbow height: vertical distance from the sitting surface to the underside of the elbow	Male	184	232	274	27	Kroemer et al., 1994
		Female	176	220	264	27	Kroemer et al., 1994
16	Thigh height: vertical distance from the sitting surface to highest top of thigh surface	Male	149	168	190	13	Kroemer et al., 1994
		Female	140	159	180	12	Kroemer et al., 1994
17	Seated stomach depth: horizontal depth of trunk at the level of abdominal extension	Male	229	299	374	45	Sanders, 1983
		Female	195	247	309	48	Sanders, 1983
18	Knee height (sitting): vertical distance from the floor to the upper surface of the knee	Male	514	559	606	28	Kroemer et al., 1994
		Female	474	515	560	26	Kroemer et al., 1994
19	Popliteal height (sitting): vertical distance from the floor to the underside of the knee	Male	395	434	476	25	Kroemer et al., 1994
		Female	351	389	429	24	Kroemer et al., 1994
20	Forward thumb tip reach: horizontal distance from the back of the shoulder blade to thumb tip with arm raised at shoulder level	Male	739	801	867	39	Kroemer et al., 1994
		Female	677	735	797	36	Kroemer et al., 1994
21	Buttock-to-knee distance (sitting): horizontal distance from the back of uncompressed buttock to the front of the kneecap	Male	596	616	667	30	Kroemer et al., 1994
		Female	542	589	640	30	Kroemer et al., 1994
22	Buttock-to-popliteal distance (sitting): horizontal distance from the back of uncompressed buttock to the back of the knee	Male	458	500	545	27	Kroemer et al., 1994
		Female	440	482	528	27	Kroemer et al., 1994
23	Elbow-to-fingertip distance: horizontal distance from the back of the elbow to middle-finger tip with lower arm horizontal in sitting position	Male	448	484	524	23	Kroemer et al., 1994
		Female	406	443	482	23	Kroemer et al., 1994
24	Forearm-to-forearm breadth: horizontal distance between the outermost points on the forearms in sitting position	Male	474	546	620	44	Kroemer et al., 1994
		Female	415	468	528	35	Kroemer et al., 1994
25	Hip breadth (sitting): maximum horizontal distance across the hips in the sitting position	Male	329	367	412	25	Kroemer et al., 1994
		Female	342	384	432	27	Kroemer et al., 1994
26	Foot length: distance from the back of the heel to the tip on the longest toe measured in longitudinal (forward) axis	Male	249	270	292	13	Kroemer et al., 1994
		Female	224	244	264	12	Kroemer et al., 1994
27	Shoe length: distance from back of heel to front edge of sole	Male	277	299	319	13	Sanders, 1983
		Female	241	264	286	15	Sanders, 1983
28	Foot breadth: maximum horizontal breadth across the foot perpendicular to the longitudinal axis	Male	92	101	109	53	Kroemer et al., 1994
		Female	82	90	98	49	Kroemer et al., 1994
29	Shoe breadth: maximum breadth of shoe at outside edges of sole	Male	98	107	116	11	Sanders, 1983
		Female	89	96	107	10	Sanders, 1983

#	Dimension	Gender					Source
30	Hand length: distance from the crease of the wrist to the tip of the middle finger with hand held straight and stiff	Male	179	194	211	98	Kroemer et al., 1994
		Female	165	180	197	97	
31	Hand length: distance from the crease of the wrist to the tip of the middle finger with hand held straight and stiff	Male	183	197	212	9	Garrett, 1971
		Female	165	179	193	9	
32	Hand breadth: maximum breadth across the palm of the hand at distal ends of the metacarpal bones	Male	84	90	98	4	Kroemer et al., 1994
		Female	73	79	86	4	
33	Hand breadth: maximum breadth across the palm of the hand at distal ends of the metacarpal bones	Male	83	90	97	4	Garrett, 1971
		Female	71	77	83	4	
34	Hand depth: measured at thenar pad	Male	55	62	70	5	Garrett, 1971
		Female	45	52	58	4	
35	Hand thickness: measured at metacarpale III	Male	30	33	36	2	Garrett, 1971
		Female	25	28	30	2	
36	Thumb breadth: measured at interphalangeal joint	Male	22	24	26	4	Kroemer et al., 1994
		Female	19	21	23	1	
37	Thumb breadth: measured at interphalangeal joint	Male	21	23	25	4	Garrett, 1971
		Female	17	19	21	1	
38	Digit 2 breadth: measured at interphalangeal joint	Male	17	18	20	1	Garrett, 1971
		Female	14	15	17	1	
39	Digit 3 breadth: measured at interphalangeal joint	Male	17	18	20	1	Garrett, 1971
		Female	14	15	17	1	
40	Digit 3 depth: measured at interphalangeal joint	Male	14	16	18	1	Garrett, 1971
		Female	12	13	15	1	
41	Digit 1 (thumb) length: fingertip to crotch level	Male	51	59	66	5	Garrett, 1971
		Female	47	54	61	4	
42	Digit 2 length: fingertip to crotch level	Male	68	75	82	5	Garrett, 1971
		Female	61	69	78	5	
43	Digit 3 length: fingertip to crotch level	Male	78	86	95	5	Garrett, 1971
		Female	70	78	87	5	
44	Weight (kg)	Male	58	79	99	13	Kroemer et al., 1994
		Female	39	62	85	14	
45	Weight (kg) (with shoes)	Male	69	92	120	17	Sanders, 1983
		Female	53	72	90	19	

Thus, the mean (μ) and standard deviation (σ) are the two parameters of the normal distribution (i.e., they define the location and spread of the distribution, respectively).

The cumulative distribution of $f(x)$ is denoted as $F(x)$ and is defined as follows:

$$F(x) = \int_{-\infty}^{x} f(x)\,dx = \frac{1}{s\sqrt{2p}} \int_{-\infty}^{x} e^{-(x-m)^2/2s^2}\,dx$$

Since the normal distribution is symmetrical about mean, $F(x) = 0.5$ defines the 50th percentile value. Thus, the 50th percentile value equals the mean (μ).

Percentile values are used to evaluate accommodation with respect to a given variable (x) (i.e., what percentage of the population can fit within a given value of an anthropometric variable x). For example, 95th percentile value of x will be defined at $F(x) = 0.95$. Thus, if x is the stature of individuals, then the 95th percentile value of x will mean that only 5% of the individuals in that population would be taller than that value. From Table 2.1, row 1, the 95th percentile value of stature of males is 1855 mm. This means that only 5% of U.S. adult males are taller than 1855 mm. Thus, if a door (e.g., for a classroom) needs to be designed so that 95% of the males can walk through the door opening without ducking their heads, then the door opening height must be at least 1855 mm. Generally, additional 50- to 100-mm clearance will be provided to account for the increase in accommodation height due to shoes and caps.

COMPUTATION OF PERCENTILE VALUES

Let us assume that we want to determine the percentile value of an adult male who is 1778 mm (5 ft, 10 in.) tall. From Table 2.1, row 1, the mean and standard deviation values of stature of males are 1756 and 67 mm, respectively. Next, we need to determine the value of the standardized normal variable (z) as follows:

$$z = \frac{x - m}{s} = \frac{1778 - 1756}{67} = 0.3284$$

The standardized normal variable (z) has a mean value equal to 0.0 and standard deviation equal to 1.0. The values of $F(z)$ for different values of z (generally ranging from about −3.0 to +3.0 are available in the normal distribution tables provided in textbooks of statistics. Referring to the table of cumulative normal distribution, the $F(z)$ value for $z = 0.330$ is 0.6293. (Note: For $z = 0.3284$, $F(z) = 0.6287$.)

If you are familiar with the Microsoft Excel application, you can obtain its value using NORMDIST function (Go to Insert, Function, and Select the statistical function called NORMDIST) [e.g., for our problem, NORMDIST(1778,1756,67,TRUE) = 0.628679].

Thus, the percentile value of a male with a stature of 1778 mm is 62.87, which means that 62.87% of the males will be shorter than 1778 mm or conversely, 37.13 = (100 − 62.87)% of males will be taller than 1778 mm.

The anthropometric data such as those provided in Table 2.1 can be used to come up with approximate values of various vehicle dimensions. The values obtained will be approximate because human "functional" dimensions in actual postures (which differ from the static postures of sitting and standing, as shown in Figure 2.1, used for anthropometric measurements) used in interacting with the vehicle (e.g., entering into and exiting from the occupant compartment, sitting in a vehicle, loading or unloading items in the trunk) cannot be easily predicted from the static anthropometry-based data. This is because of many reasons such as the following: human posture angles between different body segments vary among individuals in performing different tasks, the human joints are not like simple pin joints, human body tissues deflect (e.g., compression of tissues under the buttocks and back while sitting in a chair), there is a change within a person over time (e.g., slumping or leaning in the seat), and so forth.

Some examples of the use of static anthropometric dimensions for vehicle design are provided below.

1. Maximum seat cushion width can be estimated by considering the 95th percentile hip width of females. (The value is 432 mm from row 25 of Table 2.1).
2. Minimum seat cushion length can be estimated from the fifth percentile value of buttock-to-popliteal length of females. (The value is 440 mm from row 22 of Table 2.1).
3. Space above the driver's head can be estimated by considering the 99th percentile value of sitting height of males, torso angle, and top of deflected seat.
4. Interior shoulder width (W3) can be evaluated by comparing (W3/2–W20) with half shoulder width of the 95th percentile male (or 95th percentile male forearm-to-forearm breadth of 620 mm, row 24 of Table 2.1). (Note: The Society of Automotive Engineers Inc. [SAE] dimensions W3 and W20 are defined in Chapter 3. Dimension W3 in SAE J1100 [SAE, 2009] is defined as the cross-car distance between door trim panels at shoulder height; see Figure 3.13. W20 is defined as the lateral distance between the driver centerline and the vehicle centerline. See Figure 3.20).
5. Length of interior grab handles and exterior door handles can be estimated by considering the 95th percentile value of palm width without thumb. (The value is 98 mm from row 32 of Table 2.1).

To improve the accuracy of predicting key dimensions used to develop occupant package, the SAE Occupant Packaging Committee has developed a number of SAE standards (available in the SAE Handbook, SAE, 2009; (e.g., SAE standards J1516, J1517, J941, J1052, J287, J4004) based on functional anthropometric measurements of a large number of drivers seated in actual vehicles or vehicle bucks. In the functional anthropometry, the measurements of relevant dimensions of people are measured directly under the actual postures of people while using the vehicle. This avoids the problem of estimating or calculating dimensions by using the static anthropometric data (measured under the standard standing or sitting erect postures) and assuming posture angles of different body segments. The functional anthropometric tools used in the occupant packaging are covered in Chapter 3.

APPLICATIONS OF BIOMECHANICS IN VEHICLE DESIGN

Biomechanics is applied here to study and evaluate vehicle design issues in the following four problem areas:

1. Seating comfort (designing seats and their adjustment features)
2. Comfort and convenience during entry and egress (see Chapter 8)
3. Evaluating nonseated postures during loading and unloading of items in trunks or cargo areas, changing tires, servicing vehicles, refueling, and so forth (see Chapter 9)
4. Protecting occupants in impacts with interior hardware during accidents

Most biomechanical research studies reported in the technical literature have been conducted to understand how humans get injured due to (a) cumulative trauma (i.e., repetitive movements and stresses in body tissues while performing industrial tasks) and (b) accidents that subject the human body to high levels of deceleration and forces during impacts. Since impact protection (also called "crashworthiness") is generally considered a specialized field in the automobile industry, it is not covered in this book. Further, during the normal usage of vehicles, the drivers and passengers do not perform tasks that require application of forces at higher magnitudes (i.e., near maximum voluntary strength levels) and/or at higher frequencies as compared with certain industrial tasks. Thus, the problems of cumulative trauma injuries in vehicle usages are not at all as common or severe as in the industrial tasks. However, since discomfort can be associated with stresses at submaximal levels in the human body and in awkward postures, many of the biomechanical considerations and principles are useful in improving comfort and convenience problems in vehicle usages.

BASIC BIOMECHANICAL CONSIDERATIONS

Human strength depends on many factors. They include gender, age, duration of the exertion, static versus dynamic nature of exertion, anthropometry (lengths of body segments), posture (angles of various body segments), training, motivation, and so forth. A few basic biomechanical considerations can be summarized as follows:

1. Male versus female strength: Women typically have 65%–70% of the strength of men (see Figure 2.2).
2. Effect of age: The maximum force-producing capabilities (i.e., muscular strength) of adults decrease with age (about 5%–10% decrease on average every decade after about the age of 25 years; see Figure 2.2).
3. Muscular contraction: A muscle generates its strength during contraction. The maximum strength is reached at about 4 s after muscular contraction begins.
4. Endurance time and strength trade-off: The time over which a human can continuously exert force (called the "endurance time") increases with a decrease in the level of exerted force (strength). At about 15%–20% of the maximum voluntary contraction strength, a human can maintain exertion for a long period (see Figure 2.3). The shape of the endurance time curves varies depending on factors such as individual differences, particular muscles tested, work conditions, exertion rate, rest period between exertions, and training (Chaffin et al., 1999).

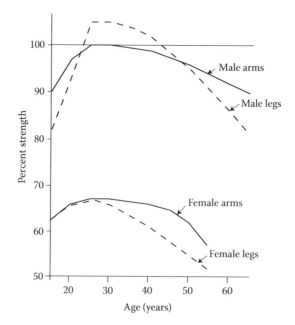

FIGURE 2.2 Strengths of arms and legs of men and women as functions of age. (Note that male strength at aged 20–22 years is set as 100%. The strength values are for isometric conditions where the muscles remain in static postures.) (Redrawn from Konz, S., and S. Johnson, *Work Design: Industrial Ergonomics*, 6th ed., Holcomb Hathaway, Scottsdale, AZ, 2004. With permission.)

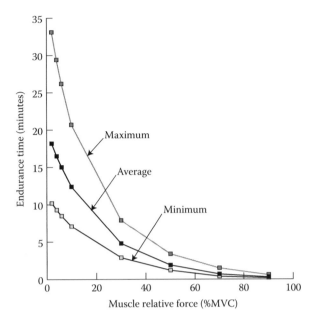

FIGURE 2.3 Trade-off between muscle relative force and endurance time. (Plotted from data presented in Chaffin, D. B., G. B. J. Andersson, and B. J. Martin: *Occupational Biomechanics.* 1999. Copyright Wiley-VCH Verlag GmbH & Co. KGaA. Reproduced with permission.)

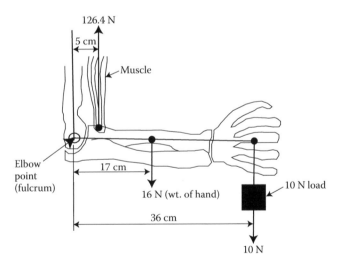

FIGURE 2.4 Upper-hand lever and force configuration while holding a weight of 10 N in the hand. (Redrawn from Chaffin, D. B., G. B. J. Andersson, and B. J. Martin: *Occupational Biomechanics.* 1999. Copyright Wiley-VCH Verlag GmbH & Co. KGaA. Reproduced with permission.)

5. Third class of lever: Most body segments related to large limb motions involve a "third class of lever system" in which the fulcrum is at one end of the lever and the external load is at the other end, with the activating force (from muscle) in between, usually close to the fulcrum. This arrangement of lever requires a larger force to be exerted by the muscle in comparison with the external load. The following example will illustrate this.

Figure 2.4 shows the lower arm held in the horizontal position by the muscle (i.e., biceps brachii) in the upper arm, with the elbow point as the fulcrum. The amount of force required in the muscle to hold a weight grasped by the hand can be calculated by computing moments around the elbow point. Let us assume that 10 N of load is held in the hand and at a distance of 36 cm (measured from the load in the hand to the elbow point). The center of gravity of the lower arm with the hand is about 17 cm from the elbow point, and the weight of the arm with the hand is 16 N. The moment pushing the hand down or clockwise around the elbow point would be ([10×36] + [16×17]) = 632 N · cm. Assuming the muscle holding the lower arm is attached 5 cm from the elbow, then the force in the muscle would be (632 / 5) = 126.4 N. Thus, in this case, to hold a load of 10 N in the hand, the reactive force in the muscle to hold the hand in equilibrium would be 12.6 times that of the load.

6. Design loads: For tasks involving large internal loads (e.g., during lifting), the job should be designed around the fifth percentile load exertion capability. Thus, 95% of the population can perform the task. For highly repetitive jobs, multiply the fifth percentile load capability by 0.15 to 0.2 (see consideration (4) above) to obtain a comfortable exertion level for accommodating a large proportion of the population over longer exertion durations.

BIOMECHANICAL CONSIDERATIONS IN SEAT DESIGN

1. Load in the L5/S1 region: One important consideration in seat design is to reduce the load on the spinal column of the seated person. Biomechanical research (Chaffin, Andersson, and Martin, 1999) has shown that during seating, the L5/S1 (the joint between the fifth lumbar vertebra and the first sacral vertebra) experiences the highest concentration of stress due to compressive force. The stresses in the L5/S1 can be reduced by providing a lumbar support that maintains the natural shape of the spinal column in the lumbar region (the natural shape of the spinal column is observed when a person is standing erect). The natural shape of the spinal column in the lumbar region is convex toward the front of the body (i.e., protruding forward, called "lordosis"). If the seatback provides the right amount of protrusion in the lumbar region at the correct height, and if the user can recline and support his or her torso on the seatback, the natural shape of the lumbar can be maintained.

2. Effect of Seatback angle and lumbar support: Figure 2.5 shows the effect of lumbar support (protrusion in centimeters) and seatback angle on the compressive force in the L5/S1 region of seated persons (Chaffin, Andersson, and Martin, 1999). The figure shows that as the seatback angle is increased from vertical (90-degree to 120-degree seatback angle with respect to the horizontal), the L5/S1 force will reduce due to transferring of the upper body (torso and head) weight into the seatback as compared with that in the spinal column. Further, the L5/S1 force can be also reduced as the amount of protrusion (i.e., the lumbar support) is increased (compare the lower four bars with the four darker bars just above in Figure 2.5).

   Based on Andersson et al. (1974a; see Figure 2.5), the load in the L5/S1 region is larger in all the seated postures (except positions 9, 10, and 11) as compared with the standing posture load of about 320 N (see seventh bar marked by a star). From Figure 2.5, we can see that the load in the L5/S1 can be reduced below this standing load of 320 N when the seatback reclined 20–30 degrees from the vertical (110–120 degrees from the horizontal) and the lumbar support of 20–50 mm is provided (Andersson and Ortengren 1974b).

3. Effect of armrests: Andersson and Ortengren (1974b) also showed that the L5/S1 load can be further reduced when the hands (lower arms) are supported on the armrests as compared with when the arms are not supported (i.e., left hanging). Thus, use of properly designed armrests will increase seating comfort by reducing load in the L5/S1 region.

OTHER SEAT DESIGN CONSIDERATIONS

Figures 2.6 through 2.9 illustrate four other occupant accommodation- and seat-comfort-related issues resulting from seat-shape- and size-related considerations.

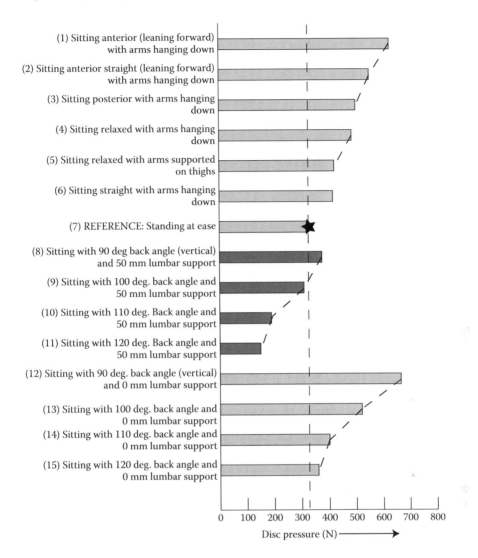

FIGURE 2.5 Effect of different sitting postures, seatback angle, and lumbar support on disc pressure at L5/S1. (Plotted from data presented in Chaffin, D. B., G. B. J. Andersson, and B. J. Martin: *Occupational Biomechanics*. 1999. Copyright Wiley-VCH Verlag GmbH & Co. KGaA. Reproduced with permission.)

4. Avoid long cushion length: Figure 2.6 (upper figure) shows that with a very long seat cushion, a seated person will leave a gap behind the user's buttocks and the lumbar region of the seatback. The seat cushion length should be shorter than the buttock-to-popliteal length of the person so that the user can support some of his or her upper body weight on the seatback and thus reduce L5/S1 load (see Figure 2.6, lower figure). Thus, if the seat cushion length is not adjustable, then it is best to design the seat cushion length for the shorter (fifth percentile) female buttock-to-popliteal length.

5. Avoid dangling feet: Figure 2.7 (upper figure) shows that if the seat cushion is too high, then the user's feet will dangle, and as a result, the pressure under the thighs will increase, creating discomfort (due to pinched veins and nerves in the back side of the knees during extended periods of driving). Thus, dangling feet should be avoided by either providing a footrest or reducing the seat height so that user's feet can be supported on the floor (see Figure 2.7, lower figure).

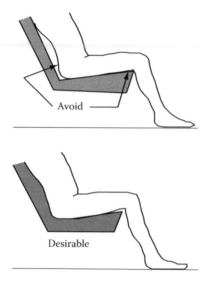

FIGURE 2.6   Avoid long cushion length.

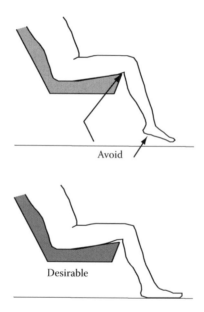

FIGURE 2.7   Avoid dangling feet.

6. Avoid curvature in the seat cushion: The curvatures in the seat cushions (see Figure 2.8), in general, should be avoided as they will put higher pressure on the body tissues surrounding the ischial tuberosites (i.e., the sitting bones—the lower protruding parts of the pelvic bones) and restrict body movements in the seat. A flatter seat cushion will allow the seat occupants to make small movements and postural changes that can increase overall seat comfort, especially during long trips (see Figure 2.8, lower figure). Provision of side bolsters in the seat cushion can increase effective curvature of the seat cushion. Therefore, tall (or heavily padded) bolsters should be avoided to improve long-term seat comfort. Increased seat curvatures and thicker bolsters also make the tasks of entry into the vehicle and exit from the vehicle more difficult (see Chapter 8).

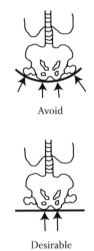

FIGURE 2.8   Avoid curvature in the seat cushion.

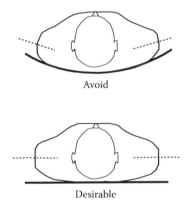

FIGURE 2.9   Avoid curvature in the seatback. (Based on Kolich, M., *Ergonomia, IJE HF*, 28(2), 125–136, 2006; Konz, S., and S. Johnson., *Work Design: Industrial Ergonomics*. 6th ed., Holcomb Hathaway, Scottsdale, AZ, 2004; from Chaffin, D. B., G. B. J. Andersson, and B. J. Martin: *Occupational Biomechanics*. 1999. Copyright Wiley-VCH Verlag GmbH & Co. KGaA. Reproduced with permission.)

7. Avoid curvature in the seatback: Provision of curved seatbacks (see Figure 2.9) or taller bolsters on the seat backs can force the driver's shoulders forward and also restrict small body movements, which, especially during long trips, can reduce comfort and increase driver fatigue. Thus, flatter seatbacks should be provided (see Figure 2.9, lower figure) to allow small body movements and increase overall seating comfort especially during longer trips. Provision of side bolsters in the seatbacks can also increase effective curvature of the seatback. Therefore, heavily padded bolsters should be avoided to improve long-term seat comfort. Increased seatback curvatures and thicker bolsters also make the tasks of entry into the vehicle and exit from the vehicle more difficult (see Chapter 8).

SEAT DESIGN CONSIDERATIONS RELATED TO DRIVER ACCOMMODATION

The following design considerations will improve the driver's accommodation and comfort (Kolich, 2006; Chaffin et al., 1999; Konz and Johnson, 2004).

1. Seat height: The seat height in vehicle package is measured using dimension H30, which is defined as the vertical height of the seating reference point from the accelerator heel point (see Chapter 3 for more details). The H30 dimension defines the driver's seated posture (defined by ankle angle, knee angle, torso angle, and seatback angle). Vehicles with overall low height (e.g., sports cars) typically have very low H30 (about 150–250 mm), whereas heavy commercial trucks have large H30 (more than 405 mm). If the seat is too high, the short driver's feet will dangle, and if the driver is unable to rest his or her heels on the vehicle floor/carpet (or on a foot rest), the driver will find the seating posture to be very uncomfortable. Therefore, based on the comfort of the fifth percentile female seated popliteal height of 351 mm, the top of the seat from the vehicle floor should not be more than about 320 mm. Power seats generally allow adjustment of the seat height so that drivers with different leg lengths can be accommodated.

   The horizontal distance between the accelerator heel point and seating reference point (defined as L53) increases as the H30 value is decreased (see Chapter 3 for more details). Thus, to minimize the horizontal space required to accommodate the driver in commercial vehicles (truck products), the seat height is increased (as compared with the passenger cars) and the driver sits more erect (seatback angle typically is more vertical, around 12–18 degrees from the vertical). In sports cars, the seatback angle will be more reclined to about 22–28 degrees from the vertical.

2. Adjustable seat: To accommodate the largest percentage of the drivers at their preferred driving posture, it is very important to allow them to adjust (1) seat height, (2) seat cushion angle, (3) seatback angle (reclining seatback), (4) height and protruding fore–aft length of the lumbar support, (5) headrest height and fore–aft location, (6) seat cushion length, (7) armrest height and length and its lateral location from the driver centerline, and (8) seat cushion and seatback bolster heights and/or angles. Power seats (which allow easy adjustments of many of the above-mentioned parameters with rocker or multifunction switches) are generally more comfortable than are nonpowered seats, especially during long trips.

3. Seat cushion length: The seat cushion length should not be longer than the driver's buttock-to-popliteal (back of knee) distance. Thus, if this length is restricted to the fifth percentile female buttock-to-popliteal distance (about 440 mm), then most drivers can use the seat and still use the back rest. Drivers with longer upper legs would prefer longer seat cushion lengths, but shorter females will not be able to use the seatback without a pillow on the seatback. Further, in case of longer seat cushion lengths, shorter females will find operation of the pedals difficult as they will be compressing the seat cushions with their thighs while depressing the pedals. Thus, an adjustable cushion length will reduce such problems and accommodate a larger percentage of the drivers.

4. Seat cushion angle: The seat cushion should slope backward by about 5–15 degrees. This will allow the user to slide back and allow the transferring of torso weight on to the seatback. Provision of an adjustable seat cushion angle will allow the user to find his or her preferred seat cushion angle.

5. Seat width: Since females have larger hip widths (breadths), the seat cushion width should be greater than 95th percentile female sitting hip width (about 432 mm; see measurement no. 25 in Table 2.1). In addition, clearance should be provided for clothing (especially thick winter coats); thus, a width of 500–525 mm at the hips can be recommended.

6. Seatback angle: The seatback angle (called A40 in SAE J1100; see Chapter 3 for more details) in automotive seating is defined by the angle of the torso line (back line) of the SAE H-point machine or the two-dimensional (manikin) template (refer to SAE standards J826 and J4002 [SAE 2009]) with respect to the vertical. The seatback angle (seat recline angle) should allow drivers to assume their preferred back angles. For passenger cars, drivers generally prefer to set the seatback angle between about 20 and 26 degrees. In trucks,

due to the higher seat height (H30), drivers prefer to sit more erect with seatback angles between about 12 and 18 degrees.

7. Seatback height: From an anthropometric accommodation viewpoint, the maximum seat-back height can be selected as the fifth percentile female acromial height, which is about 509 mm above the seat surface. However, considering the Federal Motor Vehicle Safety requirements on head restraints, the seatback height is dictated by the headrest design.

8. Lumbar area: The seat contour in the lumbar area affects the shape of the seated person's spinal column. The most important characteristic of the seat contour in the lumbar region is that it should maintain the natural curvature (bulging forward, i.e., convex, called lordosis) of the spinal column in the lower back region of the seated person. An adjustable lumbar support that allows setting its height (i.e., up and down adjustment) and protrusion location (i.e., fore–aft adjustment) would allow accommodation of different individuals while maintaining their natural lordosis.

9. Lateral location of the seat: The dimension W20-1 defines the lateral distance between the vehicle centerline and the driver's seating reference point. It should be designed so that the driver will have sufficient elbow clearance from the driver's door trim panel between the shoulder-and-elbow heights. This lateral distance from the driver centerline to the door trim panel should be larger than half of 95th percentile elbow-to-elbow width of males plus elbow clearance to avoid elbows rubbing against the door trim panel while grasping the steering wheel.

10. Armrest height: A properly designed and adjusted armrest can reduce the load on the driver's spinal column and thus increase the perception of comfort and reduce driver fatigue. The preferred height of the armrest will depend on the lateral location of the armrest from the driver centerline. Since it is difficult to position an armrest that can be perceived to be optimal by most drivers, the armrest height and lateral distance from the driver centerline should be adjustable. If the armrests are provided on both sides (i.e., on the door trim panel and on the seat or on the center console), both the armrests should be at the same height to reduce discomfort (due to leaning on one side).

11. Bolster height: The bolsters on the sides of the seat cushion and seat back can provide the driver feeling of sitting "snug or cuddled" (like in a contoured seat) in the seat and provide a sense of stability and security while negotiating curves and driving on winding roads. The bolsters restrict the seated person's movements in the seat, and, therefore, especially on long trips, such seats will be perceived to be less comfortable. (Smaller postural movements can increase the comfort of seated persons especially during longer trips.) The taller bolsters on the seatback may also move the minimum reach distance to controls and door handles more forward (due to forward shifting of driver's elbows when touching the bolsters; see Chapter 5 for the minimum reach envelopes). Further, taller bolsters will increase the difficulty in "sliding" on the seat during entry and egress.

12. Padding: Cushioning/padding is desirable because it reduces pressure by increasing support area (Konz and Johnson 2004). Seats should be covered with padded material to allow a deflection of about 25 mm and distribute the pressure under the buttocks and thighs. In general, the seat should be designed to allow higher pressure under the ischial tuberosities (i.e., the sitting bones—the lower protruding parts of the pelvic bones) and gradually decrease in outward directions. For long-term comfort, the pressure on the body tissues should not be constant. Changes in the pressures (due to deliberate massaging actions or postural movements) will reduce discomfort and fatigue. The padding also helps in reducing discomfort caused by vehicle body vibrations under dynamic driving conditions.

13. Seat track length: The locations of hip points of different drivers as they adjust the seat fore and aft define the length of the seat track. The foremost and rearmost hip points on the seat track define the seat track length. It should be long enough and placed at a horizontal

distance from the ball of foot on the accelerator pedal of 2.5 percentile to 97.5 percentile hip point locations (defined as $X_{2.5}$ and $X_{97.5}$ in SAE J1517 and J4004). Based on the SAE J4004, a seat track length of about 240 mm would be needed to accommodate 95% of the drivers in passenger cars (see Chapter 3 for more details).

RECENT ADVANCES IN DIGITAL MANIKINS

A number of three-dimensional digital human models are available to aid in the design process. These models can be configured to represent individual men and women and in different percentile dimensions for different populations. Many of the models have built-in human motion, posture simulation, and biomechanical strength as well as percentile exertion prediction capabilities. A review of these models and limitations is presented in Chapter 17.

REFERENCES

Andersson, G. B. J., R. Ortengren, A. Nachemson, and G. Elfstrom. 1974a. Lumbar disc pressure and myoelectric back muscle activity during sitting: I. Studies on an experimental chair. Scand. *J. Rehab. Med.*, 3, 104–114.

Andersson, G. B. J., and R. Ortengren. 1974b. Lumbar disc pressure and myoelectric back muscle activity during sitting: II. Studies on an experimental chair. Scand. *J. Rehab. Med.*, 3, 122–127.

Chaffin, D. B., G. B. J. Andersson, and B. J. Martin. 1999. *Occupational Biomechanics*. New York: John Wiley & Sons Inc.

Garrett, J. W. 1971. The adult human hand: Some anthropometric and biomechanical considerations. *Human Factors*, 13(2), 117–131.

Jurgens, H., I. Aune, and U. Pieper. 1990. *International Data on Anthropometry*. Geneva, Switzerland: ILO.

Kolich, M. 2006. Applying axiomatic design principles to automobile seat comfort evaluation. *Ergonomia, IJE & HF*, 28(2), 125–136.

Kolich, M. 2009. Repeatability, reproducibility, and validity of a new method for characterizing lumbar support in automotive seating. *Human Factors: The Journal of the Human Factors and Ergonomics Society*, 51(2), 193–207.

Konz, S., and S. Johnson. 2004. *Work Design: Industrial Ergonomics*. 6th ed., Scottsdale, AZ: Holcomb Hathaway.

Kroemer, K. H. E., H. B. Kroemer, and K. E. Kroemer-Elbert. 1994. *Ergonomics: How to Design for Ease and Efficiency*. Englewood Cliffs, NJ: Prentice Hall.

McDowell, M. A., C. D. Fryar, C. L. Ogden, and K. M. Flegal. 2008. *Anthropometric Reference Data for Children and Adults: United States 2003–2006*. National Health Statistics Reports, Vol. 10, accessed, June 23, 2011, http://www.cdc.gov/nchs/data/nhsr/nhsr010.pdf

Pheasant, S., and C. M. Haslegrave. 2006. *Bodyspace: Anthropometry, Ergonomics and the Design of Work*. 3rd ed. London: CRC Press, Taylor & Francis Group.

Sanders, M. S. 1983. *U.S. Truck Driver Anthropometric and Truck Work Space Data Survey*. Report CRG/TR-83/002. West Lake Village, CA: Canyon Research Group Inc.

Society of Automotive Engineers Inc. 2009. *SAE Handbook*. Warrendale, PA: Society of Automotive Engineers Inc.

3 Occupant Packaging

WHAT IS VEHICLE PACKAGING?

"Packaging" is a term used in the automobile industry to describe the activities involved in locating various systems (e.g., powertrain system, climate-control system, fuel system) and components (including occupants) in the vehicle space. Thus, it is about space allocation for various vehicle systems (i.e., hardware), accommodating "people" (i.e., the driver and the passengers) and providing storage spaces for various items (e.g., suitcases, boxes, golf bags) that people store in their vehicles.

The term "packaging" was used in the industry because the task of the package engineering is essentially "bringing in systems and components" produced by others (e.g., different suppliers) and fitting them into the vehicle space so that they will function properly to satisfy customers and users of the vehicle.

OCCUPANT PACKAGE OR SEATING PACKAGE LAYOUT

The occupant package includes drawings and three-dimensional graphic representations of the occupant compartment with the position of the occupants (key reference points, e.g., accelerator heel point [AHP], seating reference point [SgRP]), manikins (e.g., from Society of Automotive Engineers Inc. [SAE] standards J826 and J4002; note: all SAE standards are available in the *SAE Handbook* [SAE 2009]), occupant packaging tools (e.g., tools and procedures provided in the SAE standards such as J1517, J4004, J941, J1052, J287, and J1050), primary vehicle controls (steering wheel, pedals, and gear shifter), and some vehicle body and trim components (e.g., seats, instrument panels, center console, door trim panels, mirrors).

The seating package layout is thus a drawing or a three-dimensional model shown in computer-aided design (CAD) applications (e.g., CATIA, IDEAS, and ALIAS) showing locations and positioning of the driver and all other occupants (mostly in the form of manikins), eyellipses (drivers' eye locations specified in SAE standard J941), various reach, clearance and visibility zones (e.g., hand reach (HR) envelopes, head clearance contours, and fields of view), and other relevant vehicle details (e.g., steering wheel, floor, pedals, seats, arm rests, gear shifter, parking brake, mirrors, hard points, fiducial marks/points, eye points, sight lines) and dimensions. Figure 3.1 illustrates a side view of a vehicle package drawing with the above-described occupant packaging details.

It should be noted that most of the occupant packaging tools and practices used in the automotive industry were developed by ergonomics engineers by working through various subcommittees of the SAE Human Factors Committee such as the Human Accommodation and Design Devices Subcommittee. Since the SAE practices are followed in the automotive industry during the vehicle design process, the following part of the chapter will describe key dimensions, reference points, and procedures specified in a number of relevant SAE standards.

DEVELOPING THE OCCUPANT PACKAGE: DESIGN CONSIDERATIONS

In developing a vehicle package, a number of design considerations related to the functioning of various vehicle systems and interfaces between the systems and occupant comfort, convenience, and safety issues are considered. The occupant packaging considerations can be grouped into the following areas:

1. Entry and egress space: Location of the seats, seat shape, clearances required during entry and exit with various vehicle components (i.e., space available for movements of head,

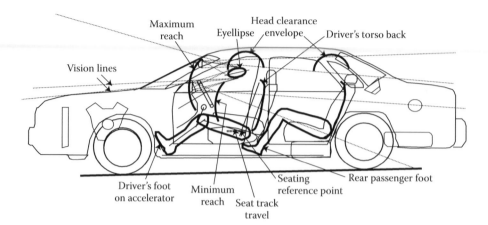

FIGURE 3.1 Illustration of a vehicle package layout.

   torso, knees, thighs, feet, hands, and torso), walk-through in the center (in vans or multi-passenger vehicles), locations of grasp handles, and so forth.
2. Comfortable seated posture: Seat height and leg space, head and shoulder room, torso angle (between torso and upper leg), neck angle (between head and torso), knee angle (between upper leg and lower leg), ankle angle (between foot and lower leg), lengths and widths of seat cushion, seat back, head rests, stresses (forces and pressures) in the spinal column, shape of seat support surfaces in the lumbar region and thigh/buttocks region, and so forth with respect to the steering wheel and pedal locations.
3. Operating controls (hand and foot controls): Locations of controls and displays; head, eye, and ear positions (for acquiring information); body movements and postures (hand, foot, head, and torso) during reaching, grasping, and operating controls; natural versus awkward postures; and use of other in-vehicle items (e.g., cup holder, map pockets, entertainment and information systems).
4. Visibility of interior and exterior areas: Eye locations; movements of eyes, head, neck, and torso during gathering of visual information from the road and inside the vehicle (e.g., visibility of displays); and available fields of view (obstructions caused by vehicle structures and components and in-direct fields from mirrors).
5. Storage spaces: Providing convenient and safe storage spaces to accommodate items brought into the vehicle during trips.
6. Vehicle service: Providing convenient access and space for performing vehicle service and maintenance tasks (e.g., refueling, checking engine oil, replacing bulbs, flat tires).

The challenge of the occupant package engineer is to assure that the largest percentage of the user population is accommodated in performing all tasks involved in the above areas during vehicle usages—while driving and not driving. The following portion of this chapter will cover issues associated with accommodating the driver (primarily area 2 above). Other areas are covered in later chapters.

SEQUENCE IN DEVELOPMENT OF VEHICLE PACKAGE

ADVANCED VEHICLE DESIGN STAGE

In many automotive companies, the advanced design departments are given the responsibility to develop new vehicle concepts. The concept development generally begins with the brainstorming activities of a multidisciplinary team involving market researchers, designers (industrial designers in the styling/design studios), engineers, and researchers who attempt to predict trends in future designs (e.g., fashions, shapes, and features in luxury products), technologies (e.g., materials, electronics, manufacturing processes), economy (e.g., energy costs and availability), markets (e.g., consumer desires and expectations in different markets and countries), government regulations (e.g., safety, fuel economy, and emissions requirements), manufacturing capabilities (availability of manufacturing and assembly plants and equipment), customer feedback (from past vehicle models and competitors), and so forth. The team defines the vehicle type, market segments, and desired product characteristics. The description of the proposed vehicle is written in a document (sometimes called the "product assumptions") and is continuously updated as new information is gathered by the team. The designers usually take the lead in creating sketches of future product concepts. The package engineering members of the team begin creation of the vehicle layout in a CAD system, which is continuously shared with the team members. The design team members discuss many different design ideas, features, engineering issues, trade-offs between different issues, engineering feasibility, costs, and timing issues and arrive at several alternate vehicle concepts.

The vehicle concepts are illustrated by creating sketches, three-dimensional computer models with different level of details from wire-frames to fully rendered vehicles (showing color, texture, reflections, shadows, etc.) that can be shown in realistic images (static or dynamic) on backgrounds of roadways and in the showroom environments, and/or physical bucks or models (e.g., full-size vehicle models with representation of interior and exterior surfaces). The vehicle concepts are generally shown to representative samples of prospective customers in market research clinics at one or more sites at different geographic locations in the selected markets. Competitive products are also included in the market research clinics to understand how the proposed concepts will be perceived by the customers in relation to selected leading competitive products. The customer reactions and responses to the product concepts, and their characteristics are documented.

The product planners also prepare a business plan for the proposed vehicle. A comprehensive business plan generally includes sections involving vehicle description, market and time schedule of producing the vehicle, proposed manufacturing facilities (e.g., where to build—in an existing plant or a new plant, in what state or country), corporate product plans (i.e., how this product fits in the overall corporation plan to produce other products and product lines), supplier capabilities (e.g., who would supply major systems, subsystems, and key components), and estimates of cash flows based on a number of assumptions (e.g., sales volumes, costs of building tools and manufacturing facilities, cost of capital, and competitors and their projected vehicle development and introduction plans).

The above outputs, that is, the product concepts, market research findings, and the business plan of the proposed vehicle, are presented to the higher management of the company to decide if the proposed product concept should be accepted.

DEVELOPMENT OF THE "ACCEPTED" VEHICLE CONCEPT

Figure 3.2 presents a flow diagram showing different tasks involved in occupant packaging and ergonomics evaluations. The process begins with Task 1, which involves defining the vehicle to be designed. As described in the previous section, this task involves inputs from a number of disciplines to prepare assumptions for the vehicle program. It is extremely important to first define the intended customer population, that is, who would buy and use the proposed vehicle. The characteristics, capabilities, desires, and needs of the users must be understood. The market researchers along with the ergonomics engineers and the designers must make every effort to gather information about the intended population. A representative sample of owners and users of the type of vehicle and from the intended market segment (e.g., luxury small four-door car, economy two-door hatchback, midsize luxury SUV) can be invited, and early product concepts can be shown. They can be extensively interviewed and asked to respond to a number of questions related to how well they like or dislike the product concepts and its details/features, their preferences, habits, and so forth. Their relevant

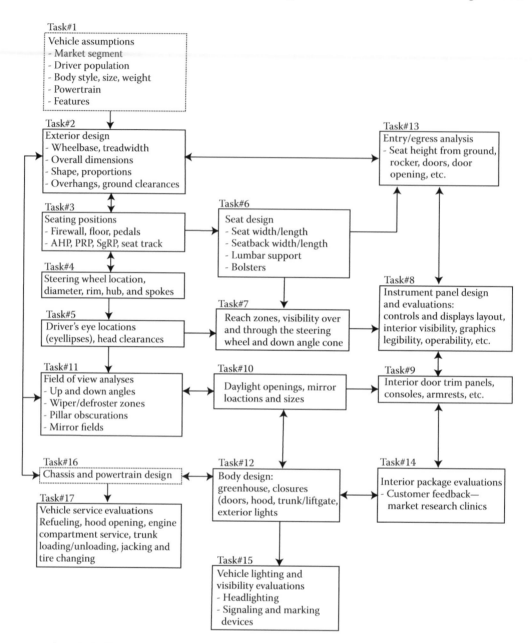

FIGURE 3.2   Flow diagram showing tasks involved in occupant packaging and ergonomic evaluations.

anthropometric dimensions can be also measured to create a database for evaluation of various vehicle dimensions. The quality function deployment (called the "QFD" in the quality management field) is an excellent tool, and it can be used at this early stage to translate the customer needs of the vehicle being designed into functional (engineering) specifications of the vehicle (Besterfield et al. 2003; Bhise et al. 2010).

The exterior design as shown in Task 2 usually leads the design process. The vehicle package engineers work concurrently by positioning the driver and the occupants in the vehicle space (Task 3) and conducting other analyses such as determining locations of primary controls (Tasks 3 and 4), determining driver's eye locations (Task 5), designing seats (Task 6), and determining maximum and minimum reach zones and visible areas (Task 7) to develop instrument panels, door trim panels, and consoles (Tasks 8 and 9). Tasks 10 and 11 are conducted to assure that the driver can obtain the fields of view needed to safely drive the vehicle. The mechanical body design (Task 12) and packaging of chassis and powertrain components (Task 16) are accomplished simultaneously by other engineering analyses departments.

At this early design phase, many analyses are performed to assure that the key vehicle parameters that define the vehicle exterior (e.g., wheelbase, tread width, overall length, width and height, overhangs, cowl point, deck point, and tumblehome) and vehicle interior (e.g., seat height, seat track length and location, and steering wheel and pedal locations) are evaluated simultaneously by involving experts from different disciplines. The key areas that link the exterior of the vehicle to the interior such as entry/egress (Task 13), fields of view, and window openings (Tasks 10 and 11) are resolved in the very early stages as the exterior and interior surfaces of the vehicle are created in the CAD models. The goal, of course, is to assure that the largest percentile values of occupant dimensions can be accommodated and functional aspects of the vehicle are not compromised. Further, the vehicle lighting design (Task 15) and illumination of lighted graphics and components (Task 8) are studied to assure that the vehicle can be used safely during nighttime.

A number of special evaluations are also conducted to assure that the drivers and the passengers can enter the vehicle and exit from the vehicle comfortably (Task 13), and extensive customer feedbacks on the interior package parameters and vehicle features are obtained (Task 14) by conducting market research clinics. Various evaluation methods used in the entire vehicle development process are summarized in Chapter 15. The next sections of this chapter will cover details related to dimensions and positioning procedures related to Tasks 3, 4, 5, and 7. The issues related to seat dimensions and designs are covered in Chapter 2. The issues related to controls and displays in Tasks 8 and 9 will be covered in Chapter 5. Chapters 6 and 7 will cover considerations related to Tasks 10, 11, and 15. The entry/exit issues will be covered in Chapter 8, and other exterior issues in Tasks 12 and 17 will be covered in Chapter 9.

DEFINITION OF KEY VEHICLE DIMENSIONS AND REFERENCE POINTS

UNITS, DIMENSIONS, AND AXES

All vehicle and occupant dimensions are measured in millimeters. The prefixes L, H, and W denote dimensions related to length (horizontal), height (vertical), and width (lateral), respectively. All angles are designated by the prefix A and are measured in degrees. (see SAE standard J1100 in the *SAE Handbook* [SAE 2009] for more details on the nomenclature and dimensions).

The three-dimensional Cartesian coordinate system used to define locations of points in the vehicle space is generally defined as follows: (a) the positive direction of the longitudinal X-axis is pointing from the front to the rear of the vehicle, (b) the positive direction of the vertical Z-axis is pointing from the ground up, (c) the positive direction of the lateral Y-axis is pointing from the left side of the vehicle to the right side, and (d) the origin of the coordinate system is located forward of the front bumper (to make all X-coordinate values positive), below the ground level (to make all Z-coordinate values positive), and at the midpoint between the vehicle width. (refer to SAE standard J182 [SAE 2009]). Figure 3.18 shows the XYZ coordinate system with its origin called the "body zero").

PACKAGE DIMENSIONS, REFERENCE POINTS, AND SEAT-TRACK-RELATED DIMENSIONS

Figure 3.3 presents a side view drawing showing important interior reference points and dimensions.

The reference points used for location of the driver and their relevant dimensions are described below.

1. The accelerator heel point (AHP) is the heel point of the driver's shoe that is on the depressed floor covering (carpet) on the vehicle floor when the driver's foot is in contact

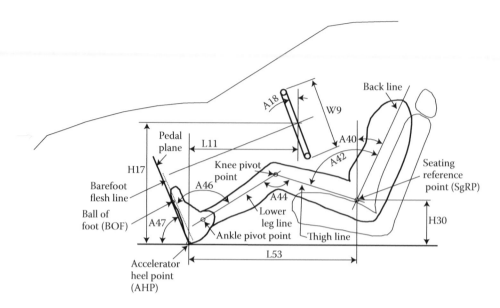

FIGURE 3.3   Interior package reference points and dimensions.

with the undepressed accelerator (gas) pedal (see Figure 3.3). SAE standard J1100 defines it as "a point on the shoe located at the intersection of the heel of shoe and the depressed floor covering, when the shoe tool (specified in SAE J826 or J4002) is properly positioned (essentially, with the ball of foot (BOF) contacting the lateral centerline of the undepressed accelerator pedal, while the bottom of shoe is maintained on the pedal plane)."

2. The pedal plane angle (A47) is defined as the angle of the accelerator pedal plane in the side view measured in degrees from the horizontal (see Figure 3.3). The pedal plane is not the plane of the accelerator pedal, but it is the plane representing the bottom of the mani- kin's shoe defined in SAE J826 or J4002. (As described later in this chapter, A47 can be computed by using equations provided in SAE J1516 or J4004. Or, it can be measured by using the manikin tools described in SAE J 826 or J4002.)

3. BOF on the accelerator pedal is the point on the top portion of the driver's foot that is normally in contact with the accelerator pedal. The BOF is located 200 mm from the AHP measured along the pedal plane (SAE J4004, SAE 2009).

4. The pedal reference point (PRP) is on the accelerator pedal lateral centerline where the BOF contacts the pedal when the shoe is properly positioned (i.e., heel of shoe at AHP and bottom of shoe on the pedal plane). SAE standard J4004 provides a procedure for locating PRP for curved and flat accelerator pedals using SAE J4002 shoe tool. If the pedal plane is based on SAE standards J826 and J1516, the BOF point should be taken as the PRP.

5. The seating reference point (SgRP) is the location of a special hip point (H-point) desig- nated by the vehicle manufacturer as a key reference point to define the seating location for each designated seating position. Thus, there is a unique SgRP for each designated seating position (e.g., the driver's seating position, front passenger's seating position, left rear pas- senger's seating position). An H-point simulates the hip joint (in the side view as a hinge point) between the torso and the thighs, and thus, it provides a reference for locating a seat- ing position. In the plan view, the H-point is located on the centerline of the occupant.

The SgRP for the driver's position is specified as follows:

a. It is designated by the vehicle manufacturer.
b. It is located near or at the rearmost point of the seat track travel.

c. The SAE (in standards J1517 or J4004) recommends that the SgRP should be placed at the 95th percentile location of the H-point distribution obtained by a seat position model (called the SgRP curve, see Figure 3.5) at an H-point height (H30 from the AHP specified by the vehicle manufacturer).

The original H-point location model was developed by Philippart et al. (1984) based on measurements of preferred sitting locations of a large number of drivers in actual vehicles with different package parameters. The sitting position of each driver was defined as the location of the driver's H-point. The H-point location was determined by the horizontal seat track position selected by the driver at the seat height (measured by H30, see Figure 3.3) in the vehicle. For any given vehicle, the H-point locations of a population of drivers can be represented by their distribution of horizontal locations. Figure 3.4 shows the distribution of the horizontal location (X) of the H-points. The 95th percentile value of H-point location distribution is generally selected as the location of the SgRP as shown in Figure 3.5. The SgRP is defined as the point located at X_{95} horizontal distance from the BOF point and H30 vertical distance from the AHP. The trajectory of X_{95} locations as a function H30 is called the SgRP curve (see Figure 3.5). The equation of the SgRP curve (provided in SAE standards J1516 and J4004) is provided in a later section of this chapter.

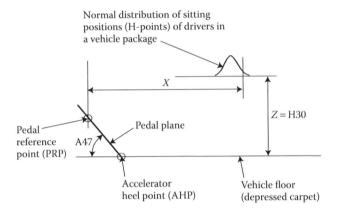

FIGURE 3.4   Distribution of horizontal location on H-points.

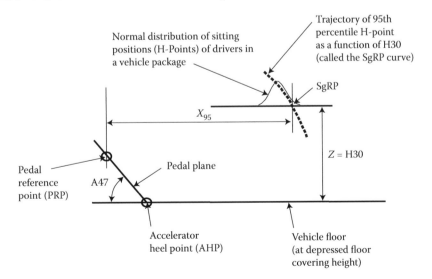

FIGURE 3.5   95th percentile H-point location curve for class A vehicles.

The driver's SgRP is the most important and basic reference point in defining the driver package. The driver's SgRP must be established early in the vehicle program and should not be changed later in the vehicle development process because of the following:

a. It determines the driver locations in the vehicle package.
b. All driver-related design and evaluation analyses are conducted with respect to this point, for example, location of eyes, interior and exterior visibility, specifications of spaces (e.g., headroom, legroom, and shoulder room), reach zones, locations of controls and displays, and door openings (for entry/exit).
c. The SgRP can be located in a physical property (i.e., an actual vehicle or a package buck) by placing the SAE H-point machine (HPM) specified in SAE standard J826 or H-point device (HPD) specified in SAE standard J4002. The HPM and HPD are three-dimensional fixtures, and they can be placed in a seat at any designated seating location to measure or verify the location of the SgRP at the seating location. The HPM is referred in the auto industry and by some seat manufacturers as "OSCAR." Since the seat is compressible and flexible, the HPM is placed on the seat and used as a development and verification tool by seat manufacturers and vehicle manufacturers to determine if the SgRP of the seat that is built and installed in an actual vehicle falls within the manufacturing tolerances from the design SgRP location. The description and procedure for location of the HPM are provided in SAE standard J826. Figure 3.6 provides a sketch of the updated HPM called the HPD. The HPD is designed with a three-segmental back pan to account for the effect of shape of the seat backs (especially in the lumbar region). SAE standard J4002 provides drawings, detailed specifications, and procedures for the use of the HPD.

The SAE HPM and the HPD (HPM in SAE standard J826; HPD in SAE standard J4002) are designed such that when they are placed on a seat, they deflect the seat somewhat like the way a real person will deflect the seat. Each device weighs 76 kg (167 lb, which is 50th percentile U.S. male weight) and has the torso contour of 50th percentile U.S. male. The devices use 95th percentile legs (10th and 50th percentile leg lengths are also available).

6. The seat track length is defined as the horizontal distance between the foremost and rearmost location of the H-point of the seated drivers. To accommodate 95% of the driver population with 50:50 male-to-female ratio, the foremost point and the rearmost points can be defined by determining 2.5 and 97.5 percentile H-point locations from the BOF. The computation procedures for determining different percentile values are specified in SAE standards J1517 and J4004. SAE standard J1517 has now been replaced by

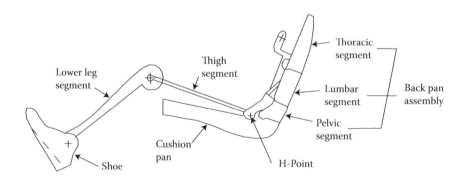

FIGURE 3.6 SAE H-point device. (Reproduced from the Society of Automotive Engineers Inc., *SAE Handbook*, Society of Automotive Engineers Inc., Warrendale, PA. 2009. With permission.)

SAE standard J4004 standard, and the SAE recommends that J4004 should be used to determine the seat track length and the accommodation levels for the U.S. driving population. It should be noted that since the introduction of SAE standards J4002, J4003, and J4004, the package engineering community within various automotive companies is slowly transitioning from the old (J826, J1516, and J1517) procedures to the revised (J4002, J4003, and J4004) procedures. Therefore, relevant information from both the procedures is provided below.

Figure 3.7 shows the original seat position location model developed by Philippart et al. (1984) and included in SAE standard J1517. SAE standard J1517 was developed by measuring actual seated positions of a large number of drivers in vehicles with different H30 values (after they had driven the vehicles and adjusted the seat location at their preferred position; Philippart et al., 1984). The H-point location model, thus, is based on functional anthropometric data (i.e., real drivers seated in actual vehicles at their preferred driving posture). SAE standard J1517 entitled "Driver Selected Seat Position" provides statistical prediction equations for seven percentile values ranging from 2.5 to 97.5 of H-point locations in the vehicle space. The 2.5- and 97.5-percentile H-point location prediction equations are generally used to establish seat track travel to accommodate 95% of drivers. The equations are quadratic functions of H30 for class A vehicles (passenger cars and light trucks) and linear functions of H30 for class B vehicles (medium and heavy commercial trucks). The class A vehicle equations are based on 50:50 male-to-female ratio. Figure 3.7 presents seven percentile curves of H-point locations obtained from equations presented in SAE J1517 for 50:50 male-to-female ratio for class A vehicles. The 95th percentile curve shown in Figure 3.7 is called the SgRP curve. The class B vehicle driver selected seat position lines are specified in SAE J1517 for 50:50, 75:25, and 90:10 to 95:5 male-to-female ratios.

SAE standard J4004 presents an H-point location procedure based on the more recent work by Flannagen, Schneider, and Manary (1996, 1998) for class A vehicles. The recommended seat track lengths to accommodate different percentage of drivers are presented in Figure 3.8. The horizontal (X) locations of the front and rear locations of the seat track are specified with respect to a reference location called X_{ref}. This reference distance is measured aft of the PRP. The X_{ref} is a linear function

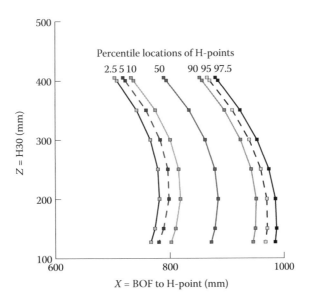

FIGURE 3.7 H-point location curves for 2.5 to 97.5 percentile H-points as functions of H30 for class A vehicles. (Drawn from equations provided in SAE standard J1517 in *SAE Handbook*, 2009.)

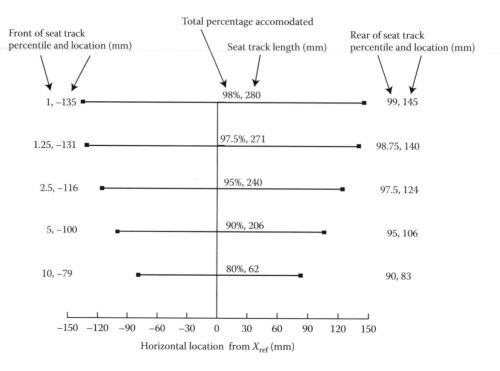

FIGURE 3.8 Recommended seat track lengths. (Drawn from data provided in SAE standard J4004 in *SAE Handbook*, 2009.)

of H30, steering wheel location (L6), and type of transmission (with or without a clutch pedal). SAE standard J4004 suggests that until the year 2017, the BOF and AHP determined according to SAE standard J1517 may be used in lieu of the PRP cited in J4004 document. However, SAE standard J4004 should be used to determine the seat track length and the accommodation levels for the U.S. driving population.

The equations illustrating the above-described two procedures are provided in a later section of this chapter.

INTERIOR DIMENSIONS

A number of interior package dimensions shown in Figure 3.3 are described in this section. The dimensions are defined using the nomenclature specified in SAE standard J1100.

1. AHP to SgRP location: The horizontal and the vertical distances between the AHP and the SgRP are defined as L53 and H30, respectively (see Figure 3.3).
2. Posture angles: The driver's posture is defined by the angles of the HPM or the HPD. The angles shown in Figure 3.3 are defined as follows:
   a. Torso angle (A40). It is the angle between the torso line (also called the backline) and the vertical. It is also called the seat back angle or back angle.
   b. Hip angle (A42). It is the angle between the thigh line and the torso line.
   c. Knee angle (A44). It is the angle between the thigh line and the lower leg line. It is measured on the right leg (on the accelerator pedal).
   d. Ankle angle (A46). It is the angle between the (lower) leg line and the bare-foot flesh line, measured on the right leg.
   e. Pedal plane angle (A47). It is the angle between the accelerator pedal plane and the horizontal.

3. Steering wheel: The center of the steering is specified by locating its center by dimensions L11 and H17 in the side view. The steering wheel center is located on the top plane of the steering wheel rim (see Figure 3.3). The lateral distance between the center of the steering wheel and the vehicle centerline is defined as W7. The diameter of the steering wheel is defined as W9. The angle of the steering wheel plane with respect to the vertical is defined as A18 (see Figure 3.3).

4. Entrance height (H11): It is the vertical distance from the driver's SgRP to the upper trimmed body opening (see Figure 3.9). The trimmed body opening is defined as the vehicle body opening with all plastic trim (covering) components installed. This dimension is used to evaluate head clearance as the driver enters the vehicle and slides over the seat during entry and egress.

5. Belt height (H25): It is the vertical distance between the driver's SgRP and the bottom of the side window daylight opening at the SgRP X-plane (plane perpendicular to the longitudinal X-axis and passing through the SgRP; see Figure 3.10). The belt height is important to determine the driver's visibility to the sides. It is especially important in tall vehicles such as heavy trucks and buses to evaluate if the driver can see vehicles in the adjacent lanes, especially on the right-hand side. The belt height is also an important exterior styling characteristic (e.g., some luxury sedans have high belt height from the ground as compared with their overall vehicle height).

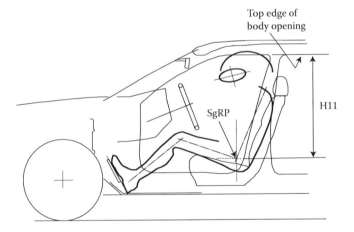

FIGURE 3.9 Entrance height (H11).

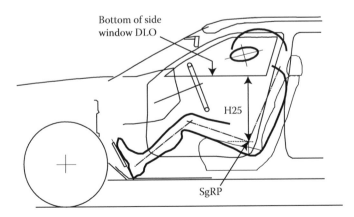

FIGURE 3.10 Belt height (H25).

6. Effective headroom (H61): It is the distance along a line 8 degrees rear of the vertical from the SgRP to the headlining, plus 102 mm (to account for SgRP to bottom of buttocks distance; see Figure 3.11). It is one of the commonly reported interior dimensions and is usually included in vehicle brochures and websites.

7. Leg room (L33): It is the maximum distance along a line from the ankle pivot center to the farthest H-point in the travel path, plus 254 mm (to account for the ankle point to accelerator pedal distance), measured with the right foot on the undepressed accelerator pedal (see Figure 3.12). It is also one of the commonly reported interior dimensions and is usually included in vehicle brochures and websites.

8. Shoulder room (W3; minimum cross-car width at beltline zone): It is the minimum cross-car distance between the trimmed doors within the measurement zone. The measurement zone lies between the beltline and 254 mm above SgRP, in the X-plane through SgRP (see Figure 3.13. It shows a cross-sectional front view of the vehicle.) It is also one of the commonly reported interior dimensions and is usually included in vehicle brochures and websites.

9. Elbow room (W31; cross-car width at armrest): It is the cross-car distance between the trimmed doors, measured in the X-plane through the SgRP, at a height of 30 mm above the highest point on the flat surface of the armrest. If no armrest is provided, it is measured at 180 mm above the SgRP (see Figure 3.14).

10. Hip room (W5; minimum cross-car width at SgRP zone): It is the minimum cross-car distance between the trimmed doors within the measurement zone. The measurement zone extends 25 mm below and 76 mm above SgRP, and 76 mm fore and aft of the SgRP (see Figure 3.15).

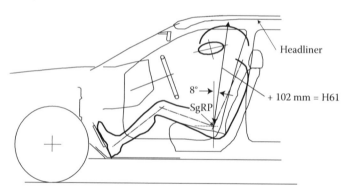

FIGURE 3.11   Effective head room (H61).

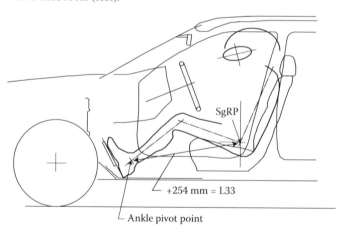

FIGURE 3.12   Leg room (L33).

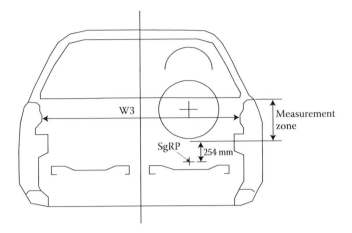

FIGURE 3.13 Shoulder room (W3).

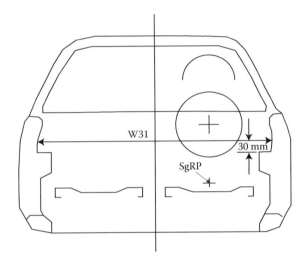

FIGURE 3.14 Elbow room (W31).

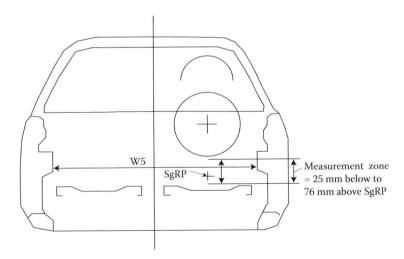

FIGURE 3.15 Hip room (W5).

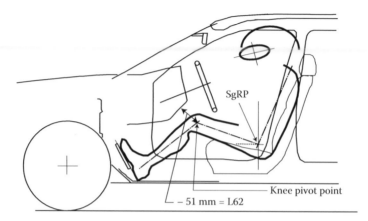

FIGURE 3.16   Knee clearance (L62).

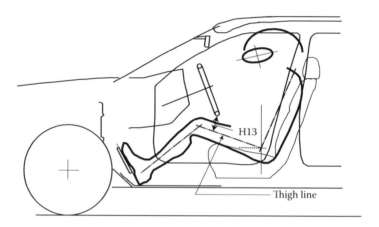

FIGURE 3.17   Thigh room (H13).

11. Knee clearance (L62; minimum knee clearance—front): It is the minimum distance between the right leg K-point (knee pivot point) and the nearest interference, minus 51 mm (to account for the knee point to front of the knee distance) measured in the side view, on the same Y-plane as the K-point, with the heel of shoe at FRP (floor reference point; see Figure 3.16).

12. Thigh Room (H13; steering wheel to thigh line): It is the minimum distance from the bottom of the steering wheel rim to the thigh line (see Figure 3.17).

DRIVER PACKAGE DEVELOPMENT PROCEDURES

In this section, we cover basic steps involved in positioning the driver, determining the seat track length, positioning eyellipse and head clearance envelopes, determining maximum and minimum reach envelopes, and positioning the steering wheel.

1. Determine H30 = height of the SgRP from the AHP.

   The H30 value is usually selected by the package engineer based on the type of vehicle to be designed. The H30 dimension is one of the dimensions used in the SAE standards to define class A vehicles (passenger cars and light trucks) and class B vehicles (medium and heavy trucks). The values of H30 for class A vehicles range between 127 and 405 mm. It should be noted that smaller values of H30 will allow lower roof height (measured from the vehicle floor) and will require longer horizontal space (dimension L53 and X_{95}) to accommodate the driver—like in a sports car. Conversely, if a large value of H30 is selected, the taller cab height and shorter horizontal space (dimension L53 and X_{95}) will be required to accommodate the driver. The class B vehicles (medium and heavy trucks) will have large values of H30 (typically 350 mm and above) so that less horizontal cab space is used to accommodate the driver, and thus, longer longitudinal space is available for the cargo area.

The BOF-to-SgRP dimension is usually determined by computing the X_{95} value (i.e., 95% of the drivers will have their H-point forward of the SgRP; measured in mm) from the following equation given in SAE J1517. (This equation is called the SgRP curve in SAE J4004.)

$$X_{95} = 913.7 + 0.672316z - 0.00195530z^2$$

where z = H30 in millimeters.

2. Determine pedal plane angle (A47).

The value of the pedal plane angle in degrees is obtained by using the following equation from SAE standard J1516.

$$A47 = 78.96 - 0.15z - 0.0173z^2$$

where z = H30 in centimeters (note: this z value is in centimeters—for the above equation only). In SAE standard J4004, the pedal plane angle is defined as alpha (α), where

$\alpha = 77 - 0.08$ (H30) (degrees from horizontal) (note: H30 is specified in millimeters).

3. The vertical height (H) between the BOF and AHP can be computed as follows:

$$H = 203 \times \sin(A47)$$

It should be noted that distance between AHP to BOF is specified as 203 mm in SAE standard J1517 and 200 mm in SAE standard J4004.

4. The horizontal length (L) between the BOF and AHP can be computed as follows:

$$L = 203 \times \cos(A47)$$

5. The horizontal distance between the AHP and SgRP (L53) can be computed as follows:

$$L53 = X_{95} - L$$

6. The seat track length is defined by the total horizontal distance of the fore and aft movement of the H-point (for a seat that does not have vertical movement of the H-point). The foremost H-point and rearmost H-point on the seat track are defined by the vehicle manufacturer. To accommodate 95% of the drivers (with 50% males and 50% females), the foremost point is defined as at $X_{2.5}$ horizontal distance from the rearward of the BOF and the rearmost point is defined as at $X_{97.5}$ horizontal distance from the rearward of the BOF. SAE standard J1517 defines $X_{2.5}$ and $X_{97.5}$ distances as follows:

$$X_{2.5} = 687.1 + 0.895336z - 0.00210494z^2$$
$$X_{97.5} = 936.6 + 0.613879z - 0.00186247z^2$$

where z = H30 in millimeters.

$$TL23 = X_{95} - X_{2.5}$$
$$= \text{horizontal distance between the SgRP and the foremost H-point}$$

$$TL2 = X_{97.5} - X_{95}$$

= horizontal distance between the SgRP and the rearmost H-point

Total seat track length to accommodate 95% of the drivers = TL1

where TL1 = TL23 + TL2 = $X_{97.5} - X_{2.5}$.

If SAE standard J4004 is used to locate the seat track, then the X distance of the H-point reference point aft the PRP is computed as follows:

$$X_{ref} = 718 - 0.24(\text{H}30) + 0.41\,(\text{L}6) - 18.2t$$

where L6 is the horizontal distance from the PRP to the steering wheel center (see Figure 3.18) and t is the transmission type ($t = 1$ if clutch pedal is present and $t = 0$ if no clutch pedal is present).

The foremost and rearmost points on the seat track are obtained from data presented in Figure 3.8. It should be noted that the X-axis of Figure 3.8 presents distances of the foremost and rearmost points with respect to X_{ref}. From Figure 3.8, for 95% accommodation, the TL1 would be 240 mm.

7. The seat back angle (or what is also called the torso angle) is defined by dimension A40 (measured in degrees with respect to the vertical). With the reclinable seat back feature, a driver can adjust the angle to his or her preferred seat back angle. The seat back angle in the 1960s and 1970s was defined as 24 or 25 degrees by many manufacturers (due to bench seats that were not reclinable). However, with the reclinable seat back features, most drivers prefer to sit more upright with angles of about 18–22 degrees in most passenger cars and about 15–18 degrees for pickups and SUVs. The seat back angles selected by class B (medium and large commercial trucks) drivers are generally more upright—about 10–15 degrees.

8. The driver's eyes are located in the vehicle space by positioning eyellipses in the CAD model (or a drawing) of the vehicle package. The "eyellipse" is a concocted word created by the SAE by joining the two words "eye" and "ellipse" (using only one "e" in the middle for the joint word). The eyellipse is a statistical representation of the locations of drivers' eyes used in visibility analyses.

   SAE standard J941 defines these eyellipses, which are actually two ellipsoidal surfaces (one for each eye) in three dimensions (they look like two footballs fused together at average interocular distance of 65 mm; see Figure 3.18 in plan view and rear view). The eyellipses are defined based on the tangent cutoff principle, that is, any tangent drawn to the ellipse in two dimensions (or a tangent plane to an ellipsoid in three dimensions) divides the population of eyes above and below the tangent in proportions defined by the percentile value of the eyellipse. Sight lines are constructed as tangents to the ellipsoids.

   SAE standard 941 has defined four eyellipsoids by combinations of two percentile values (95th and 99th) and two seat track lengths (shorter than 133 mm and greater than 133 mm). The eyellipsoids are defined by the lengths of their three axes (X, Y, and Z directions; shown in Figure 3.18 as EX, EY, and EZ). The values of EX, EY, and EZ for the 95th percentile eyellipse with TL23 > 133 mm are 206.4, 60.3, and 93.4 mm, respectively. (The values of EX, EY, and EZ for other combinations for percentile and seat track travel are available in SAE standard J941.) The eyellipses are located by specifying X, Y, and Z coordinates of their centroids. The ellipsoids are also tilted downward in the forward direction by $\beta = 12$ degrees (i.e., the horizontal axes of the ellipsoids are rotated counterclockwise by 12 degrees; see Figure 3.18).

   The coordinates of the left and right eyellipse centroids [(X_c, Y_{cl}, Z_c) and (X_c, Y_{cr}, Z_c), respectively] with respect to the body zero are defined in SAE standard J941 as follows (see Figure 3.18):

$$X_c = \text{L}1 + 664 + 0.587\,(\text{L}6) - 0.178(\text{H}30) - 12.5t$$

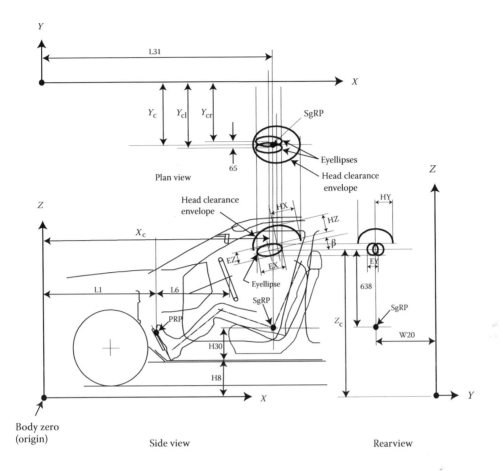

FIGURE 3.18 Location of eyellipses and head clearance envelope.

$$Y_{cl} = W20 - 32.5$$
$$Y_{cr} = W20 + 32.5$$
$$Z_c = 638 + H30 + H8$$

where (L1, W1, H1) = coordinates of the PRP, L6 = horizontal distance between the BOF (or PRP) and the steering wheel center, and $t = 0$ for vehicle equipped with automatic transmission and $t = 1$ for vehicle with clutch pedal (manual transmission). [Note: The SgRP coordinates with respect to the body zero are (L31, W20, H8 + H30). See Figure 3.18. L1 = L31 − X_{95}.]

9. The eyes of tall and short driver on the 95th percentile eyellipse are located at 46.7 mm (half of EZ = 93.4 mm) above and below the eyellipse centroid. By taking into account that the eyellipses are tilted 12 degrees forward, the height of the 46.7 mm can be adjusted to 46.7/cos 12 or 47.74 mm.

10. The head clearance envelopes are defined in SAE standard J1052 (see Figure 3.18). They were developed to provide clearance for the driver's hair on the top, front, and side of the head. They are defined as ellipsoidal surfaces (above the centroid only) in three dimensions with specified dimensions of three axes from the centroid. The dimensions are shown in Figure 3.18 as HX, HY, and HZ. The values of HX, HY, and HZ for the 99th percentile head clearance ellipsoid are 246.04, 166.79, and 151 mm, respectively, for seat track lengths more than 133 mm.

The head clearance envelopes are also defined as tangent cutoff ellipsoids, and clearances from vehicle surfaces such as the roof, header, or roof rails can be measured by determining amount of movements (in the three directions defined by the vehicle coordinate system) of the head clearance envelope needed to touch different interior surfaces. The centriod of the head clearance contour is (x_h, y_h, z_h) distance from the cyclopean centroid (midpoint of the left and right centroids) of the eyellipse. For seat track travel (TL23) greater than 133 mm, the values of (x_h, y_h, z_h) coordinates in millimeters are (90.6, 0.0, 52.6).

SAE standard J1052 provides four head clearance ellipsoids for the combinations of two percentile values (95th and 99th) and two seat track lengths (below 133 mm and above 133 mm). In addition, to accommodate horizontal head shift of occupants seated in the outboard (toward the side glass) locations, the standard requires an additional lateral shift of 23 mm of the ellipsoid on the outboard side. The ellipsoids are also tilted downward in the counterclockwise direction by 12 degrees.

11. The maximum hand reach data are provided in SAE standard J287. The reach distances are based on the controls reach studies conducted by the SAE (Hammond and Roe, 1972; Hammond, Mauer, and Razgunas, 1975). In these studies, each subject was asked to sit in an automotive buck at his preferred seat track position with respect to the steering wheel and the pedals. The subject was then asked to grasp each knob (like the old push–pull head lamp switch knob) with three fingers and slide the knob (mounted at the end of the horizontally sliding bar) as far forward as he could comfortably reach at each of the vertical and lateral bar locations (see Figure 3.19). The experimenters were looking for the maximum rather than the preferred reach distances. SAE standard J287 provides tables that present horizontal distances forward from an HR reference plane at combinations of different lateral and vertical locations.

The HR plane is a vertical plane, and it is located perpendicular to the longitudinal axis of the vehicle. The location of the HR plane from the AHP is established by computing the value of $[786 - 99G]$, where G = general package factor. If the above-computed value of HR is greater than L53, then the HR plane is located at the SgRP. The G value is computed by using the following formula in SAE standard J287 FEB2007:

$$G = 0.00327 \ (H30) + 0.00285 \ (H17) - 3.21$$

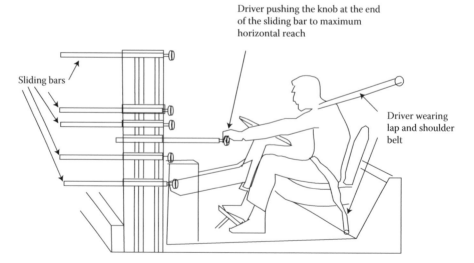

Driver pushing the knob at the end of the sliding bar to maximum horizontal reach

Sliding bars

Driver wearing lap and shoulder belt

FIGURE 3.19 Maximum hand reach study buck. (The buck shown in the above picture was configured to represent a heavy truck package.)

where H17 = height of the center of the steering wheel (on the plane placed on the driver's side the steering wheel rim) from the AHP (see Figure 3.20).

The values of G vary from −1.3 (for a sports car package) to +1.3 (for a heavy truck package).

The reach tables are provided for combinations of the three variables: (a) type of restraints used by the driver (unrestrained = lap belt only; restrained = lap and shoulder belt), (b) G value, and (c) male-to-female population mix. Figure 3.20 presents a side and plan view showing the reach contours.

The reach contours actually generate two complex surfaces, one for each hand, in the three dimensions. Figure 3.21 presents cross sections of the reach surfaces at different lateral locations for the left hand (top figure) and right hand (bottom figure) from a reach table (table 4 from SAE standard J287).

To account for differences in reach distances obtained by an extended finger (e.g., reaching to a push button with extended single finger) or full grasp (all fingers grasping a control—like a gear knob on a floor shift), 50 mm is added or subtracted, respectively, from the value obtained from the tables provided in SAE standard J287.

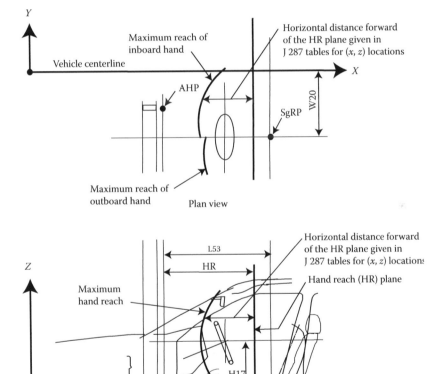

FIGURE 3.20 Plan and side views showing the HR plane and horizontal distances forward of the HR plane. (Provided in tables of SAE standard J287.)

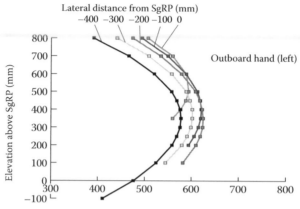

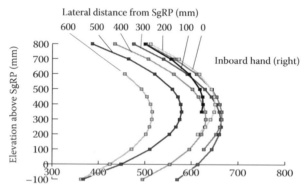

FIGURE 3.21 Maximum horizontal reach. (Plotted from data in table 4 of SAE standard J287, February 2007.)

12. The minimum hand reach is the shortest distance (i.e., closest to the driver) that a short driver seated at the foremost point on the seat track (i.e., her H-point located at the forward-most point of the seat track) will be comfortable in reaching for a control. A side view of the hemispherical minimum reach envelopes is shown in Figure 3.22. The drawing procedure for the minimum comfortable reach envelopes is covered in Chapter 5.

13. The steering wheel location is constrained by the maximum and minimum reach enve-lopes, visibility of the roadway, and thigh clearance (see Figure 3.22). The steering wheel should be placed rearward of the maximum reach (SAE standard J287) and forward of the minimum reach envelopes. The sight line (or the visibility) over the top of the steer-ing wheel rim from the short driver's (fifth percentile) eye point should allow the driver to view the road surface. The ground intercept distance of about 6–21 m (20–70 ft) in front of the front bumper is generally considered acceptable. The thigh clearance between the bottom of the steering wheel and the top of seat should allow accommodation of at least a 95th percentile thigh thickness during entry and egress.

In addition to meeting the above requirements illustrated in Figure 3.22, the nominal location of the steering wheel is also determined by benchmarking steering wheel loca-tions of other vehicles (e.g., superimposing steering wheel locations of other vehicles using common SgRP and/or BOF) and by using subjective assessment techniques in vehicle bucks (see Chapter 12). Further, use of a tilt and telescopic steering column would allow most drivers to adjust the steering wheels to their preferred positions.

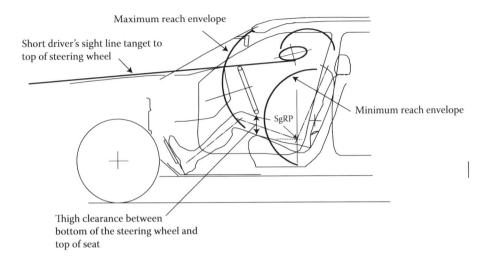

Maximum reach envelope

Short driver's sight line tanget to top of steering wheel

Minimum reach envelope

SgRP

Thigh clearance between bottom of the steering wheel and top of seat

FIGURE 3.22 Considerations related to location of the steering wheel.

14. SgRP couple distance (L50) is the longitudinal distance between the SgRPs of adjacent rows.

$$L50\text{-}1 = \text{SgRP couple distance between the front to second rows}$$

$$L50\text{-}2 = \text{SgRP couple distance between the second to third rows}$$

OTHER ISSUES AND DIMENSIONS

Other package- and ergonomics-design-related issues such as entry/exit, field of view, opening of the hood and servicing of the engine, opening of the trunk (or liftgate), and loading and unloading of items, and so forth are covered in subsequent chapters. An Excel-based spreadsheet program is provided at the publisher's website for the readers to better understand the various inputs and calculate the resulting package dimensions. The program can be also used to set up different driver packages, analyze an existing package, or conduct sensitivity analyses by changing combinations of different input parameters and studying the resulting driver packages.

REFERENCES

Besterfield, D. H., C. Besterfield-Michna, G. H. Besterfield, and M. Besterfield-Scare. 2003. *Total Quality Management*. Upper Saddle River, NJ: Prentice Hall.

Bhise, V. D., H. Dandekar, A. Gupta, and U. Sharma. 2010. *Development of a Driver Interface Concept for Efficient Electric Vehicle Usage*. SAE Paper 2010-01-1040. Presented at the 2010 SAE World Congress, Detroit, MI.

Flannagen, C. C., L. W. Schneider, and M. A. Manary. 1996. *Development of a Seating Accommodation Model*. SAE Technical Paper 960479. Warrendale, PA: Society of Automotive Engineers Inc.

Flannagen, C. C., L. W. Schneider, and M. A. Manary. 1998. *An Improved Seating Accommodation Model with Application to Different User Populations*. SAE Technical Paper 980651. Warrendale, PA: Society of Automotive Engineers Inc.

Hammond, D.C., D. E. Mauer, and L. Razgunas. 1975. *Controls Reach: The Hand Reach of Drivers*. SAE Paper 750357. Warrendale, PA: Society of Automotive Engineers Inc.

Hammond, D. C., and R. W. Roe. 1972. *SAE Controls Reach Study*. SAE Paper 720199. Warrendale, PA: Society of Automotive Engineers Inc.

Philippart, N. L., R. W. Roe, A. J. Arnold, and T. J. Kuechenmeister. 1984. *Driver Selected Seat Position Model*. SAE Technical Paper 840508. Warrendale, PA: Society of Automotive Engineers Inc.

Society of Automotive Engineers Inc. 2009. *SAE Handbook*. Warrendale, PA: Society of Automotive Engineers Inc.

4 Driver Information Acquisition and Processing

INTRODUCTION

Driving a vehicle is an information-processing activity. During driving, the driver continuously acquires information from various senses (vision, hearing, tactile, vestibular, kinesthetic, and olfactory), processes the acquired information, makes decisions and takes appropriate control actions to maintain vehicle motion on the roadway, and navigates to an intended destination. Vision is essential for driving. It is estimated that during driving, a driver receives over 90% of the inputs from his or her eyes. Therefore, in this chapter, we will begin with understanding the structure of the human eye and the capabilities of the human visual system (visual capabilities are also called human visual functions). The acquired visual information is sent to the brain, and the brain processes the information along with the information stored in the memory to make numerous decisions.

It is important to understand that most driver failures occur due to failure in obtaining the necessary information in the right amounts at the right time and the right place. When a driver is asked to describe how he or she got involved in an accident, the most common type of responses are the following: "I did not see the target (a pedestrian, a car, a curve, a sign, etc.)" or "I did not realize that the other vehicle was approaching so fast, or I misunderstood the situation." Thus, the vehicle designer should constantly think about designing the vehicle to reduce the chances of driver information-processing failures and errors.

On many occasions, the driver may find that he or she has too many tasks to do within a very short time interval due to traffic, roadway situations, state of his or her vehicle, and/or other non-driving tasks or distractions (e.g., answering a cell phone). Understanding the various demands placed on the driver and how the driver should prioritize and time share between different tasks are also areas of great importance to the vehicle designers.

IMPORTANCE OF TIME

Understanding the amount of time that the driver needs to perform different tasks is probably the most important concept in designing the driver–vehicle interface. Most drivers take about 0.5–1.2 s to read speed from an analog speedometer with a moving pointer on a fixed scale (Rockwell et al., 1973). To view the objects in a driver's side-view mirror, most drivers will make about 0.8- to 2.0 s glances. In operating more complex devices such as radios and climate controls, the drivers typically make two to four glances; and each glance takes about 1.0 s in performing tasks such as selecting a radio station, changing temperature, or changing fan speed (Bhise, 2002; Jackson et al., 2002).

A vehicle traveling on the highway at 100 km/h (62 mph) is equivalent to traveling 28 m (90 ft) per second. Thus, when a driver takes time away from the forward scene to make a 1-s glance, the vehicle travels 28 m on the roadway. If the driver takes a glance for more than 2.5 s time away from the roadway to perform other tasks, the driver will have difficulty in maintaining his or her vehicle within the lane. And if the driver takes more than 4.0 s away from the road, he or she is almost guaranteed to drift outside the driving lane (Senders et al., 1966).

Thus, it is important to design equipment inside the vehicle that drivers can use with glances no longer than about 1.5 s, and the total number of glances away from the road should be as few as possible. The time that the driver takes to perform a task depends on the complexity of the task (e.g., number of items to search and read from a display, number of sequential actions or steps to perform, number of decisions to make, number of hand and/or finger motions to make to operate a control) and the capabilities of the driver to obtain the necessary information, make decisions, and execute the necessary responses.

The simplest model of human information processing is based on a series of four steps that involve (1) acquiring information available from the sensors (e.g., primarily visual receptors inside the eyes), (2) processing the sensed information to understand the situation, (3) selecting what to do (i.e., selecting a response), and (4) executing the response. To perform each of the steps, the driver will need time, and the total time taken to complete all the four steps will thus depend on the complexity associated in each of the steps. The four-step model and other information-processing issues are covered later in this chapter.

UNDERSTANDING DRIVER VISION CONSIDERATIONS

Since the drivers obtain most of their information visually, we will begin with the structure of the human eye. The structure of the eye will provide a basic understanding of the visual information that can be available for processing.

STRUCTURE OF THE HUMAN EYE

The human eye is like a camera. It has a lens, an adjustable iris (diaphragm with an adjustable diameter aperture in the middle to allow the light inside), and a surface where the image is formed. The sensor surface is called the retina in the eye (similar to the film/image surface in a camera; see Figure 4.1).

The retina contains photosensitive receptors. The receptors are of two types: cones and rods. The rod and cone receptors have different sensitivity to light. The cone receptors are sensitive under daytime lighting conditions (called photopic vision), and they also provide color vision. The rod receptors provide vision under dark visual conditions, called scotopic vision. The mesopic vision is when both the rods and cones are active, that is, when the day and night visions overlap under

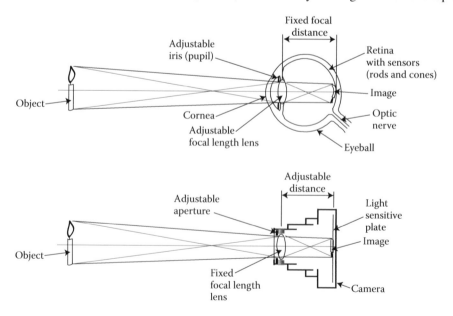

FIGURE 4.1   Comparison of the human eye with a camera.

dusk or dawn and under most night-driving conditions with low-beam headlamps. Under very dark scotopic conditions, vision is only possible with the rod receptors. Under scotopic vision with the rod receptors, a human cannot perceive any color, and the vision consists of different shades of gray from white to black. Thus, the photopic, mesopic, and scotopic vision are defined in terms of level of luminance (physical brightness measured in candela per square meter) as follows (Konz and Johnson, 2004):

Photopic vision: cones only—daytime vision: above 3 cd/m^2
Mesopic vision: both cones and rods function: 10^{-3}–3 cd/m^2
Scotopic vision: rods only—nighttime vision: 10^{-6}–10^{-3} cd/m^2

There are about 7 million cones in the fovea in the center of the visual field. The center of the visual field is defined by the visual axis of the eye, and the fovea is a region of about 1.0–1.8 degrees centered at the visual axis. The fovea is packed with all cone receptors, and it covers a field approximately of the size of a penny held at an arm's length. There are about 125 million rods scattered over the entire area of the retina except the fovea (which does not contain any rods). The distribution of the rods and cones is presented in Figure 4.2.

It is important to note that the density of the cones in the center (at the fovea) is the highest, about 140,000 cones/mm^2 of the retinal surface. The cone density decreases rapidly as the angular distance from the visual axis (called the perimetric angle or the eccentricity angle) increases (see Figure 4.2). At 10-degree and 30-degree perimetric angles, the cone density is about 10,000 and 5,000 cones/ mm^2, respectively. The visual ability to distinguish small details (called the visual acuity) is directly proportional to the cone density. Thus, to see a small detail, the driver would need to aim his or her eyes (i.e., to point the visual axis of each eye) at the detail so that the image of the detail falls on the foveae. Figure 4.2 also shows that in the outer regions of the visual field (called the peripheral visual field), there are relatively too few cones (about 5000 cones/mm^2) in the peripheral field. Therefore, in the peripheral visual field, the drivers can only detect large objects and have overall awareness of the objects but cannot see finer details.

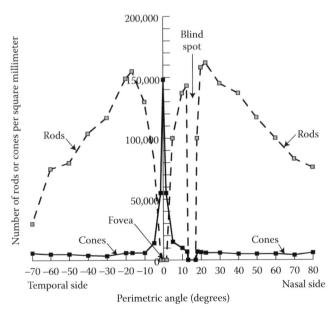

FIGURE 4.2 Distribution of rod and cone receptors in the human eye. (Redrawn from Boff, K. R., and J. E. Lincoln, *Engineering Data Compendium: Human Perception and Performance*, Vol. 1, Harry G. Armstrong Aerospace Medical Research Laboratory, Wright-Patterson Air Force Base, OH, 1988.)

There are different types of cone receptors that respond to different wavelengths of light. The cones are specialized in detecting red, green, and blue colors. Thus, there are (a) red-sensitive cones, (b) green-sensitive cones, and (c) blue-sensitive cones. The color vision characteristics and color perception are thus related to relative sensitivities of the cones and also to the presence or absence (in color deficient persons) of the above three types of cones.

The very high concentration of cones (about 140,000/mm² along the visual axis) in the foveal region provides high-visual-resolution capability (i.e., provides clearest/sharpest vision) and also higher speed of information transfer to the brain during photopic and mesopic ambient lighting conditions. Thus, to see small visual details (e.g., reading signs or graphic elements in displays), the driver will move his or her eyes and fixate (i.e., aim axes of both eyes) on the visual detail and dwell at that location for a short period, typically about 100–00 ms.

If an image on the retina is perfectly stationary for over a second (i.e., the image is stabilized), the image fades away (Yarbus, 1967). Therefore, human eyes normally make a series of rapid fixations (i.e., dwell or fixate on a visual detail for about 200–300 ms, and then the eyes move very rapidly [or make very small movements called microsaccades within less than a few milliseconds] to capture another image to refresh and continue vision).

VISUAL INFORMATION ACQUISITION IN DRIVING

During driving, the driver continuously moves his or her eyes in a series of rapid jerky movements. The eyes dwell, that is, fixate or remain steady, for a brief period typically 0.2–0.3 s and sometimes as long as 1 s, and then move (i.e., make an eye movement called the saccade) to make the next fixation, and so on. It is only during a fixation that the eyes receive information. If the fixation remains longer than about 1–3 s (i.e., when the image on the retina remains constant for more than 1 s), the image fades away. Thus, the driver has to continually make eye movements to refresh or to maintain vision and also to gather information from different parts of the visual fields. Many of the fixations made to maintain vision are involuntary and unconscious. The eye movements made on the driving scene can be measured by an eye movement measurement system (generally called an eye-marker system).

The speed of eye travel between fixations is very fast—from about 200 degrees/s for a 5-degree eye movement to about 450 degrees/s for about 20-degree eye movement (Yarbus, 1967). The eye movement time T (in seconds) for an eye movement of magnitude α (degrees) is estimated to be $T = 0.021\ \alpha^{2/5}$ (Yarbus, 1967). Thus, 5- and 20-degree eye movements take about 40 and 70 ms, respectively.

The fovea, because of its greater temporal and spatial resolution capabilities, can also acquire information faster than in the peripheral parts of the visual field. Thus, during driving, the driver moves his or her line of sight or the visual axes toward details in the road scene that he or she needs to recognize or use as a reference while obtaining information from extrafoveal (outside the fovea) regions.

ACCOMMODATION

The lens in the eye can change its power (i.e., its convexity) to focus a sharp image on the retina and to clearly see objects located at different distances. The process of changing focusing power is called the accommodation. A person with a normal range of accommodation can focus on objects located at a long distance away (far distance, i.e., infinity) to objects as close as about 90 mm (near distance) from his or her eyes. The human accommodation ability decreases with age due to hardening of the lens. After about 45 years of age, the closest (near) distance at which a person can focus (or clearly see small details) increases beyond his or her hand-reach distance (about 800 mm). Thus, for an older person to see objects at closer distances, reading glasses are required. Reading glasses are generally made to allow the reader to see objects (primarily reading material) at near distances of about 350–400 mm. Bifocal lenses are generally designed so that the person can read at closer distances from the lower part of the bifocal lenses, and the upper part of the lenses are used for far distance viewing.

Therefore, it is important to realize that if a display is provided at a viewing distance greater than the near reading distance (about 400 mm), older people will not be able to clearly see the display with the lower part of their bifocal lenses. The traditional instrument clusters (i.e., speedometer and other gauges) are generally located at about 800–900 mm from the driver's eyes, and they cannot be clearly viewed from the lower or the upper parts of the bifocals, unless the driver uses trifocal or continuously changing focal distance lenses. The header-mounted displays (mounted directly above the windshield) can especially cause reading difficulties to older drivers because to read the displays located higher and at close distances through the lower parts of their bifocal lenses, the drivers will need to tilt their heads up by a large angle. The convex mirrors located within hand-reach distances (e.g., on the driver's door) are also difficult for older drivers to use because the images of far away objects in the convex mirror are located (i.e., focused) very close to the mirror. The head-up displays are designed to focus display images at larger distances (beyond the front bumper of the vehicle) and are, therefore, useful for the older drivers.

INFORMATION PROCESSING

The visual images available to the driver from his or her eyes are processed by the brain. During the information processing, relevant details from the visual images are extracted (i.e., recognized and placed in the short-term memory) and used along with the information retrieved from the long-term memory (stored information from learning, practice, and experience) to interpret the present situation, and a decision is made on whether a response action is needed at that time. If a response is needed, then the information is further processed to decide on what to do (i.e., to select a response). Based on the selected response, signals are sent to appropriate muscles to generate movements to execute the response action (e.g., a hand movement to operate a control).

SOME INFORMATION-PROCESSING ISSUES AND CONSIDERATIONS

In the cognitive sciences, a number of studies have been conducted to understand how humans process information and make decisions (Fitts and Posner, 1967; Kantowitz and Sorkin, 1983; Sanders and McCormick, 1993; Wickens et al., 1997). Some of the challenging questions and issues studied are presented below:

1. Is the human a single- or a parallel-channel information processor? Can a human process information coming from different sources or sensors simultaneously or is the information queued up for a single processor to handle, one chunk at a time? In the early stages of learning or when a driver is under stress, most of the conscious information processing will be performed under the single-channel mode. As the driver learns and acquires skills, many simpler tasks can be performed simultaneously.
2. What is a human's channel capacity? At what rate can information be processed? Hick's Law and Hick–Hyman Law described in the next section allow us to estimate human information-processing rates.
3. Why is it important to measure reaction times? The duration of the reaction time allows us to determine the amount of information processed by the brain to decide on an output action.
4. What is the role of attention in information processing? If the driver is not attentive or not paying attention, the information available at the sensors (or the sensory level) may not be processed by the brain. Thus, an inattentive or distracted driver will fail to understand what is going on at a given moment.

5. How does a human operator time share between tasks? Normally, the driver performs many tasks while driving and shares resources (e.g., sensors located in different parts of visual fields or different sense modalities) between different tasks (e.g., driving and listening to the radio or talking to other passengers). Due to limitations in capabilities of the information-processing resources, the driver generally attends to the important tasks and disregards other lesser priority tasks.

6. How does the driver select what he or she should attend to? The driver must be processing information available from different sources to decide which resource to concentrate on and what to disregard.

7. How did the driver get to a destination while he or she was preoccupied with something else? While the brain is actively processing some other information, the driver's preattentive (or unconscious) information-processing activities (if they are highly learned) can take over and perform the driving tasks without the driver being fully aware of the situation.

To understand human information-processing capabilities, we will review a few simple but important human information-processing models and concepts in the next sections.

HUMAN INFORMATION-PROCESSING MODELS

1. Basic Four-Stage Model of Human Information Processing

A simple four-stage serial information-processing model is presented in Figure 4.3. The model shows that the information from an input stimulus is processed through four stages, namely, (1) stimulus encoding and decoding, (2) central processing, (3) response selection, and (4) response execution. The model shows that an input stimulus (or an event) in the first stage is received at the sensor (e.g., eye). For example, the stimulus may be a traffic light that turns red. Some relevant features or cues (the traffic signal device and its signal color) from the image are sent to the central processor in the brain, which processes the information further to interpret the stimulus (i.e., decode the meaning of red signal) and understands the situation (i.e., the red traffic signal will require the driver to stop). The information about the situation is further processed in the third stage to decide on the response to be selected (e.g., the driver decides to stop the vehicle instead of going through the intersection). Finally, in the fourth stage, the response is executed (i.e., information or instructions are sent to the appropriate muscles to make movements such as moving the right foot from the gas pedal to the brake pedal to stop the vehicle).

The above four-stage model shows that the information is processed serially (i.e., the processing in the sequence of four stages must occur before the response output occurs). Further, the processing of information in a preceding stage must be completed before the next stage can begin processing. This model, thus, suggests that the response time to react to the input stimulus (i.e., time taken between the occurrence of the stimulus and when the human operator initiates an output or a response movement) will be equal to the sum of the time taken by each of the four stages described above.

The concept that the information is processed in a series of stages was supported by the following two simple experiments reported in the early information-processing literature: (a) Donders' Subtraction Principle and (b) Psychological Refractory Period.

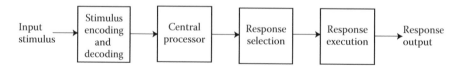

FIGURE 4.3   Four-stage serial information-processing model.

2. Donders' Subtraction Principle

This principle is based on the additive (and subtractive) nature of the components of reaction time. To prove that information is processed in a series of steps (or in a serial manner), Donders, a Dutch physiologist, conducted three reaction time experiments (Kantowitz and Sorkin, 1983). The three experiments, called Experiments A, B, and C, are illustrated in Figure 4.4.

In Experiment A, Donders provided a stimulus light (S_1) and a response key (R_1). He asked his subjects to press the response key R_1 as soon as the stimulus light S_1 was turned on. The response time measured, between when the stimulus S_1 came on and when the response R_1 key was pressed, was called RA. The situation in this experiment is the simplest (i.e., there is only one stimulus and one response). Thus, the reaction time in such a situation is called the simple reaction time.

In Experiment B, he provided two stimulus lights (S_1 and S_2) and two response keys (R_1 and R_2). He asked his subjects to respond as quickly as possible and press the response key R_1 if the stimulus light S_1 came on and press the key R_2 if the stimulus light S_2 came on. The response time measured between when the occurrence of a stimulus (S_1 or S_2) and the response (R_1 or R_2) was called RB. The situation in this experiment is more complex as after a stimulus occurs, the subject needs to detect and recognize which stimulus has occurred and also needs to select which key to press. Thus, the reaction time RB will be larger than RA because the subject makes two additional decisions.

In Experiment C, he provided two stimulus lights (S_1 and S_2) and two response keys (R_1 and R_2). He asked his subjects to respond as quickly as possible and press the response key R_1 only if the stimulus light S_1 came on and do nothing if the light S_2 came on. The response time measured between when the occurrence of a stimulus and the correct response was called RC. In this situation, the subject needed to recognize the stimulus and press the key R_1 if the S_1 came on.

From subsequent analysis of the reaction times data, Donders found that RB > RC > RA. This result proved that information processing occurred in a sequential manner because of the following additive nature of the information-processing time.

RA = simple reaction time
RB = RA + stimulus recognition time + response selection time
RC = RA + stimulus recognition time

3. Psychological Refractory Period

In another set of reaction time experiments (Fitts and Posner, 1967; Welford, 1952), subjects were asked to observe two successive signals which were presented within a short period

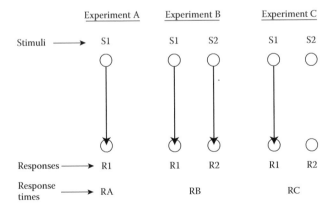

FIGURE 4.4   Donders' three experiments.

(called the intersignal interval = ΔT) between the two signals. The subjects were asked to respond (by pressing a key) as soon as the second signal was detected. The intersignal interval was varied in different trials between about 100 ms and a few seconds. The reaction time to the second signal was plotted against the intersignal interval time (ΔT; see Figure 4.5). The plot showed that the reaction to the second signal was observed to be much longer when ΔT was shorter than about 500 ms. The region of shorter values of ΔT, where the reaction time increased is known in the literature as the "psychological refractory period." It is called "refractory" in the sense that the brain waits to process the information about the second signal until the information about the first signal is processed. This phenomenon, thus, supports the serial information-processing theory because the information from both the signals cannot be processed simultaneously (or in parallel processing).

4. Simple Reaction Time

When there is only one stimulus and a subject is asked to respond using only one response key as soon as that stimulus occurs, the situation is considered to be simple (like the Experiment A above). The durations of simple reaction time typically range between about 100 and 300 ms for fully alerted subjects.

Simple reaction time (minimal reaction time) includes (Wargo, 1967):

   a. receptor delay of about 1–38 ms (depending on type of receptor, e.g., mechanical receptors in the inner ear respond more quickly than receptors in the retina as they work on photochemical reactions. [Van Cott and Kinkade, 1972])
   b. afferent transmission delay of about 2–100 ms (the delay time depends on the signal transmission distance between the sensor and the brain)
   c. central processor delay of about 70–100 ms
   d. efferent transmission delay of about 10–20 ms
   e. muscle latency and activation delay of about 30–70 ms.

The total of the above times is 113–328 ms. If we assume that a driver is traveling at 100 km/h (28 m/s), then during a simple reaction time of about 300 ms, the vehicle will travel about 8.4 m. In addition to the above simple reaction time, more time is generally required to complete a response action, such as moving a hand to press a button or moving a foot to operate a pedal.

5. Choice Reaction Time

When a driver is in a situation where he or she has a choice to select a response among several possible responses and make the selected response, then the situation is called the choice reaction time situation. The time taken from the instant when the response was

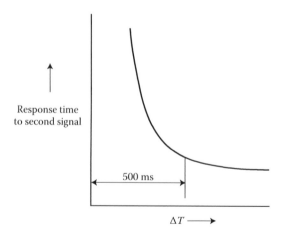

FIGURE 4.5 Reaction time to second signal as a function intersignal interval (ΔT).

requested to the instant when the selected response was made is called the choice reaction time. For example, while approaching an intersection and when the signal light turns yellow, the driver has the following three choices: (a) to decelerate and stop before the intersection, (b) to continue driving without changing speed, or (c) to accelerate. Thus, the time taken by the driver to decide on an action after the signal turned yellow can be called the choice reaction time for the situation.

The relationship between the number of choices and the reaction time has been studied by many researchers. Hick's Law of Information Processing is probably the most popular model describing this relationship. Hick's Law of Human Information Processing (Hick, 1952) is stated as follows:

$$RT = a + b \log_2 N$$

where RT = reaction (or response) time (measured in seconds; the time elapsed between
occurrence of a stimulus and the human response)
a = stimulus detection and decoding time (s)
$1/b$ = rate of human information processing of central processor measured in bits/s.
N = number of equally likely stimuli (number of choices)

Hick's Law of Information Processing, thus, states that the reaction time will increase in proportional to the logarithm to the base 2 of the number of equally likely choices (N) available to the human operator. Hick developed the logarithm to the base 2 transformation of N based on the concept of information developed by Shannon and Weaver (1949). According to this concept, the information (H) measured in "bits" is a measure of uncertainty. The value of information is directly related to the amount of uncertainty reduced. Thus, the amount of information (H) in a choice reaction situation that a decision maker processes by associating to an event (choice) among a set of N equally likely events (i.e., each of the N events occurs with equal probability of $p = 1/N$) can be defined as follows:

$$H = \log_2 \left(\frac{P_A}{P_B} \right)$$

where P_A = probability at the receiver of the event after the information is received
P_B = probability at the receiver of the event before the information is received

In the case when all the events are equally likely, $H = \log_2(1/p) = \log_2 N$.

It should be noted that P_A in the above ratio of probabilities is 1.0. This means that after the information is received, the subject is certain about what has occurred. Thus, the probability of what event (or choice) has occurred is 1.0. The probability of the occurrence of the event before the information was received (P_B) was $p = 1/N$ (as the receiver considered the event to be equally likely among the possible N events).

An Excel-based computer program to measure reaction times in choice reaction time situations ranging from $N = 1$ to 8 is available at the publisher's website (see Appendix 3). The task involves the subject pressing the number key on the computer keyboard corresponding to the number (stimulus) displayed on the screen as soon as the number appears on the screen. The number is randomly selected by the program within a selected set (choices) of numbers (N). The reader can use the program and plot his or her data (reaction times on the y-axis and $\log_2 N$ values on the x-axis) to verify Hick's Law by fitting a straight line to the data. The value of the intercept of the fitted line on the y-axis represents 'a' and 'b' is the slope of the line. The values of the constants a and b in the above Hick's Law expression typically vary between 0.1 to 0.5 s/bit and 0.1 to 0.25 s/bit, respectively, for alerted and practiced subjects.

The assumption of all choices being equally likely is applicable to situations like rolling an unbiased six-sided die or a two-sided coin. In the real world, the possible choices have different probabilities. For example while approaching an intersection, a driver may have three choices: do nothing (continue driving at the same speed), brake, or accelerate. The probabilities of the three choices will be different and will depend on other variables such as the distance of the vehicle from the intersection, the traffic signal color, and the driver's urgency to reach his or her intended destination.

Thus, if we assume that there are N events and an ith event occurs with probability p_i such that the sum of all the probabilities, $p_1 - p_N$, is equal to 1.0, then the information in the situation can be computed as follows:

$$H = \sum_{i=1}^{N} p_i \log_2 \left(\frac{1}{p_i} \right)$$

Thus, taking into account the unequal probabilities of the N possible choice events, Hick's Law was modified by Hyman in 1953. The modified law is known in the literature as Hick–Hyman Law (Hyman, 1953). Hick–Hyman Law is stated as follows:

$$RT = a + b \sum_{i=1}^{N} p_i \log_2 \left(\frac{1}{p_i} \right)$$

To get a feel of the amount of information in different situations, Table 4.1 shows calculations of information under two situations: Situation 1: when all the events are equally likely and Situation 2: when one of the N events occurs with probability equal to .9 (i.e., it occurs 90% of the time) and the remaining $N - 1$ events are equally likely (i.e., each of the $N - 1$ events occurs with the probability $(1.0 - .9)/(N - 1) = 0.1/(N - 1)$). The amount of information in Situations 1 and 2 are called H_1 and H_2, respectively.

Table 4.1 shows the values of H_1 and H_2 for N ranging from 1 to 16. For any row of the table with $N > 1$, the value of H_2 is substantially lower than the value of H_1. Thus, the choice reaction time in Situation 2 will be shorter than in Situation 1 (assuming the same values of the constants a and b in the above expression). The data in Table 4.1 also illustrate that in most decision-making situations, experienced people (experts) will have shorter reaction times than will inexperienced people. Because for an experienced individual, all the possible numbers of choices are not equally likely and due to his or her experience, he or she will select the most likely choice (i.e., a choice with the highest probability of occurrence), which will result in a lower reaction time than a novice who will treat all choices to be equally likely.

6. Factors Affecting Reaction Time

Human reaction time is influenced by many other factors (which also affect the constants a and b in the above expressions). A partial list including some of the more commonly known factors is given below:

a. Type of sensor or sense modality (e.g., mechanical sensors in human ear have shorter delay times than photochemical sensors in the human eye)
b. Stimulus discriminability or conspicuity with respect to the background or other stimuli (signal-to-noise ratio or clutter)
c. Number of features, complexity, and size of feature elements in the stimuli
d. Amount of search the human operator conducts (e.g., size of search set)
e. Amount of information processed (uncertainty and number of choices and their occurrence probability)
f. Amount of memory search

TABLE 4.1

Computation of Information in Two Situations

	Situation 1: Using Hick's Law			Situation 2: Using Hick–Hyman Law		
N	$p = 1/N$	$H_1 = \log_2 (1/p)$	p_1	p_i for $i > 1$	H_2
1	1.000	0.000	1.0	0.000	0.000
2	0.500	1.000	0.9	0.100	0.469
3	0.333	1.585	0.9	0.050	0.569
4	0.250	2.000	0.9	0.033	0.627
5	0.200	2.322	0.9	0.025	0.669
6	0.167	2.585	0.9	0.020	0.701
7	0.143	2.807	0.9	0.017	0.727
8	0.125	3.000	0.9	0.014	0.750
9	0.111	3.170	0.9	0.013	0.769
10	0.100	3.322	0.9	0.011	0.786
11	0.091	3.459	0.9	0.010	0.801
12	0.083	3.585	0.9	0.009	0.815
13	0.077	3.700	0.9	0.008	0.827
14	0.071	3.807	0.9	0.008	0.839
15	0.067	3.907	0.9	0.007	0.850
16	0.063	4.000	0.9	0.007	0.860

Note:

H_1 = Information when all outcomes are equally likely.

H_2 = Information when the first event occurs with $p_1 = 0.9$ and all other events ($i = 2 - N$) occur with $p_i = (1 - .9)/(N - 1)$ is computed as follows: $0.9 \log_2(0.9) + (N - 1)$ $p_i \times \log_2 (1/p_i)$.

g. Stimulus–response compatibility (e.g., how similar is the mapping or association of the stimuli to the responses)

h. Alertness of the subject

i. Motivation of the subject

j. Expectancy (how expected, or known from past experience is the event, in terms of when and where it could occur)

k. Mental workload (other tasks that the subject is time sharing at that time)

l. Psychological stress (e.g., emotional state of the subject)

m. Physiological stress (e.g., tired, fatigued, or in an environment affecting bodily functions)

n. Practice (how familiar or skilled is the subject to the situation)

o. Subject's age (older subjects are usually slower and more variable)

7. Fitts' Law of Hand Motion

After learning about Hick's Law, Fitts developed the law of human movements, which is known as Fitts' Law (Fitts, 1954; Fitts and Posner, 1967). Fitts' Law states that the amount of time taken to make a movement to reach a target (usually a hand movement with visual feedback, e.g., to move hand from the steering wheel to touch a button in a radio) is proportional to a variable called the index of difficulty (ID). Fitts' Law is stated as follows:

$$MT = c + [d \times ID]$$

where MT = movement time (measured in seconds; it is the time taken to reach to a target of width W located at distance A)

c = muscle preparation time (s)

d = constant (equivalent to the inverse of rate of human information processing of central processor)

ID = index of difficulty = $\log_2(2A/W)$

A = movement distance

W = target width (denotes accuracy of movement)

Thus, to reduce the time taken to make a movement, the ratio of A/W must be reduced. For example, to reduce the difficulty and time taken to reach to a button (target) on a car radio, the button can be located close to the location from where the driver's right hand is coming from (e.g., 2 o'clock position on the steering wheel) or the button can be made larger or both of the above can be altered to reduce A/W.

The Fitts' Law can be also easily verified by using targets and hand movement procedure presented in Appendix 4.

HUMAN MEMORY

The human memory system can be considered to have three types of memory storage subsystems, namely, sensory memory, working memory, and long-term memory. Figure 4.6 shows how the three memory types work in terms of the flow of information.

Sensory memory: The human sensory receptors (e.g., sensors in eyes, ears, skin) can store sensed information for a brief period. The sensory memory subsystems that hold information from the eyes and ears are called the iconic storage and echoic storage, respectively. The sensed information is stored for a brief time (about 1 s for iconic memory and up to 3–5 s for the echoic memory), after which it must enter the working memory or be lost (Wickens et al., 1997).

Sensory memory can be experienced as follows. If a subject is shown a 3 × 3 matrix of nine numbers in a very short period (e.g., exposure of less than 100 ms—usually through a tachistoscope) and then asked to recall a number at a particular location in the matrix, the subject can usually recall the number correctly even about a second after the matrix is turned off. This suggests that the image of the matrix is available in the iconic storage and the information from the storage can be retrieved after the original stimulus is withdrawn. Similarly, after a verbal message is announced and when questioned immediately after the message, most people can retrieve (from their echoic memory) what they just heard within a few seconds after the original message was presented.

Working memory: Working memory is at the center of the human memory system. It is also called the short-term memory. The information from the sensory subsystem is sent to the working memory. The information generally cannot be placed in the long-term memory or retrieved from long-term memory without passing through working memory.

Working memory has a limited capacity. About five to nine items (the magic number 7 plus or minus 2) can be stored in the working memory (Miller, 1956). Rehearsal is the most common control process to maintain information in the working memory. If rehearsal is stopped, the information can be lost from the working memory. Thus, it takes effort (e.g., continual refreshing) to maintain information in the working memory.

A demonstration of the limited capacity of the working memory can be easily given. For example, read a string of 10 one-digit numbers to a person one at a time (so that the numbers are not connected or associated as two or more digit numbers) in a random order; then, after all the numbers

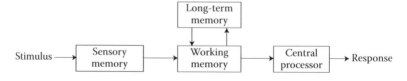

FIGURE 4.6 Three types of human memory subsystems.

are read, ask the person to recall the numbers (or write them down). It will be found that most people will not be able to recall all the numbers. But if only 5 one-digit numbers are presented in a random order, most people will correctly recall all the numbers. The working memory of most people can easily handle five (i.e., 7 minus 2) digits (or items) but cannot handle more than nine (i.e., 7 plus 2) digits (or items) in the working memory.

These limitations of the working memory are very important in designing equipment where people need to remember a number of items (such as lists, keys, preset buttons, or functions) for a short period in their working memory while performing a task. If the working memory is overloaded, people will make errors or they may not be able to perform the task without other aids (e.g., labels, written sheets, or visual screens).

Long-term memory: The information in long-term memory can be stored for a very long time, usually through the entire life span of an individual. In contrast with the working memory, no effort or capacity is required to maintain information in long-term memory. Once the information has been transferred to long-term memory from working memory, it is there forever (though it may be difficult to retrieve). No rehearsal is needed to maintain items in long-term memory. Most memory models assume that information in the long-term memory is coded (i.e., stored with some flags or cues) according to some "meaning." Many times people find that it is difficult to remember where an item was stored, but once they remember the context (or the flag) under which the information was stored, the information can be retrieved very quickly even after many years of inactivity.

GENERIC MODEL OF HUMAN INFORMATION PROCESSING WITH THE THREE MEMORY SYSTEMS

A recent and more complete model of human information processing was provided by Wickens et al. (1998; see Figure 4.7). This model includes the four-stage model of information processing along with the working and long-term memory used between the middle two stages involving perception and response selection. It also shows that attention is required for the last three stages,

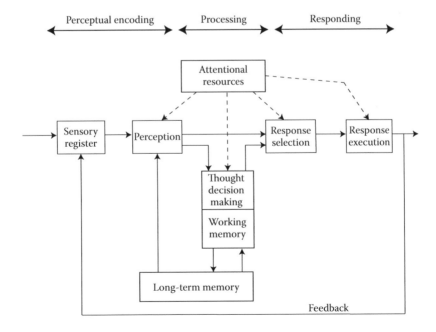

FIGURE 4.7 A more complete model of human information processing. (Redrawn from Wickens, C., S. E. Gordon, and Y. Liu, *An Introduction to Human Factors Engineering*, Pearson Prentice Hall, Upper Saddle River, NJ, 1998. With permission.)

namely, perception, response selection, and response execution. The model also includes a feedback loop to receive information about the executed response into the sensory register.

HUMAN ERRORS

Human operators will make errors in operating equipment due to a number of reasons. We are interested here in understanding how a vehicle can be designed to minimize possibilities of driver errors. Most human errors occur due to information-processing failures. Treat (1980) studied more than 13,500 police-reported accidents and found that in more than 70% of the accidents, human errors were identified as the definite causes of accidents. The definite causes were defined as the accidents in which the in-depth accident investigating multidisciplinary team members were over 95% confident that the accidents were caused by human factors (e.g., recognition errors, decision errors, performance errors).

DEFINITION OF AN ERROR

1. An act involving an unintentional deviation from the truth or accuracy (G. & C. Merriam Company, 1980)
2. An out-of-tolerance action
3. Inconsistent with normal, programmed behavioral pattern, and that differs from prescribed procedures

In general, when an operator's capabilities are overloaded (or exceeded) or when the operator is inattentive, an error is possible (i.e., when any of the links in an information-processing model is broken). The likelihood of human error will increase when human factors principles or guidelines are violated, that is, when the equipment is not designed with proper consideration of operator characteristics, capabilities, and limitations. Humans will always make errors because they are not machines. The basic error rate data available in the literature (Gertman and Blackman, 2001) show that even for the best "human-factored" equipment, error rates of the order of 1 error in 10,000 operations to 1 error in 1,000 operations are quite common.

TYPES OF HUMAN ERRORS

Human errors can be classified as follows:

1. Detection error—failure to detect a signal or a target (e.g., a driver fails to see a pedestrian). The detection error can occur due to a number of reasons such as if the signal was weak in relation to its background (on noise), the signal violated driver's expectancy (spatial or temporal), or the driver was not attentive or distracted.
2. Discrimination error—failure to discriminate between signals (e.g., stop lamps were perceived as tail lamps).
3. Interpretation error—failure to recognize a situation, a signal, a hazard, a scale, and so forth (e.g., a tachometer reading was interpreted as the speed reading).
4. Omission error—failure (or forgetting) to perform a required action (e.g., forgetting to look in the side-view mirror before changing lane).
5. Commission error—performing a function that should not have been performed. A commission error can involve
   a. Extraneous act error—introducing a step or a task that should not have been performed (e.g., changed radio station while turning on the windshield defrost function)
   b. Sequential error—performing a step or a task out of sequence (e.g., changed radio station band after selecting a radio preset station).
   c. Time error—failing to perform within an allotted time or excessive time spent (e.g., long looks, more than a single look while using a control or a display).

6. Substitution error—using or substituting another item (control or display) instead of the desired one (e.g., pressed accelerator pedal instead of the brake pedal).
7. Reversal error—reversing the direction of activation or interpreting a displayed signal in opposite direction (e.g., increased temperature instead of decreasing).
8. Inadequate response error—error in judgment or estimation of signal magnitude, distance, speed, and so forth or insufficient movement or insufficient force applied during control activation (e.g., insufficient brake pedal movement during stopping).
9. Legibility error—error related to not being able to read a display (due to factors such as small font size, insufficient light, excessive glare, parallax).
10. Recovered error—an error has occurred, but the operator could correct the error after some elapsed time. (Note: Many errors made by human operators are recovered [corrected], and thus undesired events [e.g., accidents] are avoided.)
11. Unrecovered error—an error that the operator fails to correct (or will not or cannot correct).

UNDERSTANDING HUMAN ERRORS WITH THE SORE MODEL

A number of different models have been developed to predict human errors (Leiden et al., 2001). One simple model called the SORE model is described here. The SORE model stands for stimulus, operator, response, and environment (see Figure 4.8). The model shows that the operator responds after detecting a stimulus, and the response is seen by the operator in the environment as feedback. In general, when an operator's capabilities are overloaded or exceeded, an error is possible (i.e., when any of the links in the SORE information-processing model is broken). For example, an error could occur if (a) the S–O link is broken (i.e., stimulus is not detected by the operator), (b) the O–R link is broken (e.g., the operator does not understand the stimulus, and thus, he or she cannot make a response), (c) the R–E link is broken (e.g., the response is not seen or heard in the environment), and (d) the E–O link is broken (e.g., the operator does not receive feedback from the activation of a control).

PSYCHOPHYSICS

Psychophysics is a science that relates physical characteristics of a stimulus (e.g., visual detail) to psychological response (i.e., how a person perceives and reacts to the stimulus). Some typical psychophysical problems are (a) detection of a signal (or a visual target or a change in the pointer position in a gauge), (b) just noticeable difference between two stimuli (e.g., smallest difference in brightness between two tail lamps that can be recognized by an observer), (c) equality (when two signals are perceptually equal in magnitude, e.g., determining if two stop lamps on the back of a car are equally bright), (d) magnitude estimation (e.g., estimating speed of a vehicle or distance between two vehicles), (e) interval estimation (e.g., determining the time [interval] required to reach a particular location from a given vehicle speed), (f) production of magnitude of a stimuli (e.g., pushing the brake pedal hard enough to produce required stopping distance), and (g) rating on a scale (e.g., providing a rating on ease or difficulty using a 10-point scale, where 1 = very difficult and 10 = very easy).

It should be noted that any one of the above psychophysical problems involves information processing and decision making. Thus, application of psychophysics will help in improving functionality and customer satisfaction of a product. (Note: Chapter 10 on Automotive Craftsmanship and Chapter 15 on Vehicle Evaluation Methods cover several applications on perception of quality and subjective scaling related to this topic.)

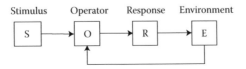

FIGURE 4.8 The SORE model for human error analysis.

VISUAL CAPABILITIES

Psychophysical methods are used to measure different visual capabilities. Some commonly considered visual capabilities of the drivers are

1. Detection of objects (e.g., visibility of visual targets)
2. Perception and identification of colors
3. Differences (or change) in luminance
4. Equality of appearance (e.g., if two trim parts "match," i.e., the appearance of the two trim parts is equal by considering their color, brightness, and texture)
5. Difference (or change) in colors (color differentiation or change in color coordinates needed to perceive a color difference)
6. Recognition of details or object recognition (e.g., reading letters or Snellen visual acuity)
7. Depth perception (i.e., ability to determine if two visual objects are at the same or different viewing distances)
8. Accommodation (i.e., ability of eyes to focus at different viewing distances)

VISUAL CONTRAST THRESHOLDS

One of the basic human visual capabilities (or visual functions) is based on the ability of humans to perceive differences in luminance (physical brightness; measured in foot Lambert or candela per square meter). Perception of visual contrast is a basic function that is needed to detect visual objects or to recognize visual details. For example, a circular object (target) can be seen against a background because of a difference in the luminance of the target against its background.

The visual contrast of a target against its background is defined here as $C = \dfrac{|L_t - L_b|}{L_b}$,

where L_t = luminance of the target and L_b = luminance of the background.

Figure 4.9 presents visual threshold contrast curves as functions of the adapting luminance and target size. To compute if a target is detectable (i.e., visible), values of the following variables are needed: (a) contrast (C) of the target against its background as shown above, (b) adaptation luminance (it is the luminance level to which the observer's eyes are adapted), and (c) target size (it is the

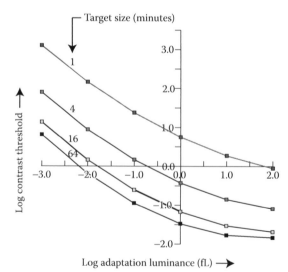

FIGURE 4.9   Visual contrast threshold curves for 1/30th second target exposure used for driver visibility and legibility modeling.

angle subtended by the target at the observer's eyes computed in minutes). The target is considered to be visible if the value of the contrast (C) at the abscissa value of the adaptation luminance falls above a threshold contrast curve corresponding to the computed target size (interpolated curve for the target size) while looking for the target.

The visual contrast threshold curves developed originally by Blackwell (1952) and modeled by Bhise et al. (1977) are presented in Figure 4.9. The log values used in the abscissa and ordinate of the graph are to the base value of 10. It should be noted that the contrast threshold curves presented in Figure 4.9 were obtained by Blackwell under 1/30th of second exposure. The 1/30th of the second exposure duration is much smaller than the typical eye fixation durations (about 1/3 s). However, since Bhise et al. (1977) found that the higher contrast thresholds for 1/30th s (as compared with at 1/3rd of second exposure) predicted visibility distances to stand-up targets more accurately under the more difficult actual dynamic driving conditions, the Blackwell's contrast thresholds obtained using 1/30th of second were used for modeling driver's night visibility of targets on the roadway and legibility of displays. The visibility and legibility prediction models using the above threshold curves are presented in Chapter 12.

VISUAL ACUITY

Visual acuity is the ability of a person to recognize a visual detail. The most common reference to visual acuity is based on the ability to see and recognize letters. Visual acuity scores are described by ratios such as 20/20, 20/40, and 20/200. The ratio is based on two distances. The numerator denotes the distance in feet from which the smallest size of letters on a visual acuity chart can be correctly read by the subject. The denominator denotes the maximum distance in feet from which the same row of letters can be correctly read by a person with normal vision. Thus, 20/40 means that the subject could read the smallest letters from 20 ft or 6.1 m (this reference distance is set for far-vision test in an optometrist's office), which a person with normal vision could have read correctly from 40 ft (12.2 m). The 20/20 means normal vision, that is, the subject's visual acuity is as good as the person with normal vision. On the other hand, a score of 20/400 generally means legally blind as the person could read the letters from 20 ft (6.1 m), which a person with normal vision could read from a distance of 400 ft (122 m).

Near visual acuity is measured at a reference distance of 14 in. (0.36 m; the shorter distance from which printed text is read by most people under most reading situations). For near acuity, the scores are also presented in ratios 20/20, 20/40, and so forth like the far acuity scores, but the numbers do not indicate actual viewing distances. However, the ratios do indicate the ability of the reader to read the letters in relation to the distance from which a person with normal vision can read the letters located at the near distances.

In a standard visual acuity test, six visual acuity tests are conducted under combinations of two distances and three viewing conditions. The two distances are 14 in. (near) and 20 ft (far or 6 m if using a metric test), and the three viewing conditions are left eye only (right eye occluded), right eye only (left eye occluded), and both eyes open.

The minimum visual requirements for a driver's license in most states in the United States are as follows: (a) minimum uncorrected (driving without glasses or contacts) far visual acuity score of 20/40 to qualify for an unrestricted driver's license and (b) minimum corrected (with glasses or contact lenses) far visual acuity of 20/50 to qualify for a restricted license (i.e., allowed to drive with corrective lenses).

For visual acuity tests, different targets have been used, for example, letters, checkerboard patterns, gratings, Landolt "C" rings at various orientations, alignment of two lines (Vernier acuity), and so forth. Figure 4.10 illustrates a visual acuity test using the letter "E" as the test object where the critical visual detail is assumed to be the middle stroke (of width h). In this test, the subject is shown letters of different sizes (representing different viewing distances). The visual acuity is defined by the smallest letter that the subject can correctly identify. The visual acuity is measured by the reciprocal of the angular size of the smallest visual detail correctly identified

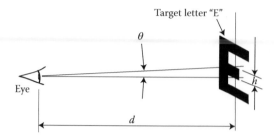

FIGURE 4.10   Visual acuity measurement (defined by 1/angle measured in minutes subtended at the stroke of width $h = 1/\theta$ at the observer's eye).

by the subject (shown as angle θ measured in minutes) in Figure 4.10. The visual acuity is, thus, equal to $1/([\tan^{-1}(h/d)]/60)$. (Note: $\tan^{-1}(h/d)$ is measured in degrees.)

DRIVER'S VISUAL FIELDS

The normal human visual field with both eyes is called the ambinocular field, and it extends about 95 degrees to the left and 95 degrees to the right, about 60 degrees upward, and about 75 degrees downward with respect to the forward line of sight (i.e., axes of both eyes pointed straight ahead). The minimum visual field requirements for a driver's license in most states in the United States are that a person with two functional eyes must have a field of vision of 140 degrees (horizontal). A person with one functional eye must have a field of vision of 105 degrees (horizontal).

Due to the distribution of the cones (refer to Figure 4.2), the vision is most detailed in the foveal vision (about 1 degree radius). The visual field outside the fovea is called the extrafoveal vision. It can be approximately categorized as the central visual field (up to about 30 degrees eccentricity angle) and the peripheral visual field (beyond 30 degrees eccentricity angle). During driving, the foveal visual field provides the most detailed vision to read details such as highway signs. The central field provides visual information (i.e., presence and locations) on most targets such as roadway, other vehicles, and traffic control devices. The peripheral vision provides awareness of larger targets (e.g., vehicles in the adjacent lanes) and provides information on moving targets and motion cues.

The smallest sizes of targets that can be seen in the visual field of a single (right) eye are shown in Figure 4.11. The figure shows the fields for detecting targets of 1, 10, and 20 min (angular size of the targets) for the right eye.

Figure 4.12 shows the extent of the field of perception of different colors in right eye. The size of the field varies depending on the color. It is largest for yellow and blue and smallest for green.

Figure 4.13 shows that, for both males and females, the total available visual field narrows with age. Thus, older drivers will not be able to detect targets in far peripheral fields as far outward as compared with younger drivers.

OCCLUSION STUDIES

The information-processing capacity of a driver must be greater than the amount of information that the driver needs to process while driving. The amount of information that a driver needs to process changes continually during driving due to changes in road geometry, road background (scenery), and traffic situations.

Some researchers have attempted to measure the driver's visual information-processing needs by using an occlusion device, which, as the name suggests, occludes (i.e., blocks) the driver's vision for a preset amount of time. Senders et al. (1966) developed a helmet with a movable shutter, which could continually cycle between open or closed positions at preset intervals (such as open for 1 s and closed for 2 s and repeat the open or closed cycles) over the entire driving course, and measured the maximum speed at which the drivers were willing to drive. Other researchers have done studies

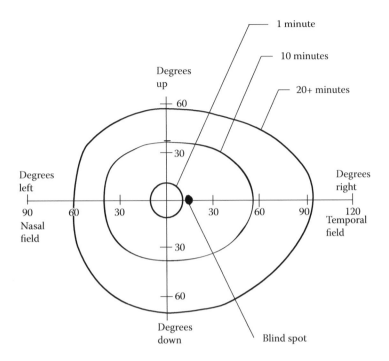

FIGURE 4.11 The normal achromatic monocular visual field of targets of various sizes for right eye. (Redrawn from Boff, K. R., and J. E., Lincoln, *Engineering Data Compendium: Human Perception and Performance*, Vol. 1, Harry G. Armstrong Aerospace Medical Research Laboratory, Wright-Patterson Air Force Base, OH, 1988.)

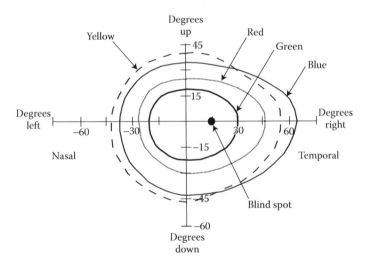

FIGURE 4.12 Right eye's field showing the extent of fields for different colors. (Redrawn from Boff, K. R., and J. E. Lincoln, *Engineering Data Compendium: Human Perception and Performance*, Vol. 1, Harry G. Armstrong Aerospace Medical Research Laboratory, Wright-Patterson Air Force Base, OH, 1988.)

by simply asking the drivers to keep their eyes closed as much as they could and open them only as needed to safely drive (i.e., maintain the car in the driving lane) at an instructed speed.

The relationship between the average amount of time the subjects were willing to keep their eyes closed (i.e., the eyes-off-the-road time) and the driving speed obtained from such studies is presented in Figure 4.14.

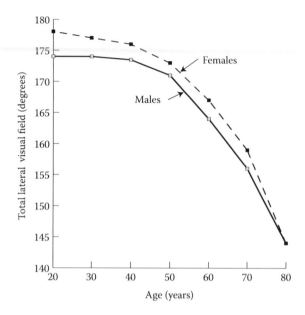

FIGURE 4.13 Effect of age on size of lateral visual field. (Redrawn from Boff, K. R., and J. E. Lincoln, *Engineering Data Compendium: Human Perception and Performance*, Vol. 1, Harry G. Armstrong Aerospace Medical Research Laboratory, Wright-Patterson Air Force Base, OH, 1988.)

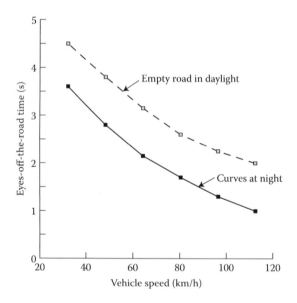

FIGURE 4.14 Relationship between eyes-off-the-road time (occlusion time) as a function of vehicle speed.

Figure 4.14 shows that on an empty road, the drivers were willing to keep their eyes closed about 2 s on average at 113 km/h (70 mph) and about 4 s at 40 km/h (25 mph). However, when the road environment was more demanding, such as driving on a winding curvy road at night, they closed their eyes on average for only 1 s at 113 km/h (70 mph). The data presented in Figure 4.14 thus provide us limits on time away from the road that must be considered while designing driver interfaces. This issue is further covered in Chapter 5 in designing and evaluating driver controls and displays and in Chapter 14 on driver workload measurement.

INFORMATION ACQUIRED THROUGH OTHER SENSORY MODALITIES

While driving, the driver also obtains information from other sensory modalities. The information obtained through perception of sound, vibrations, vehicle movements, touch, and smell also can provide useful information in performing many in-vehicle tasks. The information can also have negative effects such as annoyance, fatigue, and interference or masking of other useful information. For example, engine and road noises provide the driver with information on vehicle speed. However, too many or too loud noises can induce fatigue and disrupt conversations with other occupants or reception of audio programs.

Human factors engineers work with other engineers in specialized functions such as acoustics and sound engineering; noise, vibrations, and harshness; and interior trim and materials to make sure that the vehicle provides the necessary cues of the right type and magnitude and enhances the pleasing perceptions (e.g., sound of the exhaust that conveys engine power, sound of door closing that provides the feeling of "solidness" or "solid build quality," tactile feedback received from "crisp" detent feel while operating electrical switches, smell of "genuine" leather from the seats).

HUMAN AUDITION AND SOUND MEASUREMENTS

An understanding of human auditory capabilities and limitations provides knowledge for auditory/sound design and evaluation issues related to interior quietness, auditory display designs (e.g., beeps, buzzers, and auditory warnings), voice controls, and perception of sound quality.

Some basic information and issues in consideration of auditory systems are given below.

1. The human auditory capabilities are specified by measuring loudness thresholds as a function of the sound frequency (measured in hertz, equal to cycles/second). The loudness is measured in decibels and defined as follows:

$$\text{Loudness (dB)} = -10 \log_{10} \left(\frac{P}{P_0} \right)^2$$

   where P = pressure in sound pressure wave (measured in Newton per square meter)
   P_0 = lowest perceptible sound pressure (measured in Newton per square meter)

2. The human ear can sense sounds of frequencies between about 20 and 20,000 Hz. The sound in the human ear is sensed by hair-like sensors inside the cochlea (inner ear). There are about 30,000 hair-like sensors in each inner ear (Konz and Johnson, 2004).
3. The human hearing thresholds vary with sound frequency. The region of best sensitivity (or the lowest loudness thresholds) is within the frequency range of about 500–2000 Hz.
4. The human hearing thresholds are measured by instruments called audiometers, which can measure the loudness of a just-perceptible signal as a function of the signal frequency. The sound levels in work environments are measured by instruments called sound level meters. Since the perception of loudness is affected by the sound frequency, most sound level meters use a weighting filter, which converts the incoming sounds (with a combination of different frequencies) and provides a weighted loudness value that is equivalent to the loudness perceived by a listener with normal hearing. The A-weighting filter, used most commonly, is applicable to sound levels around 70 dB. Sound loudness values measured using the A-weighting filter are designated with a unit label called dBA.
5. Human hearing thresholds increase as people get older. The hearing threshold increases are more pronounced in the higher frequency region (over 5000 Hz; Sanders and McCormick, 1993; Van Cott and Kinkade, 1972).
6. The phon is another measure of loudness. It is based on the loudness of the 1000-Hz reference signal that will be perceived to be equivalent in loudness to the loudness of the signal being measured.

7. The just-noticeable differences in loudness and just-noticeable differences in frequencies of humans decrease with an increase in loudness level (at which the listener's ears are adapted to; Van Cott and Kinkade, 1972). Thus, most people will increase the volume of music to enjoy thorough perception of small differences (or changes—increases or decreases) in sound loudness or sound frequencies (or pitch).

8. In general, the loudness of a sound source experienced by a human decreases according to the inverse square law (i.e., the sound loudness decreases proportional to an inverse of the square of the distance between the source and the listener). The higher frequency sounds attenuate faster than the low-frequency sounds as compared with the attenuation predicted by the inverse square law (Van Cott and Kinkade, 1972). Thus, to improve detection of a signal over farther distances, the designer of sound alarm systems should select sound signals of frequencies below 1000 Hz.

9. The sound loudness in modern passenger vehicles with windows closed ranges between about 65 and 70 dBA at 98 km/h (60 mph).

10. For a warning signal to be heard inside the passenger compartment by most occupants, its loudness should be at least about 15 dB more than the overall interior noise level. The detection probability of a sound signal will also improve as the frequency of the signal differs from the frequencies of the noise.

11. Noises of higher frequencies (over 2000 Hz) are perceived as "tinny" (cheap and harsh like a tin can) as compared with noise at lower frequencies (below 1000 Hz). Thus, to provide perception of "solidness" from interior components (e.g., sounds from switches, latches, door closing), the designers should reduce or remove higher frequency sounds created during their operations.

OTHER SENSORY INFORMATION

During driving, a driver also experiences vibrations. The vibrations in vehicles are perceived through three types of sensory systems with overlapping frequencies. The three sensitivity ranges to vibrations are (1) 0 to 1–2 Hz sensed through vestibular system in the brain (as motion sickness and whole body vibrations), (2) 2 to 20–30 Hz sensed through biomechanical body resonances, and (3) >20 Hz sensed through somesthetic mechanoreceptors such as proprioceptive receptors located in muscles, tendons, and exteroceptive receptors located in cutaneous tissues. The human perception and response to vibrations depend on frequency, amplitude, direction of vibration, duration of vibration, mass of the body segments, posture level of muscle contraction, and fatigue and presence of body supports (Chaffin et al., 1999). Vehicle vibrations over 20 Hz are generally perceived as sound or noise, and frequencies below 20 Hz are perceived as vibrations. The vehicle suspension and seat design engineers evaluate ride and vibrations experienced by the occupants. Some vibrations are also perceived through a driver's contacts with the vehicle controls such as the steering wheel, pedals, and gear shift knobs. Excessive and prolonged vibrations in vehicles are generally associated with higher complaints of driver discomfort and fatigue (especially in heavy truck products). The perception of quality and craftsmanship is also associated with other sensory perceptions related to vibrations, tactile feel of interior touch areas (e.g., smoothness, softness), and smells from emissions from interior materials. Some of these issues are covered in Chapter 10.

APPLICATIONS OF INFORMATION PROCESSING FOR VEHICLE DESIGN

SOME DESIGN GUIDELINES BASED ON DRIVER INFORMATION-PROCESSING CAPABILITIES

Based on the information presented in this chapter and other studies reported in the literature related to the information-processing abilities of the drivers, many design guidelines can be developed to reduce the driver's time needed to use displays and operate controls. Some important basic guidelines used in designing controls and displays are provided below.

1. Design controls and displays such that they can be used in short (1-s) glances. Minimize or eliminate driver control/display interface configurations that will require multiple glances (or operational steps) away from the forward road scene. Avoid functions buried in menus or hidden features that will require multiple control actions and glances to operate.

2. Place important and frequently used displays close to the driver's normal line of sight while looking forward. Place controls and displays in expected locations and at locations that will involve minimal head, torso, and hand/wrist movements. (Note: SAE standard J1138 provides recommended locations for primary and secondary vehicle controls [SAE, 2009].)

3. Avoid locating controls and displays in areas that are not visible to drivers from their normal eye locations (see driver eyellipses in SAE standard J941 and in Chapter 3). The steering wheel, stalks, levers, and driver's hands (hand positions just prior to usage of the control or the display) can cause visual obstructions. Also, avoid obscurations or masking due to veiling glare, specular glare, or reflections into display surfaces or lenses due to the sunlight or other external lights.

4. All displays must be legible to drivers (especially older drivers) from the farthest eye locations under day, dusk (with and without headlamp/panel lights activation), and in night-driving conditions (see Chapter 12 for legibility prediction).

5. In developing controls and display layouts, apply principles related to (a) importance of the function, (b) frequency of use, (c) sequence of use, (d) functional grouping, and (e) controls and display association (see Chapter 5). Further, prioritization within the above principles is generally needed to avoid crowding of controls and displays in regions close to the steering wheel.

6. Place controls at locations that are comfortable to reach, grasp, and operate (see Chapter 3 for maximum reach envelopes). Also, avoid locating controls that require awkward hand/wrist deviations. Controls located too close to the torso generally require severe wrist deviations (e.g., chicken winging)—especially if the driver cannot move his or her elbows away from the torso (see minimum reach zones in Chapter 5).

7. Tiny controls that require more precise hand movements are difficult to locate and reach. (Note: Fitts' Law covered earlier in this chapter directly relates hand movement time to index of difficulty.)

8. Direction of movements of controls and displays should conform to population stereotypes (see SAE standard J1139 in SAE, 2009).

9. All controls should provide a distinct, perceptible feedback (visual, auditory, or tactile) to inform the driver about the completion of control activations.

10. All controls and displays must be labeled (provide identification and/or setting label) by use of accepted wording or symbols (refer to Federal Motor Vehicle Safety Standard 101 [National Highway Traffic Safety Administration, 2010]; SAE standard J1138 [SAE, 2009]).

11. Avoid similar-looking controls or displays. Provide cues/coding (due to shape, size, color, texture, sound, force feel, etc.) to discriminate between different controls or display details to reduce confusion between similar-looking items.

12. Some features such as head-up displays, voice displays, and voice-activated controls will require redundant (or alternate) methods of use due to their possible unavailability or undesirability under some driving environments (e.g., head-up displays under bright roadway background) and to some types of drivers (e.g., speech- or hearing-impaired drivers cannot use voice controls/displays).

CONCLUDING COMMENTS

The automobile manufacturers are currently facing a difficult challenge of determining what features should be provided for the drivers to use while driving. The rapidly advancing sensor, information, communication, display, and lighting technologies are allowing development of new devices that can provide the driver with information at high speeds and in large quantities. These technologies also have the potential to perform many tasks performed normally by drivers. Thus, if these systems are developed such that they provide just the right type of information and/or control in the right amounts and at the right time and place, then there is a great potential to improve driver convenience and safety. Additional information on future in-vehicle devices is provided in Chapter 17.

Many traffic safety experts and organizations are concerned that these new devices can overload information-processing capabilities of drivers and thus distract drivers from the primary driving tasks. The benefits that these new devices can offer are also substantial. For example, while most drivers realize the additional risk of using cellular phones while driving, there are a number of benefits (e.g., keeping in touch with home/office and a feeling of security—knowing that you can call for help immediately) that may well preclude their prohibition. Similar emotional arguments for access to the Internet in cars are also currently being made. Additional information on driver workload measurement is provided in Chapter 14.

The human factors and safety engineering community must, therefore, come with performance requirements and design guidelines to assure that useful and safe in-vehicle devices can be developed. Additional issues related to this area are covered in Chapter 5 and Part II of this book.

REFERENCES

Bhise, V. D. 2002. *Designing Future Automotive In-Vehicle Devices: Issues and Guidelines.* Proceedings of the Annual Meeting of the Transportation Research Board, Washington, DC.

Bhise, V. D., E. I. Farber, and P. B. McMahan. 1977. Predicting target detection distance with headlights. *Transportation Research Record*, 611, Washington, DC: Transportation Research Board.

Blackwell, H. R. 1952. Brightness discrimination data for the specification of quantity of illumination. *Illuminating Engineering*, 47(11), 602–609.

Boff, K. R., and J. E. Lincoln. 1988. *Engineering Data Compendium: Human Perception and Performance.* Vol. 1. Wright-Patterson Air Force Base, OH: Harry G. Armstrong Aerospace Medical Research Laboratory.

Chaffin, D. B., G. B. J. Andersson, and B. J. Martin. 1999. *Occupational Biomechanics.* New York: John Wiley & Sons Inc.

Fitts, P. M. 1954. The information capacity of the human motor system in controlling amplitude of movement. *Journal of Experimental Psychology*, 47, 381–391.

Fitts, P. M., and M. I. Posner. 1967. *Human Performance.* Belmont, CA: Books/Cole Publishing Company.

G. & C. Merriam Company. 1980. *Webster's New Collegiate Dictionary.* Springfield, MA: G. & C. Merriam Company.

Gertman, D. L., and H. S. Blackman. 2001. *Human Reliability and Safety Analysis Data Handbook.* New York: John Wiley.

Hick, W. E. 1952. On the rate of gain of information. *Quarterly Journal of Experimental Psychology*, 4, 11–26.

Hyman, R. 1953. Stimulus information as a determinant of reaction time. *Journal of Experimental Psychology*, 45, 423–432.

Jackson, D., J. Murphy, and V. D. Bhise. 2002. *An Evaluation of the IVIS-DEMAnD Driver Attention Demand Model.* Paper presented at the annual congress of the Society of Automotive Engineers Inc., Detroit, MI.

Kantowitz, B. H., and R. D. Sorkin. 1983. *Human Factors: Understanding People–System Relationships.* New York: John Wiley and Sons.

Konz, S., and S. Johnson. 2004. *Work Design: Industrial Ergonomics.* 6th ed. Scottsdale, AZ: Holcomb Hathaway.

Leiden, K., K. R. Laughery, J. Keller, J. French, W. Warwick, and S. Wood. 2001. *A Review of Human Performance Models for the Prediction of Human Error.* Report prepared for NASA Ames Research Center, Moffett Field, CA by Micro Analysis & Design Inc., Boulder, CO.

Miller, G. A. 1956. The magical number seven plus or minus two: Some limits on our capacity for processing information. *Psychological Review*, 63, 81–97.

National Highway Traffic Safety Administration. 2010. *Federal Motor Vehicle Safety Standards*. Federal Register, Code of Federal Regulations, Title 49, Part 571, U.S. Department of Transportation, accessed June 23, 2011, http://www.gpo.gov/nara/cfr/waisidx_04/49cfr571_04.html.

Rockwell, T. H., V. D. Bhise, and Z. A. Nemeth. 1973. *Development of a Computer Based Tool for Evaluating Visual Field Requirements of Vehicles in Freeway Merging Situations*. Report VRI-1. Warrendale, PA: Society of Automotive Engineers Inc.

Sanders, M. S., and E. J. McCormick. 1993. *Human Factors in Engineering Design*. 7th ed. New York: McGraw-Hill Inc.

Senders, J. W., A. B. Kristofferson, W. H. Levison, C. W. Dietrich, and J. L. Ward. 1966. *The Attentional Demand of Automobile Driving*. *Highway Research Record*. Also in the Proceedings of the 46th Annual Meeting of the Highway Research Board, Washington, DC.

Shannon, C. E., and W. Weaver. 1949. *The Mathematical Theory of Communication*. Urbana, IL: University of Illinois Press.

Society of Automotive Engineers Inc. 2009. *SAE Handbook*. Warrendale, PA: Society of Automotive Engineers Inc.

Treat, J. R. 1980. *A Study of Precrash Factors Involved in Traffic Accidents*. Highway Safety Research Institute Report, University of Michigan, Ann Arbor, MI.

Van Cott, H. P., and R. G. Kinkade (eds.). 1972. *Human Engineering Guide to Equipment Design*. Sponsored by the Joint Army-Navy-Air Force Steering Committee, McGraw-Hill Inc./U.S. Government Printing Press.

Wargo, M. J. 1967. Human operator response speed, frequency and flexibility: A review and analysis. *Human Factors*, 9 (3), 221–238.

Welford, A. T. 1952. The psychological refractory period and timing of high speed performance: A review and a theory. *British Journal of Psychology*, 43, 2–19.

Wickens, C., S. E. Gordon, and Y. Liu. 1998. *An Introduction to Human Factors Engineering*. Upper Saddle River, NJ: Pearson Prentice Hall (Addison Wesley Longman Inc.).

Yarbus, A. L. 1967. *Eye Movements and Vision*. New York: Plenum Press.

5 Controls, Displays, and Interior Layouts

INTRODUCTION

The driver obtains information available from different displays (e.g., speedometer, fuel gauge, radio display, warning lights, warning sounds, symbols, labels, road views from the windshield and window openings, inside and outside mirrors, etc.) and generates outputs (e.g., moves pedals and the steering wheel, pushes buttons, turns stalks, etc.) to control the vehicle motion and/or change the states of different in-vehicle devices (e.g., change radio station). To obtain information from displays and operate controls, the driver uses various information-processing and control-activation capabilities. Depending on the levels of different capabilities and available resources, the driver may or may not make an appropriate control action. The time taken by the driver to make a control action will also depend on the amount of information that the driver will need to process and his or her information-processing capabilities. If the controls and displays are not designed for ease in performing these tasks, the driver may not be able to complete the tasks within the available time or make errors.

The controls and displays design research began with studies of pilot errors (Fitts and Jones, 1961a and 1961b). Soon after World War II, the Air Force launched a systematic study of errors made by pilots in situations where accidents and near accidents occurred. The pilots were asked to recall incidents where they almost lost an airplane or witnessed a copilot make an error in reading aircraft displays or operating controls. From the analyses of the data gathered from these critical incidents, Fitts and Jones found that practically all the pilots, regardless of experience or skill, reported making errors in using cockpit controls and instruments. They also concluded that it should be possible to eliminate or reduce most of these pilot errors by designing equipment in accordance with human requirements. Similarly, driver errors in using displays and controls can be reduced if they are designed in accord with the human engineering criteria.

Chapter 4 provided basic information on driver information processing, driver errors, and some information-processing-based guidelines for designing controls and displays. This chapter is intended to provide information on many important considerations and issues in designing and evaluating controls, displays, and their layouts.

CONTROLS AND DISPLAYS INTERFACE

The controls and displays are the interface between the human operator (the driver) and the machine (vehicle). Figure 5.1 illustrates this interface. Thus, the problem of controls and displays design is regarded as a problem of human–machine interface design.

In designing the driver interface, the vehicle designer should always keep in mind the following basic considerations:

1. Drivers will prefer to minimize their mental and physical efforts in using controls and displays.
2. People will prefer not to use what they do not understand.
3. Study the user population, user characteristics, and the variability among the users in the population. Characteristics of users (such as their age, familiarity with the equipment

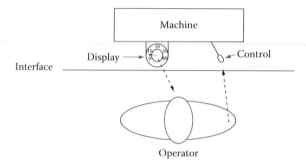

FIGURE 5.1   Controls and displays serve as the interface between the human operator and the machine.

and situations, expectations, eye and hand positions, and visual characteristics) must be considered during the design of controls and displays.

4. Study the usage conditions and driving situations when the controls and displays are used. These conditions and situations will provide an insight into the driver's informational needs, time constraints, environmental conditions (e.g., light levels), driver' tasks, and so forth needed to design controls and displays.

The following simplistic characteristics of good controls and displays are useful in understanding many basic considerations in designing a complex driver interface such as an automotive instrument panel—many of which currently have more than a hundred different controls and displays.

CHARACTERISTICS OF A GOOD CONTROL

1. A driver should be able to operate the control quickly with minimal mental and physical effort.
2. Minimal number of eye glances should be needed to complete the desired control operation (e.g., most turn signal activations are performed without looking at the turn signal stalk).
3. Any control activation should require minimal hand/finger movements (e.g., operating a control without moving hands from the steering wheel). (Note that hands-free operations or controls activated by voice commands can reduce or eliminate hand/finger movements.)

CHARACTERISTICS OF A GOOD VISUAL DISPLAY

1. The driver should be able to read and understand the display (i.e., obtain the required information) quickly with minimal mental and physical efforts.
2. The driver should be able to acquire the necessary information from a visual display in a few short eye glances.
3. The driver should not require any gross body movements (such as leaning over involving excessive head-and-torso movements) to obtain needed information. (Note that auditory displays do not require the driver to make any eye, head, or torso movements.)

The driver interface consists of controls, displays, and other items (e.g., air distribution vents and storage spaces for cups, maps, etc.). These items are placed mainly in instrument panel, center console, and door trim panels. Designing these interior components should not be approached as a number of individual problems of designing and locating each of the items in the interior space. The location and selection of these items should be considered a systems problem involving careful consideration and trade-offs between many different ergonomic and functional requirements. The drivers, through their past experience, expect certain items to be placed in certain expected locations (e.g., speedometer should be located high and in front of the driver, the turn signal control should be on the outboard stalk within finger-tip reach from the steering wheel rim, the headlamp switch should be located on the left side of the driver). The drivers also expect or prefer certain types of controls for certain control functions. The items should be also placed according to frequency of usage and their importance during vehicle usages. When preferred locations of many items cluster in the same general area (primarily around the steering wheel), the ergonomics engineer needs to prioritize and allocate spaces to different items by studying many considerations, principles, and design guidelines.

This chapter presents these issues in the following order: (1) types of controls, (2) types of displays, (3) general design considerations, (4) control design considerations, (5) display design considerations, (6) control and display location principles, (7) controls and displays evaluation methods, (8) ergonomics summary chart, and (9) some examples of controls and displays design issues.

TYPES OF CONTROLS AND DISPLAYS

There are many different types of controls and displays. A designer has to select the right types of controls and displays such that drivers can understand and associate them with their functions. The controls and displays must be designed so that they work as a system. Therefore, their layout (locations and orientations of both the displays and controls) is very important. They must be placed in expected locations and in obscuration-free areas, should be well labeled, and should move in directions expected by most drivers (i.e., they meet direction-of-motion stereotypes). They must be associated with each other in terms of characteristics such as physical, visual, and/or functional grouping. Many controls and displays are combined together. For example, many controls have displays such as identification and setting labels. Some displays have controls within them, for example, a touch screen (see Figure 5.2).

In-Vehicle Controls

The controls can be classified as follows:

1. Continuous versus Discrete: Continuous controls allow for the setting of a controlled parameter at any point within its control or movement range (or scale). Typical controls used for this application are rotary controls (e.g., volume control in audio products), slide switches, and levers. Discrete controls have detent (or preset) positions that only allow a user to set it at one of the detents. Typical examples of such controls are rotary fan speed controls (with off, low, mid, and high settings), detented slide controls, rocker switches, toggle switches, and gear shifter.

2. Push Buttons: Push button switches require the simplest kind of grasp called the contact grasp, which merely involves an extended finger to touch and apply force (usually about 1.8 to 5.3 N in low current switches) to activate a microswitch contact. They do not require the operator's fingers to grasp the control surface (as compared with a knob that requires a bending of the fingers and grasping before activating). Since many push or touch buttons can be activated with very little force, the drivers will prefer to have a rest area where the operating hand (or other fingers) can be supported close (within the hand/finger-reach distance) to the button. It should be noted that with the advances in electronics (low current circuits), the push button is probably the cheapest control to produce, and many small buttons can be mounted within small areas on the control panels and arranged in different patterns (e.g., keyboards). The operator's ability to precisely locate and activate a small button, especially following a large hand movement, can become a source of user complaint. Fitts' Law of hand motion discussed in Chapter 4 is important to reduce difficulty and movement time in reaching a push button.

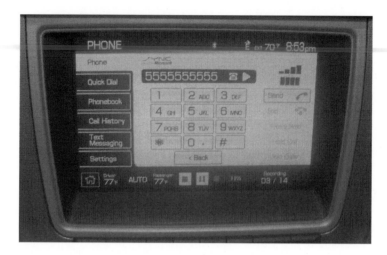

FIGURE 5.2   Touch screen with controls overlaid on the display surface.

3. Touch Screens: The touch screen displays have a control surface overlaid on the top of the display, and thus, it can be operated by finger touch without any extra input control device (see Figure 5.2). The touch areas are displayed visually, and they act as controls. The touch controls are the most direct form of the control interface as the information display and controls are on one surface. Thus, they have the potential to be intuitive, expected, and natural (like pointing a finger).

Since the touch control area is accessed by visual information about its location and size, the accuracy of activation will depend on the visibility of the touch button, finger size, application of the Fitts' Law (to reduce index of difficulty, see Chapter 4), and provision of hand/finger support. The advantages of the touch displays are as follows: (a) the input device is also the output device (generally the same display), (b) it reduces the hand and eye movements needed to find and touch/grasp the control, and (c) it eliminates finger bending and grasping motions (thus, it generally reduces activation time as compared with a control with a grasp area), and (d) no extra input device or packaging space is needed.

Some problems of the touch displays are (a) obstructions of the touch areas due to the operating finger and the hand, (b) broad finger contact does not allow for fine control movements, (c) long finger nails can cause difficulties in orienting a finger and achieving adequate skin contact, (d) lack of tactile feedback in conventional touch screens, (e) sunlight falling on the screen or reflections in the screen will reduce legibility, (f) a parallax will affect finger positioning accuracy (especially infrared-based touch systems), (g) a capacitive touch screen will not work with gloved hand or pen stylus (unless it is conductive), (h) finger touch can cause prints/smudge marks on the display surface, and (i) touch screen surface can wear and get scratched unless a hard coating or other protective materials are considered.

While touch screens have been available for many years, recent advances in resistive and capacitance technologies have increased their use in automotive applications. The resistive technology uses two layers of material separated by a gap. As pressure is applied to the surface of the film, the two layers touch, and an event is triggered. Thus, the resistive technologies actually require the user to touch and press the screen, whereas the capacitive touch technology only requires the properties of capacitive fields broken by the proximity of a finger to trigger an event. The key disadvantage for both of these technologies is that they lack force feedback typical with the conventional mechanical buttons. Since there is no mechanical actuation with the touch switch, it is difficult to determine if and when an event was triggered. However, graphic (visual), sound, and haptic (touch feel) feedback can be incorporated to reduce this deficiency. The visual feedback can be accomplished by simply changing the visual elements in the screen as soon as a finger touch is sensed. The sound feedback can be used by an internal sound device on the printed circuit board of the device, while a second option is to use the vehicle speaker system. An alternative to an audible indication is a haptic feedback upon a touch event. With this design, an actuator creates a perceptible tactile/force feedback. One method to accomplish this feedback is to use a floating structure with specified amplitude versus time vibratory movements or a crisp tactile feeling similar to a conventional switch. In general, users prefer at least a sound or haptic feedback in addition to the visual feedback. There are other advantages of touch switch sensors. These include reduced package space due to the thin switch structure and easy-to-clean surfaces with no edges or gaps. They are generally more durable due to a lack of moving parts or electrical contacts to wear out. They also can achieve a significant weight reduction due to using fewer parts.

4. Rocker Switches: Two-position rocker switches that select between two modes (e.g., on or off) are the most common switches on the automotive instrument panels. The nonprotruding portion of the rocker switch indicates the set mode (currently selected) of the switch. On the other hand, the protruding portion gives visual and tactile cues to indicate the mode available for future setting action. Society of Automotive Engineers, Inc. (SAE) standard J1139 (SAE, 2009) provides guidelines on the direction of activation of different automotive switches (including the rockers) in different mounting orientations. Some rocker switches have more than two positions. Such rocker switches are difficult to set at a required setting and are prone to setting errors. The difficulty arises because the rocker does not provide sufficient visual cues to determine its selected setting and other available settings, unless a visual display is placed next to the switch to provide the setting information.

5. Rotary Switches: Most common rotary switches have knobs that are grasped and rotated in the clockwise (to increase or turn on) or counterclockwise (to decrease or turn off) directions. The knobs can be designed in different shapes and sizes and with pointers or markings (e.g., rotary headlamp switches, thumb wheels, ring switches), and they can be mounted on different surfaces and in different orientations (e.g., rotary headlamp switch mounted at the end of a turn signal stalk). SAE standard J1139 (SAE, 2009) provides guidelines on the direction of activation of the switches in different mounting orientations.

The rotary controls can be continuous or discrete (detented). The continuous rotary can be set at any position within its range, whereas the discrete rotary can be set at one of its fixed detent positions.

Figure 5.3 shows a continuous rotary temperature control in the middle of a climate control. It incorporates a momentary push button "to turn on and off" the automatic feature of the climate control. The rotary control is also surrounded by five momentary push buttons to set different climate control modes. Each mode selection push button (except the off button) also has a small LED to indicate its status.

6. Multifunction Switches: There are many combination switches created to allow activation of many functions (e.g., a rotary switch that can be pulled or pushed, a rotary switch that can also be moved like a joystick [see Figure 5.4]; a stalk [e.g., a turn signal] control can have rotary and slide switches for wiper or light control; and push buttons can be arranged in different groups, layouts, and orientations). To indicate various available functions, visual labels and additional cues (e.g., shape, texture, color, and orientation) on the control or its associated display are generally necessary for successful implementation of such controls.

7. Programmable/Reconfigurable Switches: A programmable or reconfigurable switch can change its controlled function depending on its selected mode. For example, the BMW's i-Drive uses a multifunction rotary control mounted with its axis normal to the horizontal surface of the center console. The reconfigurable switch must provide the driver with clear

FIGURE 5.3 Rotary temperature switch with a push button in the middle (to set auto climate control) and surrounded by five push buttons to select modes for a climate control.

FIGURE 5.4 A multifunction rotary switch (in the middle) that can be rotated as well as moved in different directions like a joystick.

information on the selected function and the available settings. In general, since the driver needs to understand the "present" (or selected) mode of the switch and other possible options, the task of operating such a switch involves information gathering and processing which in turn may increase control operation times and errors in its activations.

8. Haptics Controls: They are essentially programmable switches that can change their functions and tactile feel characteristics (e.g., force-deflection profiles, feel of detents, gains and activation directions) depending on the selected mode. For the driver to understand their movement characteristics and "present" setting, the driver must get immediate tactile feedback or information through its associated visual or auditory display.

9. Voice Controls: A voice recognition system allows recognition of the driver's spoken words and sets system functions depending on the programmed functionality. Thus, in principle, the voice controls will not require a driver to make any hand or body movements (hands-free operation) to activate the controls (except to turn the voice control on). However, the voice controls may not be acceptable in all instances. Some reasons for unacceptability are as follows: (a) some drivers may not like to talk to the vehicle, (b) some may

have a temporary disability in voice generation (e.g., a sore throat), (c) noise in the vehicle may reduce accuracy of the voice recognition system, (d) delay in voice recognition, and (e) errors in voice recognition. Therefore, redundant controls and ability to turn off the voice control system should also be provided.

10. Types of Hand Controls Used in Automotive Products: A number of different types of controls are used in the automotive products. Some of the commonly used automotive controls are as follows:
    a. Momentary push button
    b. Latching push button (push button stays in latched position)
    c. Touch button
    d. Radio push buttons (selected function button remains pushed in or is highlighted)
    e. Multibutton mode selection (push buttons for different modes grouped together)
    f. Multibutton directional (a series of push buttons arranged to increase or decrease magnitude of a function, e.g., fan speed)
    g. Momentary rocker
    h. Latching rocker
    i. Hinged pull-up/push-down rocker (used commonly for power windows)
    j. Continuous rotary
    k. Detented rotary
    l. Continuous thumbwheel
    m. Detented thumbwheel
    n. Stalks (with detents and momentary switches)
    o. Stalk rotary continuous
    p. Stalk rotary detented
    q. Toggle switch
    r. Slider continuous
    s. Slider detented
    t. Levers continuous
    u. Lever detented
    v. Joystick

IN-VEHICLE DISPLAYS

The displays can be classified as follows:

1. Static versus Dynamic: A static display does not change its displayed information or characteristics with time, that is, the display content is fixed (e.g., a printed label). A dynamic display will change its characteristics, for example, a change in the graphics will indicate change in the magnitude of the displayed variable—like the speed indicated by a speedometer.

2. Quantitative versus Qualitative: A quantitative display will provide a numeric value or precise magnitude of a displayed parameter (e.g., value of instantaneous fuel economy displayed in km/L or miles per gallon). A qualitative display will show a category or change in category of information but not its precise value (magnitude or quantity). It will only indicate direction or category, for example, a qualitative temperature gauge will show if the temperature is within its normal range (or green zone) or in abnormal (or red danger) zone.

3. Symbolic or Pictorial Graphics: These displays provide information by using symbols or pictorial graphics that the user can associate after recognizing their meaning, for example, a low tire pressure warning symbol (a cross-section of a tire with an exclamation mark). Such symbols or graphics have the potential to be "language independent" and thus can be understood by users from different countries. It should be noted that many symbols used in

the automotive instrument panels were created by technical personnel, and such symbols show vehicle parts related to the displayed function (e.g., the low tire pressure symbol shows the cross section of the tire, the check engine symbol shows a side view of the engine, and the brake malfunction symbol shows a drum brake). Many nontechnical drivers will find such symbols difficult to understand until they develop familiarity with the symbols.

4. Dedicated versus Programmable: A dedicated display is a display that presents only a given (assigned) function, whereas a programmable (or reconfigurable) display can change its function and format depending on the selected display mode. For example, in an audio display, if the radio mode is selected, then the display will show the selected radio station frequency. However, in the CD mode, the same display will show the selected track being played.

5. Visual, Auditory, Tactile, and Olfactory: Displays can be presented in different forms (or mechanisms) so that they can be identified by using certain sensory modalities (e.g., vision, hearing, tactile, or olfactory). Visual displays allow presentation of the most complex form of information by use of words, numerals, and graphics with many enhancing variables such as color, luminance, highlighting, size, shape, font, movements, etc. However, the disadvantage of a visual display is that the driver must take eyes away from the road to look at the display (i.e., the eye axes need to be aimed at the display) and the operator must be attentive or alert to process the information available from the display. On the other hand, the auditory information can be presented from a region not visible to the driver; thus, the auditory display can be used to alert the driver. However, the amount of information that can be provided in an auditory display should be relatively short and simple; but it can be coded by use of loudness, pitch/tone, voice, musical notes, etc. Tactile display can provide information by a change in physical characteristics such as surface shape, size of grip, texture, movement, or vibrations of the grasp/contact area that can be sensed and perceived by touch. Olfactory display is based on sensation of odor or a smelly substance emitted by the display (e.g., the smell added in the natural gas to indicate a gas leak).

6. Head-Down versus Head-Up Visual Displays: Most visual displays used in the automotive products are mounted in the instrument panel and thus require the driver to look down by making eye and head movements (thus, they are head-down displays). The time required to acquire information from a visual display involves the time required to move the eyes and head (if required for sightline changes over about 20 degrees), refocus eyes, transmit the image information to the brain, and process the information. The head-down displays should be placed no more than 30 degrees down from the driver's normal horizontal line of sight to allow monitoring of at least some aspect of the forward scene through peripheral vision while viewing the display.

The head-up displays (HUDs) are displays that are placed at higher locations such that the driver does not need to tilt his or her head down to look at the display. The HUDs are generally projected on the windshield and focused at distances farther than the windshield so that the driver can view the display with little or no refocusing and without looking away from the forward scene.

The advantages of a HUD are as follows: (a) reduction in eyes-off-the-road time, (b) elimination of refocusing and eye-axes convergence movements (older drivers who cannot focus at near distances can view the HUD clearly), and (c) more time available to be spent on the road scene.

The disadvantages of a HUD are as follows: (a) targets can be masked by the HUD image; (b) the projected image may not be seen over brighter backgrounds in sunlight and glare (however, newer laser HUDs have higher luminance levels); (c) poor optics can cause annoyance (e.g., double images); (d) attention switching, distraction, and visual clutter; (e) possibility of "cognitive capture" (i.e., capture driver's attention) under less demanding visual environments; and (f) need for additional controls to adjust image brightness, to adjust image location, and to turn the display on or off.

7. Types of Displays Used in Automotive Products: A variety of different types of displays are used in automotive products. Some of the commonly used automotive displays are the following:
   a. Static symbol or icon (used for identification)
   b. Static word label (identification, units, and setting labels)
   c. Analog display with scale(s) and pointer(s)
   d. Analog display with bars (bar display)
   e. Digital display
   f. Graphic display with pictures or camera views
   g. Changeable message display
   h. Color change indicator
   i. Programmable or a reconfigurable display (screen)
   j. Auditory displays (beeps, tones, buzzers, voice messages)

DESIGN CONSIDERATIONS, ISSUES, AND LOCATION PRINCIPLES

SOME GENERAL DESIGN CONSIDERATIONS

The following statements describe some general design considerations that every designer of any control, display, or layout must remember.

1. Minimize time required to use a control or a display. The time required to operate a control is the sum of time taken to (a) find and recognize the control, (b) access the control, and (c) operate the control. The time required to use a display is the sum of time taken to (a) find the display and (b) read and interpret the information presented in the display. Thus, each additive component of the time should be reduced.
2. "People inherently seek to conserve their energy and reduce effort." They like to minimize their physical and mental workload. (Note: Most drivers like features such as remote key fobs, steering-wheel-mounted redundant controls, power windows, power mirrors, automatic transmissions, etc., which reduce the number of movements and force exertions.)
3. Prioritize and display only limited amounts of information such that the driver does not have to reduce attention to the basic driving tasks, especially when traffic demands are critical. The information that the driver does not need should not be presented. This will avoid unnecessary information clutter that may only increase time to obtain needed information.
4. Locate visual displays close to the driver's normal line of sight so that driver's eye movement time is reduced and peripheral detection (and/or monitoring) of visual cues related to the primary driving tasks is not compromised. Locations within about 30 to 35 degrees from the normal line of sight are generally convenient due to (a) the driver's ability to make quick eye movements without the use of large head movements or head turns and (b) higher retinal sensitivity and higher information transmission speed in the visual field within and closer to the fovea (see Chapter 4).
5. Usability of a product (with controls and displays) is dependent on the number of applicable ergonomic guidelines that the product meets. (Note: The guidelines are presented in the next section.) In general, user satisfaction will increase as a product meets a greater percentage of the design guidelines. Violation of one or more key design guidelines can make a product very difficult to use.
6. "What people don't understand, does not exist." If a user does not understand how a product or its features (e.g., controls or displays) are to be used, then he or she will not be able to use the feature, that is, the user will disregard the feature or it will not exist as a usage choice in the user's mind.

7. "Hands-free may not be risk-free." Voice displays and voice-activated controls can inter-
   fere with other visual tasks especially because verbal information processing generally
   requires more conscious attention. However, simple well-designed voice displays or voice-
   activated controls that do not overburden a driver's working memory are generally superior
   to visual–manual interfaces.
8. Seamless integration of available features is important with potentially greater number of
   new features to be offered in future products (see Chapter 17). (The integration of various
   features should be based on consideration of concepts such as coordination, compatibility,
   prioritization, consistency, etc.)

CONTROL DESIGN CONSIDERATIONS

In designing a hand control, the following issues must be considered:

1. Location: Controls should be located such that the drivers can easily find and reach them.
   The location of any automotive control should be based on driver expectancy (i.e., where
   most drivers expect the control to be located in the vehicle space). Ideally, the driver should
   be able to locate the control blindly (without looking at it). However, if the control is complex (e.g., it includes displays such as setting labels or it is combined with other controls) and
   if it is used while driving, then it should be located in a visible area, and the eye movements
   (i.e., magnitude of the angle with which the line of sight needs to be moved to view the control from the straight-ahead viewing direction) should be as small as possible (preferably no
   more than 30 degrees). The location of the control should also be based on its grouping with
   other controls with similar functions (e.g., grouping of all light controls or climate controls)
   and associations with locations of other displays and controls. SAE standard J1138 recommends locations for various primary and secondary hand controls (SAE, 2009).
2. Visibility: Controls should be placed in obstruction-free areas and should not require excessive head or torso movements to view them. Control size, color, luminance, and contrast
   with the background should aid the driver in quickly finding and recognizing the control.
   In some cases, it may not be necessary for the entire control to be placed within the visible
   area. A partially visible control can be found with a slight head movement (about ±50 mm
   lateral or less). Some controls can be found and operated without looking, that is, by blind
   positioning of hands and tactile and/or shape coding of the grasp areas of the controls. SAE
   standard J1050 provides procedure to determine obscuration free areas (SAE, 2009).
3. Identification: In general, a control should have an identification label or symbol that
   should be placed such that it is visible and is in close visual proximity to the control. Some
   controls can be identified by touch by providing unique shape, texture, or tactile coding.
   When controls are placed in close association or grouping with other controls or displays,
   they can be identified by the function of the associated items. Federal Motor Vehicle Safety
   Standard 101 (National Highway Traffic Safety Administration, 2010) and SAE standard
   J1138 (SAE, 2009) provide requirements on identification symbols and labels.
4. Interpretation: A control should be designed so that its operation (i.e., how it is operated
   or moved) is easy to understand and interpret. The configuration, shape, appearance, and
   touch feel (e.g., texture and soft-rubber-like feel) of the grasp area of the control should
   provide additional cues about its operability (direction of activation) with minimal reli-
   ance on the user's memory. The shape of the knob should be designed such that it invites
   certain actions that are compatible with its shape (e.g., rotary knob with a pointer needs to
   be rotated flat ends of stalks need to be pulled or pushed).
5. Control Size: The grasp area of control should provide sufficient surface to grasp, and
   clearances should be provided for hand/finger access (see Table 2.1 for finger dimensions).
   Additional clearances for gloved hand should also be considered for operation of primary

and frequently used controls in colder climates. The target size of frequently used and safety-related controls should be designed to minimize the hand movement time by considering Fitts' Law of hand motions (see Chapter 3). In addition, crash-protection standards that limit the amount of protrusions of the control knobs in the head swing zone must be considered in designing the shapes of the controls.

6. Operability: The direction of motion of a control should meet the population stereotypes specified in SAE standard J1139 (SAE, 2009). The control must provide feedback (visual, tactile, or auditory) to convey completion of the control activation movement. The control effort (torque or force needed to operate) should be less than 20% of the 5th percentile female maximum strength of the muscles producing the control movement. Such a low effort level not only assures that most users can operate the control, but it also makes the control operation experience acceptable. The feeling of smoothness during control movements, crispness of detent feel, and reduction in free play or "slop" are also important attributes in improving the perception of quality feel in automotive switches.

7. Error-Free Operation: The control should be designed to minimize the possibility of errors during operation (see Chapter 4 for types of human errors).

8. Inadvertent Operation: Important controls such as those that control vehicle motion or driver visibility (e.g., gear shifter, light switch) should be designed to assure that their settings will not be inadvertently changed during normal and accidental movements of an operator's hands and body parts (e.g., driver's knee or elbow bumping against a switch on the door trim panel). In such cases, additional clearances, recessing, or shields around the grasp area of the control should be considered.

VISUAL DISPLAY DESIGN CONSIDERATIONS

In designing any visual display, the following issues must be considered:

1. Findability and Location: The display should be located such that the driver can easily find it with a minimum search-and-recognition time and without any body movements (e.g., head or torso movements). The location of any automotive display should be based on driver expectancy (i.e., where most drivers expect the display to be located in the vehicle space), eye movements (i.e., magnitude of the angle with which the sightline needs to be moved from the straight-ahead viewing direction; preferably less than 30 degrees), and locations of other displays and controls (i.e., association of the display with other displays and controls). SAE standard J1138 recommends locations of various primary and secondary hand controls and embedded or associated displays (SAE, 2009).

2. Visibility: The display should be placed in an obstruction-free area and should not require excessive head or head-and-torso movements to view it. The display size, color, luminance, and contrast with the background should aid the driver in searching and quickly finding the display. SAE standard J1050 provides the procedure to determine obscuration-free areas (SAE, 2009; see Chapter 6).

   Figure 5.5 shows a view of an instrument cluster through the steering wheel. The speedometer is the most frequently used display. If we assume that a driver looks at the speedometer about three times per kilometer and drives about 20,000 km per year, the speedometer usage will be 60,000 times/year. Care must be taken to locate the speedometer such that at least 95% or more drivers will see the speedometer and the instrument cluster area without any obscurations by the steering wheel rim, spokes, the hub area, and the stalk controls. To maximize the visibility of the instrument cluster, the steering wheel, the instrument panel, and the stalks should be designed as a system by continuous evaluation of the visibility during the design process.

3. Identification: The appearance and content of the display should allow the driver to identify its function and displayed information. Placement of a closely associated identification

FIGURE 5.5 Instrument cluster located within obscuration-free area between the steering wheel rim, spokes, hub, and the turn signal stalk.

label or a symbol will help the driver in identifying the display and thus reducing any unnecessary time required in interpreting the function of the display. For some displays, the setting labels (e.g., "A/C" for the climate control) or units label (e.g., "MPH" for the speedometer) can provide sufficient information to identify the display. Thus, additional identification labels for some displays would not be necessary (e.g., a clock).

4. Legibility: All displays that have letters and numerals should be legible under day, night, and dawn/dusk conditions. The viewing distance, letter size, font, height-to-stroke-width ratio, width-to-height ratio, luminance contrast, background luminance, glare illumination and angle, etc., should be considered to assure that the display is legible to at least 65-year-old drivers. The legibility can be predicted by using available models (Bhise and Hammoudeh, 2004; also, see Chapter 12).

5. Interpretability: The content of the display should be evaluated to assure that its displayed information can be correctly interpreted (not confused) and understood by most drivers. The appropriate use of display type, layout, scales/pointers, stereotypes, use of colors, coding, frame of reference, number of similar-looking displays in close proximity, etc., should be evaluated to assure interpretability of the display (see checklist in Table 5.2).

6. Reading Performance: The driver should be able to read the needed information very quickly (preferably in a short single glance). The information acquisition time (e.g., reading time) and reading errors should be evaluated.

CONTROL AND DISPLAY LOCATION PRINCIPLES

In determining locations of the hand controls and displays and their associations, the following principles must be considered:

1. Sequence of Use Principle: The controls and displays should be located in the order of the sequence of use to reduce eye and hand movements. The driver eye fixation location and location of the hand (to be used for control operation) prior to the use of the control and display should be considered for possible reduction of eye and hand movements.

2. Location Expectancy Principle: Controls and displays should be located based on the driver's expectancy of location of the controls and displays. To establish controls and displays expectancy, high-volume vehicles in the market segment for which the vehicle is planned must be studied to determine the most common locations of primary and frequently used secondary controls.

FIGURE 5.6   Functional grouping of controls in the center stack.

3. Importance Principle: Controls that are perceived by the drivers to be important should be located close to the steering wheel. Important displays should be located close to the driver's forward line of sight.

4. Frequency of Usage Principle: Locate frequently used controls close to the steering wheel. Locate frequently used displays close to the driver's forward line of sight.

5. Functional Grouping Principle: Controls and displays associated with a similar function (e.g., light controls, engine controls, climate controls, and audio controls) should be grouped and located together for ease in finding and operating.

     Figure 5.6 shows a center stack unit with over 50 different controls. To reduce driver workload and confusion, the controls are grouped in seven rows. The lower two rows have climate controls and seat-temperature controls. The audio controls are grouped in the middle rows. The top row includes buttons for the trip and fuel consumption display. The rows of controls are separated by spaces that are covered by a bezel with a different appearance (due to differences in material, texture, and color) than the appearance of the push buttons and the rotary knobs. The continuous rotary controls are used to adjust audio volume, radio tuning, and temperature controls with the "clockwise-to-increase" direction of motion convention. The frequently used on/off switches for the audio and climate control and fan controls are placed closer to the driver on the left-hand side. The least frequently used heated-seat controls are placed in the bottom of the center stack. The hazard switch (safety related) is placed in the easy-to-reach top row. The display is also placed higher in the center stack, well within a 30-degree down angle from the horizon.

6. Time-Pressure Principle: Controls should be located close to the steering wheel or in a prominent area if they are to be used quickly and cannot be used under the driver's discretion due to demand from external situations (e.g., sudden fogging of the windshield—which requires quick operation of the windshield defrost switch, unexpected or erratic maneuver by other vehicles—which may require quick use of the horn switch, the high beam switch, or the hazard switch).

METHODS TO EVALUATE CONTROLS AND DISPLAYS

Given a list of functions to be incorporated in the interior of a new vehicle, different design alternatives (created by different designers and/or for different vehicle models) can create a large number of possible layouts and configurations using different types of controls and displays. Realizing that many current luxury vehicles have over 100 different controls and displays located in the instrument panel, console, and door trim areas, literally thousands of different layouts can be generated to meet different styling concepts. In general, creating a new design for the sake of change or innovation alone does not produce a superior design because the changes must be made to support functional improvements. Thus, it is important for the ergonomics engineer to evaluate alternate designs and select the few that are ergonomically superior.

The possible methods that can be used to arrive at one or more superior designs generally involve a combination of the following methods or approaches:

1. Apply methods, tools, models, and design guidelines available in various ergonomic standards (e.g., *SAE Handbook* [SAE, 2009], company practices, other regulatory requirements (e.g., Federal Motor Vehicle Safety Standard [National Highway Traffic Safety Administration, 2010], United Nations Economic Commission for Europe [ECE] [European Commission, 2000], lessons learned from customer feedback and complaints, and warranty experience).
2. Develop and apply ergonomic checklists and summarize the results of the checklists (see Tables 5.1 and 5.2).
3. Conduct a task analysis on selected operational tasks to uncover potential driver errors and to suggest product improvements (see Chapter 8 for task analysis).
4. Conduct quick-react studies to evaluate driver/user performance and preferences on selected product issues using field tests, laboratory, or driving simulator studies (see Chapters 13 and 14).
5. Conduct systematic drive evaluations using representative subjects. In these tests, the subject should be asked to perform a set of different tasks (e.g., turn on the radio and find an FM station of your preference, set climate control to reduce heat) that will require the driver to use all the controls and displays and provide usability ratings. The subjects could also be observed (or video recorded) to measure the time taken to read the displays and operate the controls. Any errors made using the controls and displays can be also observed. After the test, the subjects could be debriefed and asked to describe problems encountered and "likes and dislikes" in using the control and displays (see Chapter 15 for additional information on vehicle evaluation methods).
6. Inclusion of other leading competitors (benchmarking vehicles) in the above methods is also very useful in establishing the superiority of candidate designs.

Space Available to Locate Hand Controls and Displays

After the driver position is established in the vehicle package by locating the seating reference point, accelerator heel point, ball of foot/pedal reference point, and the steering wheel (see Chapter 3), the vehicle package engineer should create zones to place the controls and displays. The controls and displays zones are bounded by (1) the maximum reach zone, (2) the minimum reach zone, (3) the visible zone through the steering wheel, and (4) the 35-degree down-angle zone. Figure 5.7 presents a side-view drawing showing the sections of the zones considered for locating controls and displays.

Maximum Reach Zone: The maximum reach zone is developed by applying SAE standard J287 procedure, which will define the space (in front of the driver) that 95% of drivers can reach (SAE, 2009). The left-hand reach and right-hand reach boundaries are illustrated in Figure 3.21. To establish the reach zones, the members of the SAE Human Accommodation Committee conducted studies by measuring how far forward drivers seated in different vehicle bucks could reach forward (in the X direction) for different Y and Z locations, using a three-finger pinch grip holding a 25-mm diameter rotary knob. The reach-distance data (in X direction) from a vertical plane (called the HR plane, placed perpendicular to the X axis) at different Y and Z locations are provided in tabular form in SAE standard J287. The standard provides tables for different combinations of three variables: (a) restrained or unrestrained reach, (b) male-to-female ratios of drivers, and (c) G-factor values

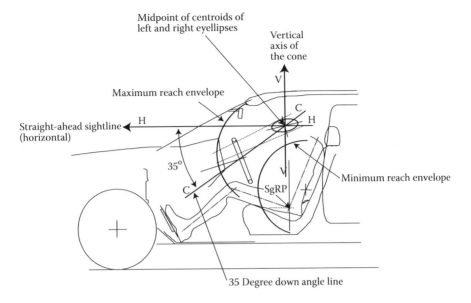

FIGURE 5.7 Maximum reach, minimum reach, 35 down angle, and visibility through the steering wheel. (The side view is shown in the vertical plane passing through the seating reference point.)

that define how the package is set up (the G-factor values range from −1.3 for a sports car with low-chair-height package to +1.3 for a heavy-truck package with high chair height). The reach distances provided in the tables are increased by 50 mm to obtain extended finger reach (to operate a push button) and decreased by 50 mm to obtain full-grasp reach (like grasping a floor-mounted shifter knob with all fingers turned inward).

Once the left- and right-hand maximum-reach surfaces (or envelopes) are placed in the drawing (or three-dimensional computer-aided design model), they set the forward boundary for locating all the hand controls that are operated during driving to assure that 95% of the drivers can reach the controls.

Minimum Reach Zones: The minimum reach zones define the closest distance (with respect to the driver's body) for locating controls. Figure 5.7 illustrates the minimum reach. The minimum reach zones are hemispherical in shape, with centers at the left- and right-elbow locations of the short driver who sits forward at the 5th percentile location in the seat track (i.e., at the 5th percentile location of the H-point defined using SAE standards J1517 or SAE 4004).

Minimum reach zones are not specified in any SAE standard. (Different vehicle manufacturers have their own internal guidelines.) They define the shortest distance (closest location from the driver) at which a hand control can be located without inconveniencing most drivers. Thus, controls should be located forward of the minimum reach zones.

A typical procedure for determining the minimum reach zones is as follows: (a) consider a short female driver sitting in the forwardmost seating position, (b) find her elbow points just touching the seat back (upper arms hanging down and elbows touching the bolsters of the seat back), (c) determine her (5th percentile female) elbow-to-knuckle length (i.e., lower arm length, which is about 400 mm). Set this dimension as the radius (R) of the hemispherical minimum reach envelopes, and (d) create two hemispherical zones of radius R, with centers at the elbow points.

Figure 5.7 shows a side view of both the maximum and minimum reach zones positioned in a vehicle package. The space between the maximum and the minimum reach zones is thus the space available to locate hand controls that are easily reached by most drivers.

Zone Defining the Visibility Through the Steering Wheel: Figure 5.7 shows the sightlines (dotted lines) of drivers from the 95th percentile eyellipses touching the steering wheel (inside of the rim and spokes and hub areas) and projecting on to a plane of the instrument cluster (which includes gauges such as the speedometer, fuel gauge, etc.). SAE standard J1050, Appendix D (SAE, 2009) provides a drafting procedure for determination of the boundary of the zone on the instrument cluster plane, which will be visible to most drivers. It is the area on the cluster plane that will be visible to at least one eye of all drivers whose sightlines from the left and right eyes are tangents to the 95th percentile left and right eyellipses, respectively.

Thirty-Five-Degree Down-Angle Cone: The 35-degree down angle zone defines the lower boundary for locating controls and displays such that the driver will not require eye movements larger than 35 degrees down in the vertical (elevation) direction (see Figure 5.7). The limit of useful peripheral/central vision is considered to be about 30–35 degrees for detecting stop lamps of lead vehicles while simultaneously looking down to view a display or a control.

Figure 5.7 shows the construction details to create a cone-shaped zone. The primary displays and controls (that cannot be blindly reached and operated) should be placed above this conical zone and close to the forward (straight ahead) line of sight. The procedure for generation of the cone is as follows: (a) find the midpoint (or cyclopean centroid) of the left and right eyellipse centroids, (b) construct 35-degree (or 30-degree) down line (shown as line CC in Figure 5.7) from the straight-ahead sightline (shown as line HH in Figure 5.7) passing through the cyclopean centroid, (c) rotate the 35-degree down line to form a cone around the vertical axis (shown as line VV in Figure 5.7) passing through the midpoint of the eyellipse centroids, and (d) locate displays and controls above the cone.

Space Available to Locate Controls and Displays: The space available to locate hand controls and displays is thus bounded by (a) space rearward of the maximum reach, (b) space forward of the minimum reach, (c) space above the 35-degree cone, and (d) visible regions seen through and around the steering wheel and the stalks.

CHECKLISTS FOR EVALUATION OF CONTROLS AND DISPLAYS

Since there are so many issues and principles to consider while designing and evaluating each control or a display, using a set of comprehensive checklists is an efficient approach used by ergonomists. Table 5.1 provides an example of such a checklist for evaluation a control. The questions in the checklist are grouped according to steps related to finding, identifying, interpreting, reaching, grasping, and operating a control. Scoring schemes can also be developed by placing appropriate weight to each question in each group for quantitative comparisons of ergonomic qualities of different controls. Scoring weights can be determined based on attributes such as importance of the control (urgency of usage), frequency of usage, and consequence of errors in not finding or using the control incorrectly.

Table 5.2 provides a similar checklist for evaluation of visual displays.

ERGONOMICS SUMMARY CHART

Table 5.3 provides an example of an ergonomics summary chart (called the "smiley faces" chart). The chart lists each control and display in different interior regions of the vehicle on the left-hand side of the table. The evaluation criteria are grouped into nine columns located in the middle of the table. The nine criteria groups are labeled as follows:

1. Visibility, obscurations, and reflections
2. Forward-vision down angle
3. Grouping, association, and expected locations
4. Identification labeling
5. Graphics legibility and illumination
6. Understandability/interpretability
7. Maximum- and minimum-reach distance
8. Control area, clearance, and grasping
9. Control movements, efforts, and operability

TABLE 5.1
Checklist for Evaluation of a Control

	No.	Question
Findability	1	Can this control be easily found?
	2	Is the control located in the expected region?
	3	Is the control visible from the normal operating posture?
	4	Are head or head-and-torso movements required to see the control?
	5	Is the control visible at night from the normal operating posture?
Identification	6	Is the control logically placed and/or grouped to facilitate its identification?
	7	Is the control properly labeled?
	8	Is the label visible?
	9	Can the label be read (legible?) from the normal operating posture?
	10	Is the label illuminated at night?
	11	Can the label be read (legible?) at night from normal operating posture?
	12	Can the control be identified by touch?
	13	Can the control be discriminated from other controls located close to it?
Interpretability	14	Can the control be confused with other controls or functions?
	15	Can an unfamiliar operator guess the operation of the control?
	16	Does the shape of the control convey/suggest activation directions?
	17	Does the control work like most other controls of that control type?
	18	Is the control grouped logically?
	19	Is the control placed within a group of controls that control the same basic function?
	20	Are there other controls within 2–3 in. that have similar visual appearance or tactile feel?
Control Location, Reach and Grasp	21	Is the control located within maximum comfortable reach distance?
	22	Can the control be reached without excessive bending/turning of operator's wrist?
	23	Is the target area of the control large enough to reach the control quickly?
	24	Can the control be reached without complex/compound hand/foot motions?
	25	Can the control be reached without torso lean?
	26	Can the control be grasped comfortably without awkward finger/hand orientations?
	27	Is there sufficient clearance while grasping the control?
	28	Is there sufficient clearance for a person with long (15 mm) fingernails?
	29	Is there sufficient clearance space for operator's hands/knuckles?
	30	Is there sufficient clearance to grasp the control with winter gloves?
	31	Is there sufficient foot clearance (if foot operated control)?
	32	Is the control located at "just-about-right" location?
	33	Is the control located too high?
	34	Is the control located too low?
	35	Is the control located too far?
	36	Is the control located too close to the driver?
	37	Is the control located too much to the left?
	38	Is the control located too much to the right?
	39	Is the control oriented to facilitate its operation?
	40	Is the control combined or integrated with other controls?
	41	Can the location of the control be changed when setting of any other control is changed?
Operability	42	Can the control be operated quickly?
	43	Can the control be operated blindly or with one short glance?
	44	Is the operation of the control part of a sequence of control operations?
	45	Can the control be operated without reading more than two words or labels?

continued

TABLE 5.1 (Continued)
Checklist for Evaluation of a Control

No.	Question
46	Can the control be operated without looking at a display screen?
47	Can the control be operated easily without excessive force/torque/effort?
48	Does the control provide visual, tactile, or sound feedback on completion of the control action?
49	Does the control provide immediate feedback (without excessive time lag)?
50	Does the control move without excessive dead space, backlash, or lag?
51	Is sufficient clearance space is provided for operating hand/foot as the control is moved through its operating movement?
52	Is regrasp required during operation of the control?
53	Can the control be moved without excessive inertia or damping?
54	Does the control direction of motion meet the direction of motion stereotypes?
55	Are more than one simultaneous movements required to operate this control?
56	Is the direction of screen/display movement related to the control movement compatible?
57	Is the magnitude of displayed movement related to the control movement "about right"?
58	Can the control be activated easily with gloved hand?
59	Can the control be operated easily by a person with long finger nails?
60	Does the surface texture/feel of the control facilitate its operation?
61	Can the operation of the control be performed with little memory capacity (five or less items)?
62	Are surfaces on the control rounded to reduce sharp corners and grasping discomfort during its operation?

TABLE 5.2
Checklist for Evaluation of a Visual Display

	No.	Question
Findability	1	Can this display be easily found?
	2	Is the display located in the expected region?
	3	Is the display visible from the normal operating posture?
	4	Are head or head-and-torso movements required to see the display?
	5	Is the display illuminated and visible at night from normal operating posture?
Identification	6	Is the display logically placed and/or grouped to facilitate its identification?
	7	Is the display properly labeled? (e.g., units shown)
	8	Is the label visible? (not obstructed or not obscured by glare/reflections)
	9	Can the label be read (legible) from the normal operating posture?
	10	Is the label illuminated at night?
	11	Can the label be read (legible) at night from normal operating posture?
	12	Can the display be identified by its appearance? (e.g., clock)
	13	Can the display be discriminated from other displays located close to it?
Interpretability	14	Can the display be confused with other displays? (e.g., similar in appearance)
	15	Can an unfamiliar operator guess the functionality of the display?
	16	Does the association of the display with a control convey its function?
	17	Does the display work like most other displays of that display type?
	18	Is the display grouped logically?
	19	Is the display placed within a group of displays or controls that have similar functions?
	20	Are there other displays within 2–3 in. that have similar visual appearance?
	21	Are any coding methods (color, shape, outlines, etc.) used to improve its comprehension?

TABLE 5.2 (Continued)
Checklist for Evaluation of a Visual Display

Display Location	22	Is the display located at a comfortable viewing distance?
	23	Is the display located close to the driver's primary line of sight (above the 35-degree down-angle cone)?
	24	Is the display area large enough to accommodate displayed information?
	25	Does the display appear cluttered?
	26	Is the display located at "just-about-right" location?
	27	Is the display located too high?
	28	Is the display located too low?
	29	Is the display located too far?
	30	Is the display located too close to the driver?
	31	Is the display located too much to the left?
	32	Is the display located too much to the right?
	33	Is the display oriented to facilitate its viewing?
Usability	34	Can the display be read quickly?
	35	If the display contains scales: Are the numerals, the scale(s) and pointer(s) easy to read? (consider: end points, progression, placement, orientation, size, and font of the numerals; scale markings: major/minor size, pointer length/width, obscuration of numerals by pointer, etc.)
	36	Can the display be read on bright sunny days with sun rays directed at the display?
	37	Is the display required? (does it serve useful function?)
	38	Is the reading of the display part of a sequence of display readings steps? (e.g., menus)
	39	Does the display provide other than visual (e.g., sound, vibrations, tactile) cues related to the displayed the information?
	40	Does the display provide immediate feedback (without excessive time lag) of a control action or change in status?
	41	Does the display change too slowly or fails to display quickly (inaction, damping or lag)?
	42	Is the display too sensitive to small changes in displayed function?
	43	Does the display direction of motion meet the direction of motion stereotypes?
	44	Are more than one simultaneous control movements are required to access this display?
	45	Is the direction of screen/display movement compatible to its related control movement?
	46	Is the magnitude of screen/display movement related to the control movement "about right"?
	47	Can the display be easily read by an older person at night?
	48	Does the background surface/texture/color of the display facilitate its readability?
	49	Can the displayed information be understood with little memory capacity (five or less items)?
	50	Are surfaces on or close to the display provide bright discomforting reflections of external or internal sources?

A 5-point rating scale (with 5 = highest score and 1 = lowest score) is used to evaluate the ergonomic guidelines in each of the above nine groups. The ratings are usually obtained from trained ergonomists (based on data obtained from three-dimensional computer-aided design model of occupant package, sitting in available interior bucks, and results from applicable design tools and models) and graphically displayed by using a graphic scale of smiley faces for each of the above nine groups for each item listed in each row. The chart provides an easy-to-view format that can be used to provide an overall ergonomics status of a vehicle interior and was found by the author to be a useful tool in various design and management review meetings. The objective of the ergonomics engineer is to convince the design team during the design review meetings to remove as many "black dots and black donuts" from the charts and increase the number of smiley faces by making the necessary design changes.

TABLE 5.3
Ergonomics Summary Chart

Ergonomics Evaluation: Vehicle X

Key

Rating				
Not Applicable ⊘	Low 1–2 ●	Mid 3 ○	High 4–5 🙂	

Control and Display Evaluation Criteria

No.	Interior Items: Controls, Displays, and Handles	Visibility, Obscurations, and Elfections	Forward-vision Down Angle	Grouping, Association, and expected Location	Identification Labeling	Graphics Legibility and Illumination	Understandability/ Interpretability	Maximum and Minimum Comfortable-Reach Distance	Control area, clearance, and grasping	Control movements, efforts, and operability	Comments: Specific Problems and Suggestions
1	Inside door handle	🙂	⊘	🙂	🙂	🙂	🙂	🙂	🙂	🙂	
2	Door pull handle	🙂	⊘	🙂	🙂	🙂	🙂	○	○	⊘	Requires some chicken-winging—should be moved forward 25–50 mm; cannot use full power grasp
3	Door lock	🙂	🙂	🙂	🙂	🙂	🙂	🙂	○	⊘	Difficult to push—touch area moves under bezel surface
4	Window controls	🙂	🙂	🙂	🙂	🙂	🙂	🙂	🙂	🙂	
5	Window lock	🙂	🙂	🙂	🙂	🙂	🙂	🙂	🙂	🙂	
6	Mirror control	🙂	🙂	🙂	🙂	🙂	🙂	🙂	🙂	🙂	
7	Turn signal stalk	🙂	🙂	🙂	🙂	🙂	🙂	🙂	🙂	🙂	
8	Wiper switch (right stalk)	🙂	🙂	🙂	🙂	🙂	🙂	🙂	🙂	🙂	

#	Item	C1	C2	C3	C4	C5	C6	C7	C8	C9	Comments
9	Ignition switch	●	⊘	☺	☺	☺	☺	☺	☺	☺	Requires head movement to see the control during key insertion
10	Cruise controls	☺	☺	☺	☺	☺	☺	☺	☺	☺	
11	Shifter	☺	⊘	☺	☺	☺	☺	☺	☺	☺	Located on console
12	Light switch (on left stalk)	☺	☺	☺	☺	☺	☺	☺	☺	☺	
13	Panel dim	☺	☺	☺	☺	☺	☺	☺	☺	☺	On left side of I/P
14	Parking brake (on console)	☺	☺	☺	☺	☺	☺	☺	☺	☺	
15	Hood release	●	⊘	☺	●	☺	☺	☺	⊘	⊘	Difficult to see from the driver's seat. Not labeled
16	Tachometer	☺	☺	☺	☺	☺	☺	☺	⊘	⊘	
17	Speedometer	☺	☺	☺	☺	☺	☺	☺	☺	☺	
18	Temperature	☺	☺	☺	☺	☺	☺	☺	⊘	⊘	
19	Fuel gauge	☺	☺	☺	☺	☺	☺	☺	⊘	⊘	
20	Oil pressure								⊘	⊘	Not on this vehicle
21	PRNDL display	☺	☺	☺	☺	☺	☺	☺	☺	☺	
22	Radio	●	●	☺	☺	☺	☺	☺	⊘	⊘	1/2–2/3 of the radio controls below 35 degrees; silver buttons on silver background difficult to locate quickly
23	Climate control	☺	●	☺	●	☺	☺	☺	☺	☺	Mode selector symbols difficult to read on silver background; low location and disassociated display

continued

TABLE 5.3 (Continued)
Ergonomics Summary Chart

#	Component	Ratings	Comments
24	Clock (top in center stack)	☺ ☺ ☺ ☺ ☺ ☺ ☺ ☺	
25	Backlight defrost	☺ ☺ ☺ ☺ ◉ ☺ ☺ ☺	Symbol difficult to read on silver background
26	Windshield defrost	☺ ☺ ☺ ☺ ◉ ☺ ☺ ☺	Symbol difficult to read on silver background
27	Traction control		Not on this vehicle
28	Parking brake	☺ ☺ ☺ ☺ ☺ ☺ ☺ ☺	On console
29	Hazard switch	☺ ☺ ☺ ☺ ☺ ☺ ☺ ☺	On center above the CD slot
30	Ash tray	☺ ☺ ☺ ☺ ☺ ☺ ☺ ●	Located too low
31	Cigarette lighter	☺ ◉ ● ☺ ☺ ☺ ☺ ●	Located too low; not enough clearance for finger grasp
32	Cup holder	☺ ☺ ☺ ☺ ☺ ☺ ☺ ⊘	Requires chicken winging; located too close and too low
33	Shifter	☺ ☺ ☺ ☺ ☺ ☺ ☺ ☺	
34	Glove box latch	☺ ☺ ☺ ☺ ☺ ☺ ☺ ☺	
35	Cruise control on/off	☺ ☺ ☺ ☺ ☺ ☺ ☺ ☺	On right spoke of S/W
36	Trunk release	☺ ☺ ☺ ☺ ☺ ☺ ☺ ◉	Difficult to see from the driver's eye point
37	Fuel fill door release	☺ ☺ ☺ ☺ ☺ ☺ ☺ ☺	On floor—left side

SOME EXAMPLES OF CONTROL AND DISPLAY DESIGN ISSUES

1. Speedometer Graphics Design

Figure 5.8 shows four analog speedometer graphic designs that were created to work with the same speedometer hardware. An ergonomics engineer was asked to evaluate the four graphic designs and recommend the best design. The top left design has the speed numerals inside the scale (tic marks), and the numerals are progression of 10's, which are easier to interpolate and read precise speed as compared with the top right and lower left designs. The top right design looks less cluttered due to the numerals presented in progression in 20's, whereas the lower left design has numerals ending is 5's, which the designer claimed may be "easier to use, because many of the speed limits on the roads are in multiples of 5 such as 25, 35, 45, 55 and 65 MPH." The lower right design was created by placing the speed numerals outside the scale such that the pointer cannot block any of the numerals.

To evaluate the designs, the ergonomics engineer created slides of the four speedometers with the pointer positioned at different random angular locations and presented the speedometers to the subjects using a 0.5-s exposure tachistoscopically. It was found that when the scales were presented in the numeric progression of 10's, the subjects could read the speeds with lower error rates and they also preferred using the speedometers (top left and bottom right). Additional information on scale design issues and progression of numbers in analog gauges can be found in the works of Van Cott and Kinkade (1972) and Sanders and McCormick (1993).

2. Power-Window Location

Figure 5.9 presents two different locations of power-window switches. The figure on the left shows the power-window rocker switches on the center console, whereas the figure on the right shows the power-window switches grouped and located on the driver's door. When drivers in an evaluation study were asked to drive the vehicles with the above two power-window switch configurations and asked to operate the switches at different times while driving, they said that the switch configuration located on the driver's door was more convenient to operate because the window switches "belong to the door" (because of their association) and they were placed higher (required only about 20- to 25-degree down-eye movements to look at them from the forward line of sight) as compared with

FIGURE 5.8   Four different speedometer designs.

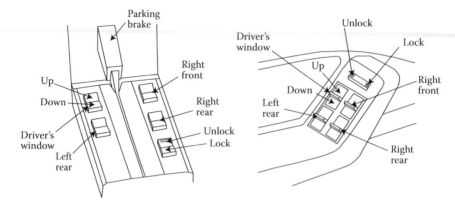

FIGURE 5.9 Two designs for power-window switches.

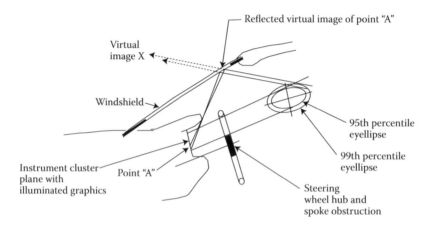

FIGURE 5.10 Sightlines analysis to predict locations of reflected images of illuminated graphics in the windshield.

the console mounted window switches, which required over 50 degrees of combined eye and head movements, which the drivers said were "buried way down in the console."

3. Annoying Reflections

Figure 5.10 shows a drawing created to analyze reflections of lighted setting labels on switches mounted on the lower part of the instrument panel. The drivers, especially those with taller eye locations, complained that at night they were able to see reflections of lighted switch labels in the windshield every time they looked at the inside rear-view mirror. The drivers complained that the reflections were "annoying" during night driving. The drawing in Figure 5.10 shows that rays from the lower part of the instrument cluster and the switches, when reflected into the windshield, could pass through the upper portions of the 95th and 99th percentile eyellipses. Thus, the tall drivers whose eyes will be located in that upper part of the eyellipses would be able to see the reflection of the lighted graphics (as virtual image, see Figure 5.10) from the lower side of the instrument panel via the windshield.

Other typical reflection problems of interior lighted components are the following: (a) Reflections of the lighted graphics of headlamp switches mounted on the left lower side of the instrument panel into the driver's side glass: These reflections in some vehicles occur along the driver's line of sight to the left outside mirror. Thus, every time the driver uses the left outside mirror, the reflections of the lighted headlamp switch can be seen. The reflections can mask some part of the driver's mirror field.

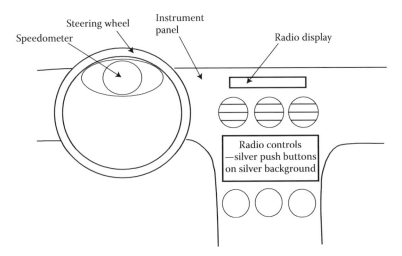

FIGURE 5.11   Disassociated radio controls with silver background.

(b) Reflections of the lighted instrument panel in the backlite (back window behind the driver) of pickup trucks: In some pick-up trucks, the drivers can see reflections of the lighted instrument panel in the backlite when the driver uses his or her inside mirror to view the rear field. Thus, an ergonomics engineer must perform reflection analyses to assure that any lighted components should not cause annoying reflections in the driver's field of view (see Chapter 6 for field-of-view evaluations).

4. Hard-to-Read Labels and Difficult-to-Operate Radio

Figure 5.11 shows a sketch of a center stack showing a high mounted radio display. The display was separated from its controls because of the three circular air registers that were placed between them. Further, the radio buttons and their surrounding background had a silver (brushed nickel) appearance. The alphanumeric labels on the radio buttons were printed in black, and thus, they appeared black during daytime. However, when the lights were turned on, the labels looked red. Because of the separation of the radio display from its controls and the difficult-to-read buttons on the radio during daytime (low visual contrast between the black printed labels and silver button background), the drivers during a drive test complained, saying that they liked the high radio display location but did not like the locations and the legibility of the radio buttons (see Chapter 12 for information on legibility evaluations).

5. Center Speedometer and Low Radio Location

Figure 5.12 shows a sketch of an instrument panel in which the instrument cluster was located high and at the center of the vehicle. The radio was located well below the 35-degree down-angle location. The initial impression of the drivers to the instrument panel was that they thought that the centrally located speedometer was very unusual, but after they drove the vehicle on a short trip, they found that they could easily look at the speedometer and monitor the road through the front windshield. The down angle to the speedometer was only about 12–15 degrees as compared with the traditional, slightly lower mounted speedometers (at about 20-degree down angle), which are viewed through the steering wheel. The radio was mounted well below the 35-degree down angle, and the drivers found it difficult to operate the radio because while looking at the radio, they could not see any of the forward road view in their peripheral vision. Additional information on the speedometer location study can be found in the work of Bhise and Dowd (2004).

6. Door Trim Panel Layout

Figure 5.13 shows two door trim panels on the driver-side door. Most drivers found that the left door was very difficult to open from the inside because the inside door opening handle was located too

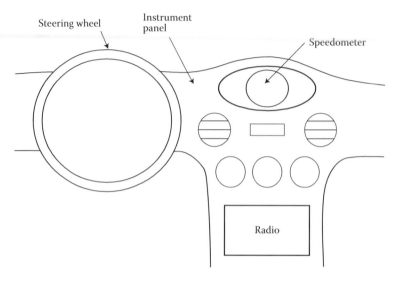

FIGURE 5.12   High center mounted instrument cluster and low mounted radio.

FIGURE 5.13   Layout of two door trim panels.

low. The door was also found to be difficult to close because the door pull cup was located too far rearward in the armrest (as it was located inside the minimum reach zone). The door trim panel in the right picture was easier to use because of the high-mounted inside door opening handle and the forward-mounted angled pull grip handle.

CONCLUDING REMARKS

A number of basic considerations in designing controls and displays were covered in this chapter. The basic considerations such as minimization of mental and physical efforts, visibility of displays, legibility and identification, reach-to-hand control, etc., are applicable to designing driver interfaces with hand controls and visual displays. With the advances in technologies related to driver information systems and entertainment systems, the content of the driver interface has been steadily increasing over time. Recent advances in reconfigurable displays , touch screens, multifunction and steering-wheel-mounted controls, voice controls, hard disks for data storage (Bhise, 2006), and Bluetooth communications have allowed even more choices in incorporating more features and their controls and displays. With these technologies, driver expectations and demands on new in-vehicle features have also increased substantially. The increases in functionality and features have led to additional issues such as interactions between different systems, priorities in displaying the status of different systems and operating their controls, increases in driver workload, and driver distractions. Part II of this book covers additional issues such as prediction of visibility and legibility, driver performance measurement, driver work load evaluation, and new technology implementation to aid the ergonomics engineer in designing future vehicles.

REFERENCES

Bhise, V. D. 2006. Incorporating hard disks in vehicles: Usages and challenges. SAE Paper 2006-01-0814. Presented at the SAE 2006 World Congress, Detroit, MI.

Bhise, V. D., and R. Hammoudeh. 2004. *A PC Based Model for Prediction of Visibility and Legibility for a Human Factors Engineer's Tool Box. Proceedings of the Human Factors and Ergonomics Society 48th Annual Meeting in*, New Orleans, LA.

Bhise, V. D., and J. Dowd. 2004. *Driving with the Traditional Viewed-Through-the-Steering Wheel Cluster vs. the Forward-Center Mounted Instrument Cluster. Proceedings of the Human Factors and Ergonomics Society 48th Annual Meeting*, New Orleans, LA.

European Commission. 2000. *The European Statements of Principles on Human Interface for Safe and Efficient In-Vehicle Information and Communication System*. In the Official Journal of European Communities, Document C(1999) 4786.

Fitts, P. M., and R. E. Jones. 1961a. Analysis of factors contributing to 460 "pilot errors" experiences in operating aircraft controls. Memorandum Report TSEAA-694-12. Aero Medical Laboratory, Air Materiel Command, Wright-Patterson Air Force Base, Dayton, OH, July 1947. In *Selected Papers on Human Factors in the Design and Use of Control Systems*, ed. H. Wallace Sanaiko. New York: Dover Publications Inc.

Fitts, P. M., and R. E. Jones. 1961b. Psychological aspects of instrument display I: Analysis of factors contributing to 270 "pilot errors" experiences in reading and interpreting aircraft instruments. Memorandum Report TSEAA-694-12A. Aero Medical Laboratory, Air Materiel Command, Wright-Patterson Air Force Base, Dayton, OH, July 1947. In *Selected Papers on Human Factors in the Design and Use of Control Systems*, ed. by H. Wallace Sanaiko. New York: Dover Publications Inc.

National Highway Traffic Safety Administration. 2010. *Federal Motor Vehicle Safety Standard*. Federal Register, Code of Federal Regulations, Title 49, Part 571, U.S. Department of Transportation, Washington, DC. http://www.gpo.gov/nara/cfr/waisidx_04/49cfr571_04.html.

Sanders, M. S., and E. J. McCormick. 1993. *Human Factors in Engineering Design*. 7th ed. McGraw-Hill Inc.

Society of Automotive Engineers Inc. 2009. *SAE Handbook*. Warrendale, PA: SAE.

Van Cott, H. P., and R. G. Kinkade (eds.). 1972. *Human Engineering Guide to Equipment Design*. Sponsored by the Joint Army-Navy-Air Force Steering Committee, McGraw-Hill Inc./U.S. Government Printing Press.

6 Field of View from Automotive Vehicles

INTRODUCTION TO FIELD OF VIEW

The objective of this chapter is to provide a background into ergonomic issues related to designing the daylight openings (called the DLOs, which include all the window openings including windshield and backlite) and other field-of-view-providing devices such as mirrors and cameras to assure that drivers can view the necessary visual details and objects in the roadway environment. This chapter will present methods used in the industry to locate various eye points in the vehicle space and draw sight lines used to measure and evaluate fields of view.

Linking Vehicle Interior to Exterior

The field-of-view analyses link the vehicle's interior design to its exterior design. The interior package provides the driver's eye locations and interior mirror. The vehicle exterior defines DLOs and exterior mirrors. Thus, the interior and exterior designs must be developed in close coordination to assure that drivers can see all the needed fields to drive their vehicles safely.

What Is Field of View?

The field of view is the extent to which the driver can see 360 degrees around the vehicle in terms of up and down (vertical or elevation) angles and left and right (horizontal or azimuth) angles of the driver's line of sight to different objects outside the vehicle. (The interior field-of-view issues related to visibility of controls and displays are covered in Chapter 5.) Some parts of the driver's visual field are obstructed due to the vehicle structure and components such as pillars, mirrors, instrument panel, steering wheel, hood, lower edges of the window openings (called the belt line), headrests.

Thus, what the driver can see while seated in a vehicle depends on the characteristics of (a) the driver, (b) the vehicle, (c) the targets (e.g., pedestrians, signs, signals, lane lines), and (d) the environment (e.g., road geometry, weather, day/night). Some details and variables associated with the above items are described below.

Driver Characteristics: The amount of visual information that a driver can obtain will depend on the following driver characteristics: eye locations in the vehicle (defined by the eyellipses in Society of Automotive Engineers Inc. [SAE] standard J941 [SAE, 2009]), visual capabilities (e.g., visual contrast thresholds, visual acuity, visual fields), visual sampling behavior (e.g., eye movements), head-turning abilities (e.g., range of comfortable head-turn angles), head movements (e.g., leaning forward, sideways, and head turning), information-processing capabilities, and driver age (which affects all driver capabilities).

Vehicle Characteristics: The vehicle characteristics related to the driver's field of view and visibility are window-opening dimensions and glazing materials (e.g., optical and installation characteristics of the glass), other components that can reduce visibility due to obscurations, glare and/or reflections of brighter objects (e.g., external light sources, high reflectance or glossy materials on vehicle surfaces), indirect vision devices (e.g., mirrors, sensors, cameras, and displays), wiping and defrosting systems, and vehicle lighting and marking systems (e.g., headlamp beam patterns, signal lamps).

Targets: The sizes, locations, and photometric characteristics of different targets and their backgrounds will affect the amount of information the driver can acquire. The targets include the roadway and traffic control devices (road geometry, lane markers, signs, signals, etc.), other vehicles (e.g., their visibility due to exterior lamps and reflectors at night), pedestrians (their size, location, movements, clothing reflectance, etc.), animals, and other roadside objects.

Environment: This will include the visual conditions due to illumination: day, night, dawn/dusk, weather (fog, snow, rain), other sources of illumination and glare (from sun, oncoming headlamps, street lighting), reflections of interior and exterior sources, and the roadway.

ORIGINS OF DATA TO SUPPORT REQUIRED FIELDS OF VIEW

The field-of-view requirements are based on information gathered from several sources that provide us an understanding of the sizes of the visual fields needed to drive vehicles safely. The information can be obtained from several sources. These include the following:

1. Targets: Objects or targets that must be seen and their locations with respect to the driver. The targets include delineation lines and traffic control devices (e.g., signs, signals) related to lane keeping, targets on possible collision course with the vehicle (e.g., pedestrians), and other vehicles in traffic. Further, information on road geometry (e.g., road widths, curvatures, grades, intersections) is needed to determine locations of the targets with respect to the driver.
2. Driver Capabilities: For example, capabilities of drivers to estimate magnitudes and changes in magnitudes of vehicle heading, distances to targets, speeds, and accelerations; driver response capabilities; and abilities to maneuver (control) vehicle.
3. Driver Feedback on Vehicle Design Features Related to Visibility: The vehicle manufacturers routinely collect feedback information from drivers and owners on vehicle features that they especially liked or disliked and problems experienced by drivers and their complaints. The visibility-related complaints such large obscurations due to vehicle pillars, smaller mirror fields, obscurations due to large headrests, mirrors, higher cowl and deck points, higher beltlines, etc., are useful in determining fields of view needed to satisfy customers.
4. Accident Experience: Studies involving analyses of accident databases have suggested relationship between accident rates and some vehicle design features under certain visibility situations (e.g., obstructions caused by wider A-pillars have been implicated to cause higher accidents in left-turning situations).
5. Past Research Studies: Results from field studies involving measurements of driver behavior, performance, and preferences while using different driver vision systems under different vehicle uses and traffic also provide information on acceptable and unacceptable sizes of driver's field of view. Examples of such studies are eye-glance measurements while using different mirror systems, effect of forward visibility on vehicle speed, distance estimation while using different mirrors, effects of plane and convex side-view mirrors on accident rates, and effect of hood visibility on parking performance.

TYPES OF FIELDS OF VIEW

The driver's fields of view can be classified based on (a) direct versus indirect fields and (b) field coverage with each eye separately, either eye or both eyes.

The direct field consists of the views that the driver sees directly by moving his or her eyes and head. These include (a) forward view (through the windshield), (b) rearview (directly looking back through the backlite [rear window]), and (c) side views (directly looking through the left and right side windows). The indirect field is what the driver views indirectly by use of imaging devices such as the inside mirror, outside mirrors, or display screens showing camera views or locations of objects detected by other sensors (e.g., blind area detection systems, backup sensors).

The monocular field is the view obtained by only one eye. Figure 6.1 shows a plan view of the human head showing the horizontal fields of view from the left and right eyes. The field of the left eye, shown as L, is the monocular field of the left eye. Similarly, R denotes the monocular field of the right eye. The ambinocular field is the sum of the fields obtained from the left and the right eye (L + R). The binocular field (B) is the common field seen by both eyes (i.e., only the overlapping portions of L and R).

Systems Consideration of 360-Degree Visibility

The direct and indirect fields that a driver can obtain while seated in a vehicle should be designed such that the driver can always obtain 360-degree visibility around his or her vehicle. Figure 6.2 shows that the driver in the subject vehicle (labeled in the figure as S) shown in the middle lane of the three-lane highway can see 360 degrees around his or her vehicle through direct, indirect, and peripheral visual fields. The fields that the driver can see directly are what he or she can see from his or her windshield and side windows by turning his or her eyes and head. The driver can also see objects in his or her indirect mirror fields shown in the figure as LMF, IMF, and RMF through the use of left outside, inside, and right outside mirrors, respectively. In addition, the figure shows peripheral direct fields (labeled as left peripheral and right peripheral fields) that the driver can see when the driver looks at the left outside or the right outside mirrors. Thus, the figure shows that the driver can see at least a part of any of the surrounding vehicles (labeled as vehicles L1 and L2 in the left adjacent lane, F as the following vehicle, and R1 and R2 as the vehicles in the right adjacent lane) in his or her direct, indirect, or peripheral fields under both daytime- and night-driving situations. Under night-driving situations, the driver can see at least one headlamp or side marker lamp of each vehicle in the left and right side adjacent lanes and at least one headlamp of the following vehicle in one of the three mirror fields or in one of his or her peripheral fields. Thus, all other vehicles around the subject vehicle can be seen either in the direct field of view or in a direct peripheral field (which was found to extend peripherally about 70 degrees from the line of sight while viewing into the left or right outside mirror [Ford Motor Company, 1973]) or in the indirect fields of the three mirrors.

The vehicle designer, thus, must assure that the direct and indirect fields (with proper aiming of the three mirrors) can provide 360-degree visibility of surrounding vehicles to most drivers (i.e., the driver can see at least a part of each of the surrounding vehicles). This means that the locations of the three mirrors, their sizes, the vehicle greenhouse (pillars and window openings), and the driver's

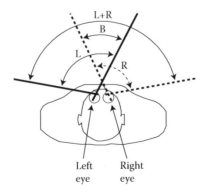

FIGURE 6.1 Left monocular (L), right monocular (R), binocular (B), and ambinocular (L + R) fields. The picture represents a plan view of the driver and horizontal fields of the left (L) and the right (R) eyes.

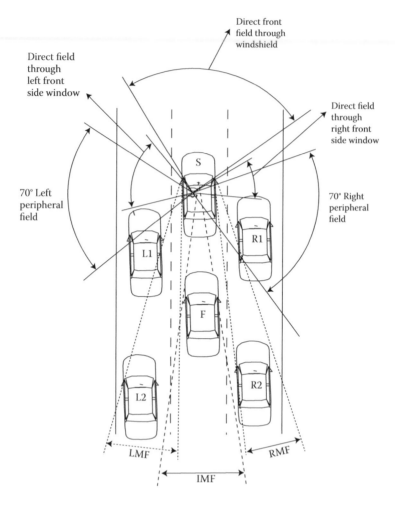

FIGURE 6.2 A 360-degree visibility from direct, indirect, and peripheral fields. (Note that the driver in vehicle S can see at least a portion of any of the vehicles on his front, sides, or rear.)

eye locations should be all designed as a system to assure that the driver can see any of the vehicles in the left and right adjacent lanes and directly behind and ahead in the driving lane.

MONOCULAR, AMBINOCULAR, AND BINOCULAR VISION

The views obtained by any one eye, sum of the fields of both eyes, and only the field common to both eyes are called monocular, ambinocular, and binocular fields, respectively. Figures 6.3 and 6.4 present photographs taken from the same vehicle to illustrate what the driver sees from his or her left and right eyes, respectively, while looking toward a left outside mirror. Since the driver receives information available from both the eyes and the brain fuses these images from both the eyes, Figure 6.5 shows superimposed views of both the monocular views.

To understand the differences between the two images and what the driver sees from the fused images, Figure 6.6 shows the outlines of the left A-pillar and the left outside mirror in Figure 6.5 from the two eyes. Figure 6.6 shows that right eye's view (shown in the dotted lines) and the left eye's view (shown in the solid lines) are different, that is, the corresponding lines of A-pillar and mirror obstructions do not fall on top of each other. Figure 6.7 shows the binocular obscuration (shaded areas) caused by the A-pillar and the outside mirror. The binocular obscuration is observed to be smaller than the

FIGURE 6.3   Monocular view from the left eye.

FIGURE 6.4   Monocular view from the right eye.

FIGURE 6.5   Superimposed view of views from the left and right eyes.

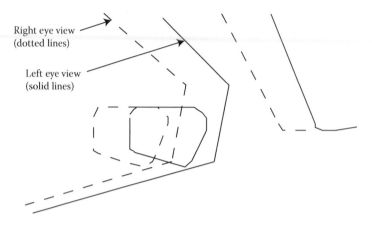

Right eye view
(dotted lines)

Left eye view
(solid lines)

FIGURE 6.6 Superimposed views from both eyes of outlines of obstructions caused by the A-pillar and the side-view mirror.

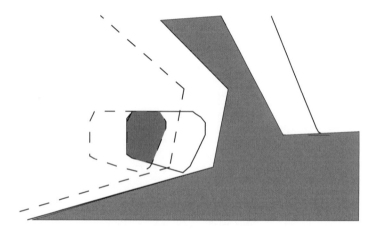

FIGURE 6.7 Binocular obstructions (shaded areas) caused by the A-pillar and the side-view mirror.

obscuration in either the left or right monocular views. This is because some portions of obscuration in the left eye are visible to the right eye and vice versa. Thus, the binocular obscuration (i.e., the field not visible to both the eyes simultaneously) is always smaller than any of the monocular obscurations.

Figure 6.8 shows another interesting effect. It shows the ambinocular outside mirror field (shaded area that covers the reflected field in the rear), which is what the driver will see as the total mirror field from the left and the right eyes together (assuming that the mirror surface is the same as the shroud that surrounds the mirror). Thus, for a two-eyed driver, the ambinocular mirror field is larger than what he or she can see with either eye. The binocular field through the mirror, where the driver can see the same field with both eyes (i.e., the overlapping field), is smaller than the field seen by either eye. Since the objects in the binocular mirror field are viewed by both eyes (and images of the object seen by each eye are slightly different), the driver gets additional information, which generally improves the perception of depth (or distances) of the objects seen in the mirror field. The portions of the ambinocular field that do not contain the binocular field are only seen by one eye. Any object seen monocularly (i.e., only with one eye) does not get the additional cues (from the other eye), and therefore, the driver's judgment of depth and location of the object seen only in the monocular field is less precise than the judgments made when the object is seen in the binocular field.

This is also an important issue in realizing that if a single camera (nonstereoscopic) is mounted outside the vehicle to replace an outside mirror, the camera view provided to the driver on a display

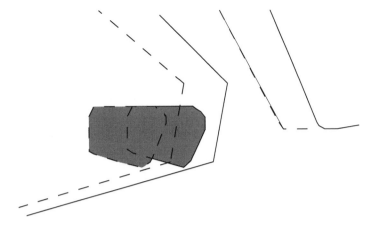

FIGURE 6.8   Ambinocular mirror field from the left and right eyes.

screen mounted inside the vehicle can only provide a monocular view. The camera view, thus, lacks the binocular cues obtained by the driver using a plane outside mirror.

The field-of-view determination procedures, therefore, must measure the monocular, ambinocular, and binocular views of the driver and evaluate (a) locations and sizes of binocular obstructions and (b) the sizes of ambinocular fields.

FORWARD-FIELD-OF-VIEW EVALUATIONS

To determine if a vehicle design will provide satisfactory forward field of view, the vehicle designer must conduct many analyses to evaluate different driver needs and requirements. Some important issues related to the forward field of view are presented below.

UP- AND DOWN-ANGLE EVALUATIONS

The up angle (A60-1) from the tall driver's eye points (95th percentile eye location) and down angle (A61-1) from the short driver's eye points (fifth percentile eye location) are generally determined by drawing tangent sight lines (in the side view) from the 95th percentile eyellipse to the top and bottom edges of the windshield (DLO, i.e., only the transparent area not covered by the black-out paint) as shown in Figure 6.9 (refer to SAE Standard J1100 for definitions of the up and down angles). The above angles are measured in the vertical plane passing through the driver centerline (i.e., through the driver's seating reference point [SgRP]) and using a mid eyellipse. The angles are measured with respect to the horizontal. A smaller up angle (A60-1) indicates that a tall driver will experience difficulty in looking at high-mounted targets (e.g., the tall driver may have to duck his or her head down to view a high-mounted traffic signal while waiting at an intersection). On the other hand, a smaller down angle (A61-1) indicates insufficient visibility for the short drivers over the cowl/hood area.

The visibility of short drivers should be also evaluated by drawing tangent sight lines over the top of the steering wheel, top of the instrument panel binnacle (most upward protruding parts), and over the hood from the short (2.5 or 5th percentile female) driver's eye locations.

VISIBILITY OF AND OVER THE HOOD

1. The visibility of the road surface (i.e., the closest longitudinal forward distance from the front bumper at which the road surface is visible, also called the ground intercept distance) is of critical concern to many drivers. The problem is worse for short drivers. In general, most drivers want and like to see the end of the hood, the vehicle corners (extremities), and the road at a

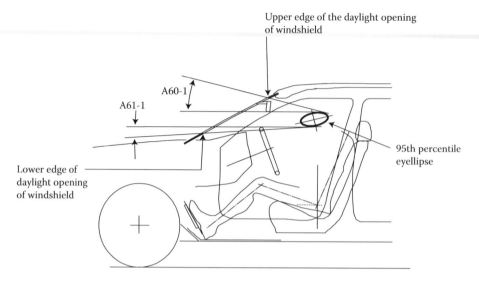

FIGURE 6.9   Tall driver up angle (A60-1) and short driver down angle (A61-1).

close distance. As more aerodynamic vehicle designs with low front ends were introduced in the United States (after the mid-1980s), many drivers who were accustomed to the long hoods with their visible front corners complained about not being able to see the ends of their hoods.
2. The view of the hood provides better perception of the vehicle heading with respect to the roadway, providing a feeling of ease in lane maintenance and while parking. (Note: Racing cars have a wide painted strip over the hood at the driver centerline to provide highly visible vehicle-heading cue in their peripheral vision).
3. Drivers in heavy trucks with long hoods experience larger obstruction of the road due to the hood. The problem can be severe if the obstruction is large enough to hide a small vehicle (e.g., bike rider, sports car) located in front of the hood. The problem often occurs when the truck is behind a small vehicle while waiting at an intersection.

COMMAND SITTING POSITION

1. The command sitting position provides the feeling of "sitting high" in the vehicle. It is opposite to the feeling of "sitting in a well" or "sitting too low" in the vehicle.
2. For command sitting position, provide (1) higher SgRP location from the ground, (2) low cowl point, (3) low beltline, (4) adequate visibility of the hood (at least 1 degrees of angle subtended by the visible part of the hood at the short driver's eyes), and (5) greater visibility of the roadway (shorter ground intercept distance from the front bumper). It should be noted that the command sitting position is one of the key positive attributes of an SUV.
3. The above command sitting feeling is also appreciated by short female drivers (with 2.5 or 5th percentile female eye height).

SHORT DRIVER PROBLEMS

Short drivers are drivers with shorter (fifth percentile and below) sitting eye heights and/or shorter (fifth percentile and below) leg lengths. The visibility problems encountered by such short drivers are the following:

1. Obstruction of the road by the steering wheel (top part of the rim) and instrument panel (or cluster binnacle, causing smaller down angle [A61-1]).

2. Unable to see any part of the hood (no visibility of the front end of the hood). (Note: Providing a raised hood ornament near the front of the hood can provide useful information in maintaining vehicle heading. Similarly, providing visibility of the corners of the hood or ends of the front fenders [via placement of "flag poles" as provided on some trucks with longer hoods] can improve ease in parking and lane maintenance.)

3. The outside side-view mirrors may obscure the forward direct view. (The upper edge of the side-view mirrors should be placed at least 20 mm below the fifth percentile female's eye point.)

4. The closest distance at which a driver can see the road (over the hood) is much longer for the short driver than are the visible distances for other drivers.

5. The shorter drivers will experience reduced rear-visibility problems during reversing or backing up (especially with a higher deck point and taller rear headrests). (Note: One check that many package engineers consider is whether a short driver can see a 1-m high target [simulating a toddler] in the rearview while backing-up in the direct rearview with the driver's head turned rearward and also while looking in the inside mirror.)

6. Since short drivers (with shorter leg lengths) sit more forward in the seat track, the driver's side A-pillar will create a larger obscuration in the forward field of view for short drivers as compared with tall drivers.

7. Short drivers require larger head-turn angles to view side-view mirrors due to their more-forward sitting position as compared with the taller drivers. This problem is more severe for short drivers with arthritis (typically older short females with a shorter range of head-turn angles).

TALL DRIVER PROBLEMS

The tall drivers are the drivers with greater (95th percentile and above) sitting eye heights and/or longer (95th percentile and above) leg lengths. The visibility problems encountered by such tall drivers are the following:

1. External objects placed at higher locations, placed above the upper sight line at up angle (A60-1 in Figure 6.9), may be obstructed from the view of tall drivers (e.g., a tall driver may have to duck his or her head down to view overhead traffic signals at intersections). The visibility near the top portion of the windshield is further limited by the shade bands and/or the black-out paint applied to the windshields.

2. The inside rearview mirror may block the tall driver's direct forward field. Therefore, the lower edge of the inside mirror should be placed at least 20 mm above the 95th percentile eye points.

3. The tall driver also sits farther from the mirrors due to a more rearward sitting position. Thus, the mirrors provide smaller fields of view to the tall drivers as compared with mirror fields of other drivers.

4. The tall driver may have more side visibility problems because of (a) more-forward B-pillar obscurations in direct side viewing (because the tall driver sits more rearward sitting position compared with other drivers) and (b) more-forward peripheral awareness zones (i.e., the peripheral zones do not extend as much rearward as for the shorter drivers [see Figure 6.2]) while using the side-view mirrors.

SUN VISOR DESIGN ISSUES

1. The lower edge of the sun visor under drop-down condition (called the sun visor dropped-down height) and its length should be designed to prevent the incidences of direct sunlight in the driver's eyes from different sun angles from the windshield and driver's side window.

2. The sun visor dropped-down height should be adjustable, and it should be capable of dropping down to accommodate the short driver's needs.

3. If the sun visor hinge mechanism becomes loose, the sun visor may accidentally swing and drop down and cause obstruction in the forward field of view. This obstruction will be more severe for taller drivers.

WIPER AND DEFROSTER REQUIREMENTS

SAE standards J902 and J903 (SAE, 2009; also Federal Motor Vehicle Safety Standards [FMVSS] 103 and 104; National Highway Traffic Safety Administration [NHTSA], 2010) provide requirements on how to establish areas in the driver's forward field that must be defogged (or defrosted) and wiped by the wipers, respectively. The requirements specify the sizes of these areas and percentages of each area that must be covered (cleaned) by the defoggers and wipers. The areas are specified by establishing four tangent planes to the 95th percentile eyellipses.

The wiper sweep areas are controlled by areas A, B, and C (SAE J903 in SAE, 2009). The wiper sweep area must be designed so that at least 80% of area A, at least 94% of area B, and at least 99% of area C must be wiped by the wipers. Areas A, B, and C are defined by drawing up, down, left, and right tangent planes to the eyellipses as shown in Figure 6.10. For passenger cars, area A is bounded by the upper tangent plane at 10-degree up angle, the bottom tangent plane at 5-degree down angle (see side view in Figure 6.10), the left vertical tangent plane at 18-degree angle to the left eye ellipse, and the right vertical tangent plane at 56-degree angle to the right eyellipse (see plan view in Figure 6.10). Similarly, the angles defining area B are 5 degrees up, 3 degrees down, 14 degrees left, and

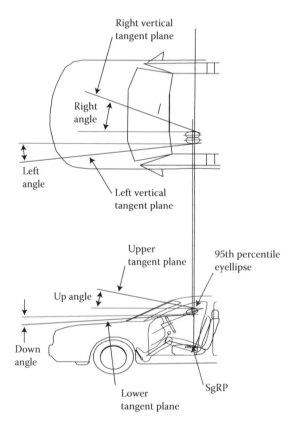

FIGURE 6.10   Plan and side views showing the four tangent planes that define the wiping areas to be covered by the wipers.

53 degrees right. The angles defining area C are 5 degrees up, 1 degree down, 10 degrees left, and 15 degrees right. The wiper area requirements for trucks (over 4500 kg), buses, and multipurpose vehicles are specified in SAE standard J198 (SAE, 2009). Since the driver in these commercial vehicles generally sits higher than that in passenger cars, the angles defining wiper areas A, B, and C are specified in SAE J198 as functions of the height of the SgRP above the ground.

OBSTRUCTIONS CAUSED BY A-PILLARS

The left and right front roof pillars (called the A-pillars), depending upon their size and shape of their cross sections at different heights with respect to the driver's eye locations, can cause binocular obstructions in the driver's direct forward field of view. The obstructions can hide targets such as pedestrians and other vehicles during certain situations. Figure 6.11 shows that during an approach and left turn through an intersection, a pedestrian crossing the street on the driver's left side and vehicles approaching from the driver's right side can be partially or completely obscured by the left and right A-pillars, respectively.

The vehicle body designers must conduct visibility analyses of such situations and minimize the obstructions caused by the pillars. The body engineers make trade-offs between (a) increasing the cross section of the pillars to meet the roof-crush requirements in FMVSS 216 (NHTSA, 2010), (b) applying padding to reduce head-impact injuries (i.e., increasing A-pillar width), and (c) reducing the binocular visual obstructions simultaneously for left- and right-side viewing . As the vehicle design progresses, there are other obstruction-related issues such as (a) thickness of the rubber seals used to secure the windshield to the pillars, (b) black-out paint applied to the glass to hide the joints and improve appearance from the exterior, and (c) manufacturing variations in pillar cross-sectional areas with respect to the SgRP (e.g., tolerances in sheet metal of the order of 4–8 mm). These three issues generally tend to increase the obstructions caused by the pillars.

SAE standard J1050 in its Appendix C (SAE, 2009) provides a procedure to measure the visual obstruction caused by the A-pillar. Figure 6.12 illustrates the binocular obstruction angle obtained by drawing sight lines tangent to the cross section of the left and right pillars.

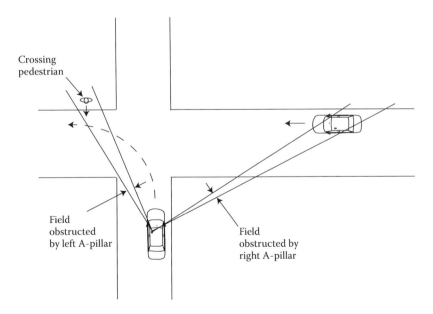

FIGURE 6.11 Obstructions caused by the left and right A-pillars during left turning at an intersection.

The obstruction angles, β_L and β_R, are determined by using the following steps:

1. The obstruction is measured in the horizontal plane passing through the eye ellipse centroids. Thus, Figure 6.12a shows the left A-pillar cross section and the 95th percentile eyellipses in the horizontal plane at the height of the centroids of the two ellipses.
2. An eye point closest to the pillar cross section is selected. Point L, thus, represents the left eye point on the left eyellipse that is closest to the pillar cross section (see Figure 6.12a). The closest eye point is selected because the obstruction angle subtended by the pillar will be largest from the closest eye point. Point R represents the right eye (notice that it is located on the right eyellipse, and it is closest to the cross section of the A-pillar), and point P represents the neck pivot of the driver. According to assumptions in SAE standard J1050, the distance between the left (L) and right (R) eyes is 65 mm, and the pivot point is 98 mm behind the midpoint of the two eyes (L and R).
3. To look in the region toward the pillar, the driver may need to make some eye movement and head turn. The maximum eye movement that most people will make without turning their head is about 30 degrees. Therefore, SAE standard J1050 allows a maximum of 30-degree eye movement in the horizontal plane. The sight lines from the left eye and the right eye are turned 30 degrees toward the pillar, and then, if required, the head is turned counterclockwise around the pivot point P until the left eye sight line is tangent on the left side of the pillar cross section. This turned head position is shown in dotted lines in Figure 6.12a. The left and right eye points at the turned head position are labeled as L_t and R_t, respectively.
4. From the right eye (R_t), a sight line tangent to the right side of the pillar cross section is drawn. This line is labeled as "sight line from right eye" in Figure 6.12a. The binocular obstruction angle of the pillar is shown in the figure as β_L, which is the angle between the "sight line from right eye" and the "sight line from left eye."
5. To compute the binocular obstruction angle caused by the right A-pillar β_R, a similar analysis should be conducted by first determining the closest right eye point (on the right eyellipse) to the cross section of the right A-pillar (see Figure 6.12b).

The FMVSS 128 (which was enacted in 1978 and later rescinded) had set 6 degrees as the criterion for maximum allowable A-pillar binocular obstruction. This requirement is still considered by many vehicle designers as an unwritten guideline for designing A-pillars. (It should be noted that the procedures for measurement of pillar obscurations in the European requirements are different from the SAE procedure described above.)

MIRROR DESIGN ISSUES

REQUIREMENTS ON MIRROR FIELDS

For vehicles sold in the United States, the inside and outside mirrors should be designed to meet the field-of-view requirements specified in FMVSS 111 (NHTSA, 2010). Figures 6.13 and 6.14 show the minimum required fields for inside and driver's side outside mirrors, respectively, for the passenger cars.

The inside plane mirror should provide at least a 20-degree horizontal field, and the vertical field should intersect the ground plane at 61 m (200 ft) or closer from the driver's SgRP to the horizon (see Figure 6.13). A procedure for determination of the mirror field of view is covered in a later section (see Figure 6.15).

The driver's side outside-plane mirror should provide (as specified in the FMVSS 111 [NHTSA, 2009]) a horizontal field of 2.4 m (8 ft) width at 10.7 m (35 ft) behind the driver at ground level, and the vertical field should cover the field from the ground line at 10.7 m (35 ft) to the horizon (see Figure 6.14).

(a) Obstruction due to left A-pillar

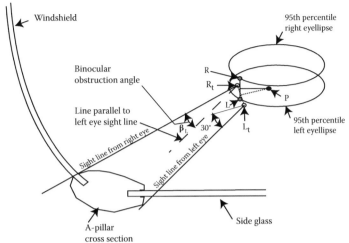

(b) Obstruction due to right A-pillar

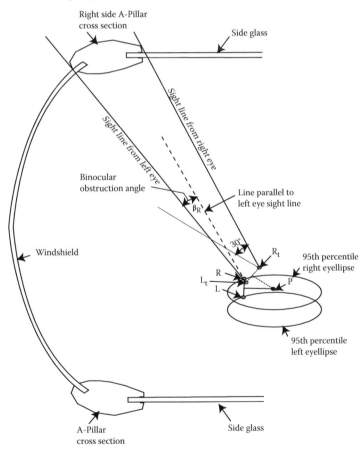

FIGURE 6.12 Determination of binocular obscuration angle due to the left (a) and right (b) A-pillars.

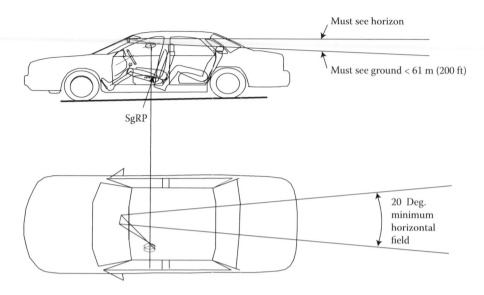

FIGURE 6.13   Inside mirror field required for passenger cars.

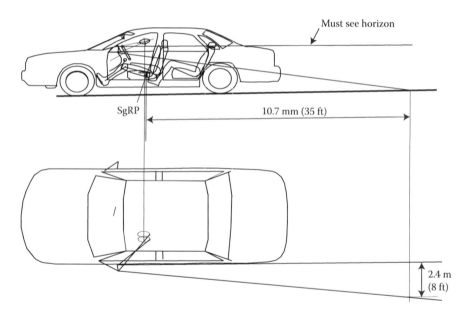

FIGURE 6.14   Driver's side mirror field required for passenger cars.

MIRROR LOCATIONS

Inside Mirror Locations

The inside mirror should be located with the following design considerations:

1. The mirror should be placed within the 95th percentile maximum reach envelope with full hand grasp using SAE standard J287 procedure.
2. The lower edge of the mirror should be located at least 20 mm above the 95th percentile driver's eye height. This assures that the mirror will not cause obstruction in the forward direct field of view for at least 95% of the drivers.

3. The mirror should be placed outside the head swing area (during frontal crash) of the driver and the front passenger (refer to FMVSS 201, NHTSA, 2010).

Outside Mirror Locations

The driver's side outside mirrors should be located with the following design considerations:

1. The driver's side outside mirror should be located such that a short driver who sits at the forwardmost location on the seat track should not require a head-turn angle of more than 60 degrees from the forward line of sight.
2. The upper edge of the mirror should be placed at least 20 mm below the fifth percentile driver's eye location to avoid obscuration in the direct side view of at least 95% of the drivers.
3. The mirror aiming mechanism should allow for a horizontal aim range large enough for a short driver to see a part of his or her vehicle and a tall driver to aim outward to reduce the blind area in the adjacent lane (to meet the 360-degree requirement illustrated in Figure 6.2).
4. In addition, to improve the aerodynamic drag and wind noise, the mirror housing design needs (reduced frontal area) should be considered along with the reduction in obscuration caused by the mirror and the left A-pillar in the driver's direct field of view (see Figure 6.7). It should be noted that improving the aerodynamic drag is a trade-off issue with the mirror field of view as it requires reduction in size of the outside mirrors.

The passenger's side outside mirror is generally located symmetrically to the driver's side outside mirror. FMVSS 111 does not require an outside passenger mirror on passenger cars or a truck if the inside mirror meets its required field of view. However, if the passenger's side outside mirror is provided, FMVSS 111 requires it to be a convex mirror with radius of curvature not less than 889 mm and not more than 1651 mm (NHTSA, 2010). FMVSS 111 also provides alternate requirements for trucks and multipurpose vehicles that cannot provide any useful field from their inside mirrors due to blockage by cargo or passenger areas.

PROCEDURE FOR DETERMINING DRIVER'S FIELD OF VIEW THROUGH MIRRORS

SAE standard J1050 presents a procedure to determine the field of view through a mirror (SAE, 2009). To determine the horizontal field of view that at least 95% of the drivers can view, one begins with a given mirror with its known location and size and the location of the 95th percentile eyellipses in the vehicle space. Figure 6.15 (see top figure) presents a plan view showing the inside mirror (on the right side of the driver) and the 95th percentile eyellipses.

The horizontal mirror field of view is determined by using the following steps:

1. The mirror surface is shown in Figure 6.15 (top figure) as MN, and the left and right eyellipses are drawn. The aim of the mirror is generally determined iteratively to assure that the ambinocular mirror field defined by the reflected sight lines from the left and the right eyes is contained within the backlite.
2. First locate the eye point that is farthest from the mirror. The farthest eye point is selected because it provides the smallest mirror field. The left eye point labeled as L on the left eyellipse is the farthest eye point from point N (which is the farthest point on the mirror from the eyellipses; see Figure 6.15 top figure).
3. The right eye, corresponding to the farthest left eye point L (65 mm apart), on the right eyellipse is identified as R. The neck pivot point for head turn corresponding to the eye points L and R is P. The pivot point P is located 98 mm behind the two eye points.

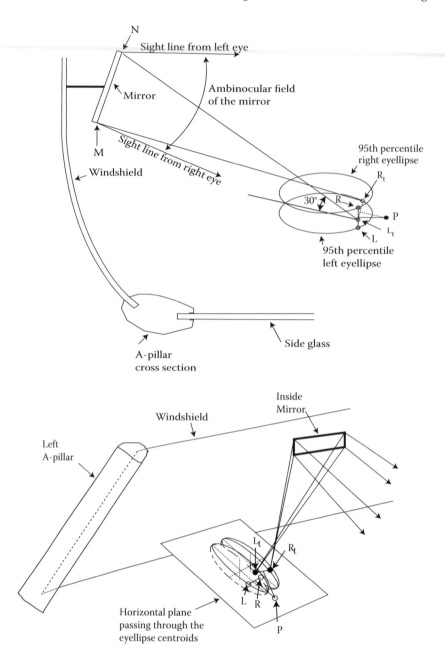

FIGURE 6.15   Determination of inside mirror field of view.

4. From the left eye L, the line of sight at 30 degrees to the right is drawn to show maximum eye turn. The head is turned clockwise around the pivot point P until the 30-degree-turned line of sight from the left eye (L) passes through point N. The left eye at the turned head position is shown as L_t.

5. The right eye at the turned head position is indicated as R_t. The line of sight from R_t is connected to the left point on the mirror M.

6. The mirror ambinocular field of view is the angle between the sight lines that are reflected (shown in the figure as "sight line from right eye" and "sight line from left eye") at points M (i.e., reflection of sight line R_tM) and N (reflection of sight line L_tN), respectively.

The horizontal field of the left outside mirror can be computed by using a similar procedure. The analysis for the left side mirror will be flipped, that is, the farthest eye point from the left mirror will be the farthest right eye point.

To compute the vertical ambinocular mirror field, the above analysis is conducted in three dimensions where instead of just reflecting the sight lines from points N and M as described above, the sight lines from r_t and l_t to all the four corners (top and bottom points on the left and right sides) of a rectangular mirror are constructed (see lower figure in Figure 6.15). The L_t and R_t turned head eye points can also be raised if the driver requires tilting head upward to maintain up angles of the sight lines up to 45 degrees maximum (see SAE standard J1050, SAE, 2009 for more details).

CONVEX AND ASPHERICAL MIRRORS

If a convex mirror is used instead of a plane mirror, the driver's field of view through the mirror will be larger. The radius of convex mirrors used in vehicles usually ranges between about 1016 and 1524 mm (40–60 in.). FMVSS 111 requires the average convex mirror radius to be between 889 mm (35 in.) and 1651 mm (65 in.) for convex mirrors mounted on the passenger's side. The field of view covered by a convex mirror increases as the mirror radius is decreased. However, the size of the images of objects in the convex mirrors will decrease (be minified) as the mirror radius is decreased. Convex mirrors with radii below about 889–1016 mm (35–40 in.) are not recommended, because the images of the objects (e.g., other vehicles in the rear field) are too small to see and estimate their distances and speeds relative to the subject vehicle, and radii greater than 1524–1651 mm (60–65 in.) are not recommended, as the images of the objects cannot be well discriminated with images seen in a plane mirror (with radius equal to infinity).

Furthermore, the images of objects in a convex mirror are located at a much closer distance to the driver's eyes than the images of the objects in the plane mirror. (The image of an object in the convex mirror is located near its focal point [which is located at half the radius of the convex mirror] behind the mirror surface.) Thus, it is more difficult for the drivers to estimate distances of objects seen in convex mirrors. Because of the minified images, drivers will tend to overestimate distances of objects (e.g., other vehicles viewed in the mirror field). Further, since older drivers cannot focus at closer distances, they will have difficulty in using convex mirrors placed on the driver's side. The difficulty is also increased due to appearance of double images for some drivers in using convex mirrors placed at closer viewing distances. Since the convex mirrors placed on the passenger's side door are at a larger distance from the driver, viewing the minified images becomes the primary problem in estimating distances of objects in the mirror.

Aspherical mirrors: An aspherical mirror is a mirror with continuously changing radius in the lateral (horizontal) direction, usually changing from a large radius (at the edge closer to the driver) to a small radius (at the edge farthest from the driver with a radius of 735 mm [30 in.] or more). The advantage of aspherical mirrors is that they provide a larger mirror fields than a plane mirror of the same size. However, many drivers experience visual strain because the images of views seen by the driver's two eyes have different sizes (levels of minification), and the drivers also experience distance-estimation difficulties with the minified images. Thus, the drivers need considerable practice to get used to the aspherical mirrors. Older drivers typically have more difficulty in getting used to aspherical mirrors than the younger drivers due to their inability to focus at closer distances and the visual strain associated with double images. FMVSS 111 currently does not require aspherical mirrors on vehicles used in the United States.

METHODS TO MEASURE FIELDS OF VIEW

The field-of-view issues described above should be analyzed to assure that a vehicle being designed will not cause visual problems when it is used in different driving situations by drivers with differing visual characteristics within the target population.

During the early design phases, as the vehicle greenhouse is being defined and the driver's eye locations have been established in the vehicle space, vehicle package engineers and ergonomics engineers should conduct a number of field-of-view analyses. The field-of-view analysis methods are generally incorporated into the computer-aided design systems used for digital representation and visualization of the vehicle. The methods essentially involve projecting the driver's sight lines to different components (such as pillars, window openings, mirrors, instrument panels, hoods, and deck surfaces) onto different projection planes such as ground plane and vertical planes placed at different distances in front, sides, and rear of the vehicle. Physical devices (e.g., sighting devices, light sources, lasers, and cameras) have also been used to conduct evaluations of physical properties (e.g., bucks, production, or prototype vehicles). However, the positioning of such devices in vehicle space with high precision is very time consuming and costly.

Early feedback on vehicle designs that reduce driver visibility by increasing obstructions (e.g., due to larger pillars, headrests, high beltlines, or smaller mirrors) should be investigated fully by using computer-aided design procedures. Questionable problems can be further evaluated by creating full-size bucks or even drivable mock-ups for market research clinics or human factors field tests. Such problems, if not fixed early, would be extremely time consuming and expensive to change during the later stages of vehicle development.

POLAR PLOTS

Creating a series of polar plots to conduct different field-of-view analyses is a very effective method for visualizing and measuring fields-of-view issues (McIssac and Bhise, 1995). A polar plot is especially useful for ergonomic analyses as it allows direct measurements of angular fields, angular location of different objects, angular sizes of different objects, and angular amplitudes of eye movements and head movements required to view different objects. It also allows incorporation of views from both eyes and thus facilitates the evaluation of monocular, ambinocular, and binocular fields and obscurations. It also simplifies the three-dimensional analysis by reducing it to a two-dimension analysis.

A polar plot involves plotting the visual field from the driver's (one or both) eye points in angular coordinates. It is equivalent to projecting the driver's view on a spherical surface with the driver's eyes at the center of the sphere. The driver's eye point is considered the origin from which sight lines are originated. Each sight line aimed at a target point can be located by determining its azimuth angle in degrees (θ) and elevation angle in degrees (Φ) with respect to the eye point (as the origin) of a coordinate system. If point P is defined by (x, y, z) as its Cartesian coordinates (with the eye point as the origin), then its polar (angular) coordinates (θ, Φ) can be computed as follows: $\theta = \tan^{-1}(y / x)$ and $\Phi = \tan^{-1}[z / (x^2 + y^2)^{0.5}]$. It should be noted that this polar plotting method does not use the distances from an eye point to any object point for the analysis. Figure 6.16 illustrates the transformation of the Cartesian location of point P to its polar location.

Figure 6.17 illustrates the polar representation of a point as it is viewed by both eyes. The figure shows that P is located at angular positions of (θ_L, Φ_L) and (θ_R, Φ_R), respectively, from the left and right eyes of an observer. Thus, the images of point P will be at different locations on the retinas of the observer's two eyes. However, the two images will be fused and perceived as a single image by the observer's brain. The figure on the right, therefore, shows locations of both the image points, namely, P_L and P_R with their respective polar coordinates (θ_L, Φ_L) and (θ_R, Φ_R), on a common origin. This technique of plotting polar locations of objects seen by each eye on a common origin point is very useful because it provides information on what the driver sees with each eye and also what can be obscured binocularly.

The above-described concepts are used in evaluating the driver's field of view as follows. First, Cartesian coordinates of all relevant objects in the driver's field of view are measured. Figure 6.18 shows a view of the DLOs of a truck-type vehicle with outside mirrors and eyellipses. The inputs required to create a plot would be Cartesian coordinates of points along all edges of the window openings and mirrors shown in Figure 6.18 and the left and right eye points represented by the

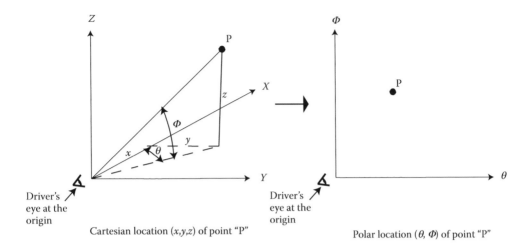

FIGURE 6.16   Transformation of Cartesian coordinates to the polar coordinates.

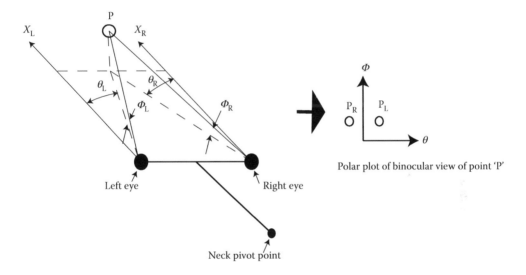

FIGURE 6.17   Illustration of viewing of point P by both eyes and its polar plot.

centroids of the left and right eyellipses, respectively. The coordinates are usually obtained from the computer-aided design model of the vehicle or by physical measurements of actual properties (e.g., exterior models or production vehicles) using computerized coordinate measurement machines. These coordinates are converted into polar coordinates from the knowledge of the coordinates of the eye points.

Figure 6.19 shows the polar plot of the driver's view (from the vehicle data shown in Figure 6.18) obtained from the driver's eyes located at the centroids of the eyellipses. The polar view in Figure 6.19 extends horizontally from over −90 to +90 degrees and vertically from −45 to +45 degrees. The plot shows the angular size of each window opening from both the left and right eyes. The left eye's view is shown in dotted lines, and the right eye's view is shown in solid lines.

The other advantage of the polar plot is that the polar coordinates of objects included in the plot provide angular locations, which can be used to directly measure driver's line-of-sight locations (i.e., combined eye movements and head turns) from the straight-ahead location (which is the origin of the polar plot, and thus, it has the polar location of [0,0]). Similarly, the size of the object shown in the

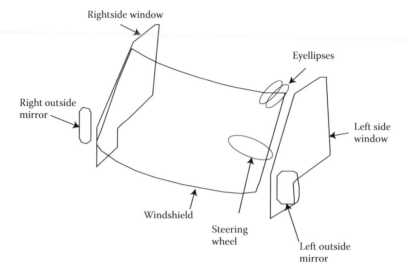

FIGURE 6.18   Data inputs for polar plotting.

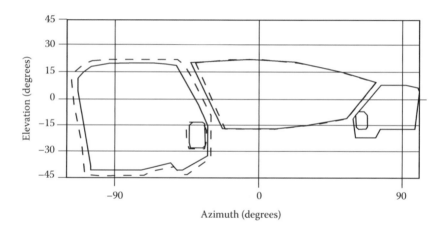

FIGURE 6.19   Polar plot of superimposed views of the window openings from the left and right eyes.

polar plot can be directly measured to determine angular sizes of monocular and binocular obstructions. Thus, the angular locations of the pillars, up and down angles, and the binocular obscurations
caused by each of the pillars can also be measured directly from the polar plot in Figure 6.19.

The roadway and many other external objects on the roadway can also be included in the polar
plots. These external objects help in understanding many visibility issues in terms of what objects
can be seen through the window openings and what objects are fully or partially obstructed by
vehicle components. Through extensive photographic measurements of objects in the driver's view,
Ford Motor Company (1973) developed targets that encompass the regions on the roadways where
different objects appeared in the photographic data. These Ford targets represent different external
objects such as overhead signs, side mounted signs, traffic signals, vehicles approaching from inter-
secting roadways, vehicles in adjacent lanes, vehicles ahead, vehicles behind the driver's car. The
targets can be placed in the polar plot to evaluate fields of view from the vehicles. McIssac and Bhise
(1995) described the use of the targets in polar plots. The article also describes the use of polar plots
in determining indirect visual fields from plane and convex mirrors by plotting virtual images of
objects seen in the mirrors and the outlines of the mirrors.

OTHER VISIBILITY ISSUES

LIGHT TRANSMISSIVITY

In the above-discussed field-of-view analyses, it was assumed that if any of the targets are located in the window openings, they will be visible to the driver. The visibility of the targets, however, depends on the visual contrast of the targets and the luminance of the background of the targets. The luminance of the targets and their backgrounds are affected by light transmission losses in the glass used for the windshield and other window openings. The light transmissivity depends on the type of glass, glass thickness, and installation (i.e., rake) angles. The light transmission losses increase with an increase in rake angle and glass thickness. The minimum light transmission (or tinting) requirements for different window openings are included in FMVSS 205(NHTSA, 2010).

OTHER VISIBILITY-DEGRADATION CAUSES

The visibility through different window openings also depends on other light effects such as (a) unwanted reflections of interior lighted components in windshields, side glass, backlites (in pickup trucks), and mirrors and (b) scattering of light due to dirty or degraded windshields and veiling glare reflections. The visibility considerations and models to predict visibility are discussed in detail in Chapter 12.

SHADE BANDS

The shade bands are usually applied on the top parts of the windshield and backlite and also in some cases on the side glass panels to reduce unwanted sun glare. The lower edges of the shade bands are generally placed above the eye locations of the tall drivers using the 95th percentile eyellipse. SAE J100 Recommended Practice provides boundaries for shade bands on glazed surfaces in class A vehicles (SAE, 2009).

PLANE AND CONVEX COMBINATION MIRRORS

A number of truck products generally use a combination of both plane and convex mirrors on the driver's side. The plane mirror is useful to view directly behind and when the reflected sight line is close to the side of the vehicle. Convex mirror provides a wider field of view and reduces blind areas. Generally, the provision of two separate mirrors is easier to use from the viewpoint of visual strain as the driver can view either the plane or convex mirror. This combination, thus, avoids the difficulties similar to those the drivers experience while using aspherical mirrors.

HEAVY-TRUCK DRIVER ISSUES

The driver's SgRP in a heavy truck is located typically at higher distances (over about 2.5 m) from the ground. The high truck driver's eye location causes the following two unique problems: (a) small cars can be hidden in front of the long hood at intersections, and (b) pedestrians, cyclists, or small vehicles in the right adjacent lane can be hidden below the beltline (of the passenger's side window). Some truck cab designs extend the belt line lower and also provide additional small window openings in the doors below the traditional side windows.

Many heavy trucks have tractor and trailer combinations. The tractor–trailer combination driver needs to view the rear tires of the tractor and the trailer during sharper turns. Therefore, they are equipped with unique plane and a convex combination side-view mirrors. The side-view mirror is called the west-coast mirror, and it typically includes a 152 mm wide × 406 mm tall (6 × 16 in.) plane mirror and a 152 × 152 mm (6 × 6 in.) convex mirror (mounted below the plane mirror).

The tall plane mirror allows the driver to view the side of the entire vehicle and the rear wheels of the tractor and trailer. The convex mirror provides a wider rearview. Both the mirrors are especially useful while negotiating wide turns in urban areas and while parking in dock areas.

CAMERAS AND DISPLAY SCREENS

A number of camera and sensor systems are available to provide the driver with additional information. Some examples of such systems are briefly described below.

1. Backup cameras and sensors are currently available on a number of vehicles to provide the driver with a visual image and/or auditory signal to alert the driver about an object in the path of a reversing vehicle. The rear-facing cameras within the vehicle body (e.g., on trunk lids, fenders, tail lamp housings) are used to provide an image of the view directly behind the vehicle.
2. Side-view cameras and sensors are also used on vehicles to provide the driver with a visual image and/or auditory signal to alert about the presence of a vehicle in an adjacent lane.
3. The display screens for camera field views in vehicles should be placed above 35-degree down-angle zone (see Figure 5.7). Some systems have placed or integrated the display in components such as inside mirrors, center stack screens, instrument clusters, or side-view mirrors.
4. Forward-facing right-side cameras (e.g., mounted on the right outside mirror) have been proposed as an aid to the driver in obtaining a better (angled) view of lead vehicles and objects ahead.

CONCLUDING REMARKS

Providing the driver with adequate views around his or her vehicle is very important and is an essential safety need. The fields of view available to the driver depend on driver positioning in the interior package of the vehicle and the integration of window openings in the exterior design of the vehicle. Therefore, during the design process, the driver's field of view must be constantly evaluated to assure that visibility requirements covered in this chapter are met to accommodate the largest percentage of drivers.

After a customer uses his or her new vehicle, the visibility problems of the vehicle can be noticed very quickly. The examples of such problems are the inside or outside mirrors are too small; obstructions caused by the steering wheel, pillars, window openings, or headrests are too large; or unwanted reflections occur during nighttime or daytime. Unfortunately, most visibility problems cannot be easily fixed without a major and expensive change in the vehicle design—which generally means waiting for the next major model change. Thus, the importance of providing the right feedback on the visibility issues during the very early stages of a vehicle program should not be underestimated.

REFERENCES

Ford Motor Company. 1973. *Field of View from Automotive Vehicles*. Report prepared under direction of L. M. Forbes, Report SP-381. Presented at the Automobile Engineering Meeting, Detroit, MI. Warrendale, PA: Society of Automotive Engineers Inc.

McIssac, E. J., and V. D. Bhise. 1995. *Automotive Field of View Analysis Using Polar Plots*. SAE Paper 950602. Warrendale, PA: Society of Automotive Engineers Inc.

National Highway Traffic Safety Administration, 2010. *Federal Motor Vehicle Safety Standards*. Federal Register, Code of Federal Regulations, Title 49, Part 571, National Highway Traffic Safety Administration, U.S. Department of Transportation, accessed June 23, 2011, http://www.gpo.gov/nara/cfr/waisidx_04/49cfr571_04.html.

Society of Automotive Engineers Inc. 2009. *SAE Handbook*. Warrendale, PA: Society of Automotive Engineers Inc.

7 Automotive Lighting

INTRODUCTION

Vehicle lighting systems are primarily safety devices because they provide visibility under night driving and convey vehicle state information to other drivers under all driving conditions. The vehicle forward lighting systems allow drivers to see the roadway, traffic control devices, route guidance signs, and targets in the roadway. The signaling and marking lamps and devices provide vehicle visibility and information on the motion characteristics of the vehicles to other drivers.

AUTOMOTIVE LIGHTING EQUIPMENT

Automotive lighting is a broad area, and it includes a number of different lamps, lighting devices, and reflex reflectors. Automotive lighting equipment can be categorized as follows:

1. Exterior lamps and lighting devices
   a. Roadway illuminating devices
      - Headlamps—low and high beams
      - Front fog lamps
      - Auxiliary headlamps
      - Cornering lamps
   b. Signaling and marking lamps
      - Parking (front), tail, stop, and turn signal lamps
      - Marking: side marker lamps, identification lamps (for trucks), and reflex reflectors
      - Backup (reversing) lamps
      - Daytime running lamps
      - Rear fog lamps
   c. Security/convenience lighting
      - Under mirror flood lamps
      - Cargo lamps (inside truck bed)
      - Running board lamps (for trucks and SUVs)
2. Interior lamps and lighting devices
   a. Illuminated displays (graphics /labels) and controls
      - Interior displays
      - Lighted labels on controls
      - Illumination lamps (or light-emitting diodes [LEDs]) for displays and controls
   b. Interior illumination
      - Dome lights
      - Map lights
      - Courtesy/convenience lamps (e.g., lamps mounted under instrument panels to illuminate floor, lamps on doors or sun visors)
3. Other lamps/lighting devices
   a. Engine, trunk, and cargo area lamps
   b. Emergency, police, and service vehicle warning lamps (e.g., blue, yellow/amber, red, or white flashing or rotating warning lamps)

Designing or evaluating a lighting system is a systems problem because the performance of a driver using the lighting system depends on the characteristics of the subject driver and other drivers, characteristics of the lighting system, and the characteristics of the driving environment (the roadway, traffic, lighting, and weather conditions).

This chapter will concentrate on headlighting and signal lighting in terms of human factors issues in their designs, photometric specifications, their effects on driver performance, and methods used for their evaluations. The visibility and glare modeling and computational issues are covered in Chapter 12.

OBJECTIVES

The objectives of this chapter are to provide an understanding into ergonomic issues related to the following:

1. Night-driving considerations involved in target detection, disability, and discomfort glare evaluation
2. Headlamp beam pattern design
3. Methods to evaluate headlamp systems
4. Signaling and marking devices
5. Signal lighting evaluation methods
6. Important research studies in vehicle lighting
7. Future technology trends and research issues to improve vehicle lighting

HEADLAMPS AND SIGNAL LAMPS: PURPOSE AND BASIC ERGONOMIC ISSUES

HEADLAMPS

The purpose of headlamps is to illuminate the roadway in front of the vehicle such that the driver can see the pavement, traffic control devices (e.g., lane lines, signs, reflectorized markers), and other targets (e.g., objects in the roadway, other vehicles, and pedestrians) far enough ahead to safely drive the vehicle at night. The low beams are designed for driving in opposed traffic to minimize blinding or discomforting drivers in oncoming vehicles. The high beams are designed to provide higher visibility under unopposed driving situations (i.e., when oncoming vehicles are absent).

The basic ergonomic issues associated with the design of headlamps are as follows:

1. Distances from the observer vehicle at which targets are detected (i.e., target visibility or detection distance) and recognized or identified (i.e., target recognition distance).
2. Effects of glare from oncoming vehicle headlamps on drivers. The glare causes two types of effects, namely, discomfort glare and disability glare. The discomfort glare reduces visual comfort and is psychological in nature. It may affect driver behavior, for example, the driver may look away from the glare source or make a dimming request to the oncoming driver to switch his or her high beam to low beam. The disability glare is physical in nature as it affects functioning of the eye and reduces visibility of objects in the driver's visual field.
3. Trade-off between visibility and glare (i.e., if the light output of headlamps is increased to improve the visibility of the driver, the increased output can increase discomfort and disability glare experienced by other drivers in oncoming vehicles). Thus, low beam patterns are designed by considering the trade-off between the subject driver's visibility and the glare effects experienced by the drivers in oncoming vehicles.

SIGNAL LAMPS

The purpose of the signal lamps and other vehicle making devices is to provide information on the presence (visibility), identification, location, orientation, and motion characteristics of a vehicle and intent of its driver (e.g., turning, stopping, backing) to other drivers.

The basic ergonomic issues associated with the design of signaling devices are as follows:

1. Distances at which a vehicle can be visible to other drivers and distances at which other drivers can correctly recognize the vehicle and determine its state or movements (e.g., discriminating a tail lamp from a stop lamp; discriminating a stop lamp from a turn signal lamp; realizing that the vehicle ahead is stopping, turning, or backing)
2. Signal lamp conspicuity or effectiveness (i.e., signal lamp size, shape, and luminance associated with the strength of the visual signal)
3. Effects of glare at night from the signal lamps on driver discomfort and disability while viewing other objects (e.g., glare from a yellow turn signal or stop lamps of a lead vehicle while waiting at an intersection at night)
4. Daytime visibility of signal lamps (e.g., stop lamp and turn signal lamp visibility when sunlight falls on its lens)
5. Trade-off between visibility and glare (e.g., visibility of the front turn signal in the presence of the low or high beam vs. the glare caused by the turn signal)

HEADLIGHTING DESIGN CONSIDERATIONS

In order to design or evaluate headlamps, one must have data on the characteristics and locations of the targets that the driver must see to safely drive the vehicle under dark nighttime conditions and in the presence of other drivers (and pedestrians) and glare sources. The headlighting design considerations are as follows:

1. The driver must be able to see the roadway at least 2 s ahead to maintain lateral control (Bhise et al., 1977b). (Note: At 100 km/h (62 mph), the vehicle will travel 56 m (180 ft) in 2 s. Therefore, the headlamp illumination must provide visibility of lane lines at least 56 m ahead to allow the driver to maintain the vehicle in the lane at 100 km/h.)
2. The driver must be able to see stationary (e.g., potholes) and moving targets (e.g., pedestrians and other vehicles) from far enough distances to avoid collisions. (Note: The visibility distance of a target should be larger than the distance that the driver will travel during his or her response time to recognize the target and to complete a maneuver, e.g., braking, to avoid a collision with the target.)
3. The driver must be able to see critical traffic control devices (i.e., delineation lines, roadside markers/reflectors, signs) from safe distances (i.e., distances within which the intended maneuver can be completed) on the highway. (Note: Some traffic control devices may be illuminated by other external light sources, or higher reflectance materials can be used to be seen under headlamp illumination.)
4. The headlamps on the observer vehicle should not discomfort other drivers (in oncoming vehicles or in leading vehicles when viewing rear through their mirrors).
5. Visual information acquisition characteristics of the drivers vary due to factors such as individual differences (e.g., due to differences in their age and visual detection thresholds, discomfort glare thresholds, and visual search behavior), state of alertness, and workload (presence of other tasks simultaneously performed) or inattention.
6. Vehicle characteristics affect the driver's visibility due to factors such as the driver's eye height and headlamp mounting height above the pavement, headlamp beam pattern, headlamp aim, vehicle loading, lamp voltage, headlamp lens cleanliness, and light transmissivity of glazing materials and their cleanliness.

7. Geometric, photometric, and traffic characteristics of roadways affect driver visibility. Some variables related to the above three characteristics are as follows: (a) geometric characteristics: road topography, lane width, road curvatures, separation distance between lanes; (b) photometric characteristics: reflectance of pavements and road shoulders, ambient luminance, and street lighting; and (c) traffic characteristics: vehicle (traffic) density, distances between cars, and ratio of cars to trucks.

8. Weather conditions (wet or snow-covered pavement; light scattering in fog, rain, and snow) also affects visibility.

9. Target characteristics (target size, shape, and target orientation; reflectance and color; and target motion) also affect visibility.

10. Other vehicles (their relative locations, headings, velocities, headlamp beam patterns, and other signaling and marking devices; driver's eye locations; and locations of headlamps) also affect driver's visibility.

Thus, headlamp design is a systems problem involving the consideration of many variables.

TARGET VISIBILITY CONSIDERATIONS

A driver must be able to perform all driving tasks safely under night-driving situations. Unfortunately, it is not possible to provide the driver the same level of visibility at night as in the daytime due to the following reasons:

1. Targets and areas required to be seen are spread in large fields around the vehicle (refer to Chapter 6 on field of view from vehicles).

2. The level of illumination from a headlamp on a target placed in front of the headlamp falls off rapidly as the distance of the target from the headlamp is increased. The level of illumination falling on the target will depend on the headlamp intensity directed at the target divided by the square of the distance between the headlamp and the target. (Note: This is the inverse square law for illumination.)

3. Target size (angle subtended by the target at the observer's eye point) decreases as the distance from the eye to the target increases. (Note: A small target requires larger visual contrast threshold for visibility as compared with a large target. The visual contrast is defined as the difference in the luminance between the target and its background divided by the luminance of the background [see Chapter 12 for more details on visibility prediction].)

4. The visual contrast required to see a target increases as the luminance of the target's background (i.e., the driver's adaptation luminance) is decreased. (Note that items (2) through (4) above are based on Blackwell's Visual Contrast Threshold Curves. See Chapters 4 and 12.)

5. Higher (more intense) headlamp output can benefit the driver in the observer vehicle, but oncoming drivers will be discomforted and disabled due to the increased glare from the observer's vehicle's headlamps. The glare is affected by the aim of headlamps (i.e., headlamp misaim) and road geometric conditions such as hilly terrain with curves and grades.

PROBLEMS WITH CURRENT HEADLIGHTING SYSTEMS

Some key problems with the current headlamp systems are as follows:

1. Insufficient visibility, especially with the low beam (e.g., insufficient seeing distances to relevant targets at highway speeds and inadequate light distribution as different light distributions are ideally needed for different night-driving situations).

2. Glare caused by headlamps (due to factors such as headlamp location, beam pattern, and headlamp misaim).

3. Variability in lamp output/beam patterns (e.g., manufacturing variations in light sources and lamp optics, fluctuating voltages, lens cleanliness, and degradations in light output due to aging).

4. Variability in headlamp aim retention (the headlamp aim can change in vehicle usage due to changes in load carried in the vehicle, changes in vehicle components and their alignments due to wear, accidents, etc.).

5. Nonuniformity (poor perception) of beam pattern on the roadway (e.g., nonuniform–splotchy or streaky appearance).

6. Lower visibility under adverse weather conditions (low visibility on wet pavements due to very low retroreflectance; attenuation and scattering of light directed above the lamp axis in rain, fog, and snow).

7. Older drivers find night driving more difficult. The visual contrast thresholds and glare effects generally increase with increase in driver age.

NEW TECHNOLOGICAL ADVANCES IN HEADLIGHTING

A number of different improvements in the headlamps systems have been made due to advances in technologies related to optics, light sources, sensors, actuators, and information technologies. Some key advances are briefly described below.

1. New light sources are more efficient, that is, they produce more light flux per unit energy input (e.g., high-intensity discharge [HID] lamps produce about 75 lm/W as compared with about 24 lm/W produced by tungsten–halogen lamps. (Note: Lumen is a unit of light flux.) The use of LEDs as a source for headlighting is just emerging with the introduction of LED headlamps in a few luxury vehicles and in aftermarket applications.

2. The ability to produce smaller headlamps with reduced frontal area, low aerodynamic drag, and smaller packaging space.

3. New optical solutions (e.g., use of projector lamps, complex reflectors, axial filament light sources, light engines with fiber optic cables [single more efficient lamp distributing light flux to different locations via fiber optic cables], multiple source lamps, etc.) to improve light distribution (e.g., providing sharper cutoffs in low beams to reduce glare experienced by oncoming drivers and to provide smoother more uniform luminance distributions on the roadway).

4. Smart headlamps (adaptive headlamps) that can change the headlamp aim and/or beam pattern as functions of velocity, road geometry, road lighting, etc.

5. Night-vision systems that provide driving scene imagery (at farther distances than conventional headlamps) on a driver display (e.g., a temperature-sensitive infrared camera provides the driver a view of the road scene in a head-up display).

SIGNAL LIGHTING DESIGN CONSIDERATIONS

Some key ergonomic considerations in designing and evaluating signal lamps are as follows:

1. Signal lamps serve as displays indicating the presence of a vehicle and provide cues or information that other observers (drivers in other vehicles, pedestrians, bicyclists, etc.) can use to determine how safely they can make their intended maneuvers.

2. Signal displays should provide needed information in the shortest possible time with minimal (or no) confusion or errors. Thus, the signal lighting system should be able to provide clear information on (a) vehicle presence (detection), (b) vehicle location (position and distance), (c) vehicle size (width, length, and height), and (d) vehicle movement (or state) of the vehicle (accelerating, decelerating, stopping, turning, etc.).

3. Signal effectiveness:
   a. Signal visibility and conspicuity: A signal lamp should be visible under all driving environments (daytime and with sunlight falling on the stop and turn signal lenses, dawn/dusk, night, rain, fog, etc.).
   b. The signal should be recognizable and interpretable with minimal confusion.
   c. The driver's response time to detect, recognize, and correctly interpret a given traffic situation indicated by vehicle signal lamps should be as short as possible (about 1 s).
   d. A signal lamp should not degrade a driver's ability to see and interpret other visual details or signals (e.g., masking of other signals, glare due to signal lamps, etc.).
4. Signal coding methods:
   a. Color or change in signal color (Note: Signal colors are defined in Society of Automotive Engineers Inc. [SAE] J578, SAE, 2009.)
   b. Intensity and/or change in intensity (e.g., required ratio of stop to tail lamp intensity)
   c. Spatial cues based on mounting location, position (mounting height, distance from outer edges of the vehicle), and separation distances between lamps on the vehicle (e.g., distance between left and right lamps)

SIGNAL LIGHTING VISIBILITY ISSUES

More specific issues related to visibility of signal lamps are briefly described below:

1. Other drivers must be able to see the subject vehicle from distances larger than their perception–reaction and maneuvering distances to safely avoid collisions.
2. The visibility of a signal lamp depends on a number of factors related to signal lamps, ambient lighting conditions, driver characteristics, and the road environmental characteristics. These characteristics are described below.

   The signal lamp characteristics that affect its visibility are luminous intensity directed at the observer, illuminated area and shape (e.g., length-to-width [aspect] ratio), beam pattern, distribution of luminance over the face on the lamp, spectral characteristics of the light emitted by the lamp (i.e., its color), and amount of dirt on the lamp that reduces its output.

   The primary driver characteristics that affect the signal visibility are driver's visual thresholds (luminous intensity for detection of point sources and visual contrast thresholds), driver's age, confidence in detection and recognition, and driver alertness.

   The ambient lighting conditions affect the visibility (or conspicuity) of the lamp. For example, a stop lamp should be visible under both the nighttime and daytime conditions. Further, when direct sunlight falls on the signal lamp lens, its on and off states must be discriminable.
3. The driver's response time to recognize a change in signal (e.g., onset of a stop lamp or initiation of a turn signal) should be very short (close to a simple reaction time of 0.3–0.5 s).
4. Driver errors in recognizing a signal correctly must be minimized (to avoid accidents resulting from not recognizing a signal or recognizing a signal too late).

PROBLEMS WITH CURRENT SIGNAL LIGHTING SYSTEMS

Since different vehicle models use different lamp designs and light sources with different technologies and manufacturing variations in lamps and light sources, there is potential to create confusion in correct recognition of signals. Issues related to such problems are presented below.

1. Confusion errors in identification of a red signal as a tail lamp, a stop lamp, or a rear turn signal

2. Recognition of status or change in vehicle motion (recognition that a vehicle is slowing, is accelerating, or has stopped)
3. Variations in lamp intensities between different brands and models of vehicles and within samples of the same vehicle make and model (i.e., manufacturing variations)
4. Equivalency of signal (strength) of lamps differing due to differences in lamp size, shape, light sources (e.g., tungsten, halogen, LED, neon, etc.), lamp color (e.g., equivalency of the red rear turn signal with the amber rear turn signal), flashing frequency, flash on versus off durations, rise time (time to achieve 85% of the minimum required intensity), etc.

New Technology Advances and Related Issues in Signal Lighting

Some issues related to introduction of new technologies in producing future signal lamps are presented below.

1. Signal lamps can be created by using different lighting technologies. For example, the light sources within a signal lamp can be created by using technologies such as tungsten or halogen filament bulbs, neon tubes, LEDs, electroluminance films, and organic LEDs. The light produced by different sources generally differs in their spectral distribution and rise time.
2. These technologies can also allow for the creation of lamps with different luminance distributions on the lamp lens. Lamps with a small single source generally cause a hot spot, that is, a brighter area centered on the lamp source, which improves its visibility as well as its effectiveness as a signal as compared with a lamp with a very uniform luminance over a larger illuminated area. Thus, a more uniform-looking LED lamp created by using many equally spaced LEDs may not be perceived to be as effective as compared with a lamp with a single light source with the same total luminous intensity (Bhise, 1981, 1983).
3. Future lamps may also be used to create large displays on the back of the vehicle. For example, such displays can convey information to other drivers by flashing words or symbols to inform the vehicle state to other drivers.
4. The output of the lamp may be affected by its temperature. For example, the light output of the LEDs reduces with increase in temperature (SAE J2650, SAE, 2009).
5. Since many of the new technology light sources do not require a metal filament (which can break with vibrations), their life is much longer than the traditional tungsten or halogen light sources. Further, such nonfilament sources can potentially reduce variability in lamp beam patterns that is typically associated in light sources with filaments.

PHOTOMETRIC MEASUREMENTS OF LAMP OUTPUTS

Light Measurement Units

The light measurement variables and their units are given below.

1. Luminous (light) flux (Φ): It is the time rate of flow of radiant light energy and is measured in lumen (lm).
2. Luminous intensity (I): It is measured in candela (cd) and defined as light flux per unit solid angle (lumen/steradian). (Note: Steradian is a unit of solid angle.)
3. Illumination or illuminance (E): It is measured in foot-candle (fc) or lux (lx).
   1 fc = 1 lm/ft^2 = 10.76 lx; lx = lm/m^2
4. Luminance (L) (physical brightness): It is measured in foot-Lambert (fL) or cd/m^2.
   1 cd/m^2 = 1 nit = 0.29 fL; cd/[ft$^2 \times \pi$] = 1 fL

5. Lamp beam pattern: Beam pattern of a lamp is generated by distribution of luminous intensity output (in candela) over a range of angles with respect to the lamp axis. The angular locations within a beam pattern are measured in degrees horizontal and degrees vertical with respect to the lamp axis.

Federal Motor Vehicle Safety Standards (FMVSS) 108 (National Highway Traffic Safety Administration [NHTSA], 2010) and SAE standards (SAE, 2009) on vehicle lighting provide minimum and maximum luminous intensity (in candela) requirements at different test point locations with respect to the lamp axis for different automotive lamps (e.g., tail lamp, stop lamp, turn signal lamp, headlamp, side marker lamp, backup lamp).

6. Total light flux output of a light source is measured in lumens. It is the integrated value of light flux emitted in all directions from the light source.

7. Illumination (E) at a distance (D) from a lamp with intensity (I) is computed as $E = I/D^2$. E is measured in foot-candle or lux, where D = distance between light source and incident surface. The distance is measured in feet for illumination measured in foot-candle (or in meters for illumination measured in lux).

8. Luminance (L) is photometric brightness. Luminance of a surface illuminated by light source is defined as $L = r \times E$, measured in foot-Lambert, where r is the reflectance of the surface (see Figure 7.1). Note: $L = (r \times E)/\pi$, where L measured in candela per meter squared and E is measured in lux.

HEADLAMP PHOTOMETRY TEST POINTS AND HEADLAMP BEAM PATTERNS

The current photometric requirements on low and high beam patterns of replaceable bulb headlamps with optical aiming described in FMVSS 108 (NHTSA, 2010) are presented in Tables 7.1 and 7.2, respectively. The requirements are specified in terms of minimum and maximum luminous intensity (specified in candela) at different angular locations (called the test points which are specified as degrees left or right and up or down with respect to the lamp axis which is parallel to the longitudinal axis of the vehicle for front and rear lamps). The differences in test point locations and their output requirements (see Figures 7.2 and 7.3), thus, create difference between the low and high beam patterns (see Figures 7.4 and 7.5).

LOW AND HIGH BEAM PATTERNS

Figures 7.4 and 7.5 show the low and high beam patterns of typical U.S. headlamps, respectively. The candlepower values shown in these diagrams were obtained from measurements made at the University of Michigan Transportation Research Institute for a sample of 20 low and 20 high beam lamps from high-volume vehicles sold in the United States (Schoettle et al., 2001).

The difference between the low and high beam patterns is primarily due to the location of the hot spot (or the highest intensity point). To limit the glare experienced by the oncoming driver, the low

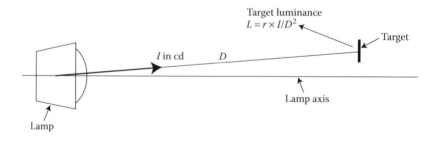

FIGURE 7.1 Luminous intensity (I) directed at a target of reflectance (r) located at a distance (D) producing target luminance (L).

TABLE 7.1
Low Beam Photometry Test Points and Minimum and Maximum Intensity Requirements

Test Point Location (Azimuth, Elevation) in Degrees	Luminous Intensity (cd)	
	Minimum	Maximum
10 up to 90 up	–	125
(–8,4) and (8,4)	64	–
(–4,2)	135	–
(1,1.5) to (3,1.5)	200	–
(1,1.5) to (90,1.5)	–	1,400
(–1.5, 1) to (–90,1)	–	700
(–1.5, 0.5) to (–90,0.5)	–	1,000
(1,0.5) to (3,0.5)	500	2,700
(0,0)	–	5,000
(–4,0)	135	–
(–8,0)	64	–
(1.3,–0.6)	10,000	–
(0,–0.86)	4,500	–
(–3.5,–0.86)	1,800	12,000
(2,–1.5)	15,000	–
(–9,–2) and (9,–2)	1,250	–
(–15,–2) and (15,–2)	1,000	–
(0,–4)	–	10,000
(4,–4)	–	12,500
(–20,–4) and (20,–4)	300	–

Source: FMVSS 108, National Highway Traffic Safety Administration, Federal Motor Vehicle Safety Standards, *Federal Register*, Code of Federal Regulations, Title 49, Part 571, U.S. Department of Transportation, 2010.

beam hot spot is directed lower by about 1.5–2.5 degrees and to the right side by about 2–3 degrees (see Figure 7.4). The high beam has its hot spot directed straight ahead along the lamp axis at (0,0) (see Figure 7.5). The intensity of high beam hot spot is generally higher than the intensity of low beam hot spot.

PAVEMENT LUMINANCE AND GLARE ILLUMINATION FROM HEADLAMPS

Figure 7.6 shows the luminance of the pavement in the middle of the driving lane resulting from a pair of low and high beams mounted on a passenger vehicle on a straight-level roadway. The figure shows that since the low beam hot spots are lower than the hot spots of the high beam, the luminance of the pavement under the low beam is higher at closer distances on the pavement as compared with the high beam. The curves presented in Figure 7.6 were generated by using the beam patterns presented in Figures 7.4 and 7.5 and by assuming 0.05 pavement retroreflectance and 0.003 cd/m^2 (0.001 fL) ambient luminance.

Figure 7.7 presents the luminance of the pavement on the left and right edges of a 3.66-m-wide (12-ft-wide) straight-level driving lane for the low and high beams mounted in a passenger vehicle. The two curves for the low beam show that the pavement in the right-hand side of the driving lane will be considerably brighter than the pavement at the corresponding distance on the left side of the

TABLE 7.2

High Beam Photometry Test Points and Minimum and Maximum Intensity Requirements

Test Point Location (Azimuth, Elevation) in Degrees	Luminous Intensity (cd)	
	Minimum	Maximum
(0,2)	1,500	–
(−3,1) and (3,1)	5,000	–
(0,0)	40,000	70,000
(−3,0) and (3,0)	15,000	–
(−6,0) and (6,0)	5,000	–
(−9,0) and (9,0)	3,000	–
(−12,0) and (12,0)	1,500	–
(0,−1.5)	5,000	–
(−9,−1.5) and (9,−1.5)	2,000	–
(0,−2.5)	2,500	–
(−12,−2.5) and (12,−2.5)	1,000	–
(0,−4)	–	5,000

Source: FMVSS 108, National Highway Traffic Safety Administration, Federal Motor Vehicle Safety Standards, *Federal Register*, Code of Federal Regulations, Title 49, Part 571, U.S. Department of Transportation, 2010.

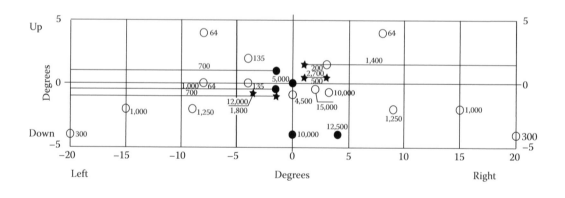

○ = Minimum   ● = Maximum   ★ = Minimum and maximum

FIGURE 7.2 Low beam photometry test point locations with respect to the lamp axis. (The numbers next to each test point symbol present the minimum, maximum, or both intensity requirement values in candela at the test point. All test points falling on a given horizontal straight line have the same intensity requirement shown above the line.)

driving lane. The figure also shows that the pavement will be brighter under high beam beyond 100 m as compared with that under the low beam. The above differences in brightness are due to less intense and lower and rightward location of the hot spot in the low beam pattern as compared with the high beam pattern (see Figures 7.4 and 7.5).

Figure 7.8 presents the amount of illumination falling into the oncoming driver's eyes from properly aimed high and low beams on a two-lane straight-level road as functions of separation distance

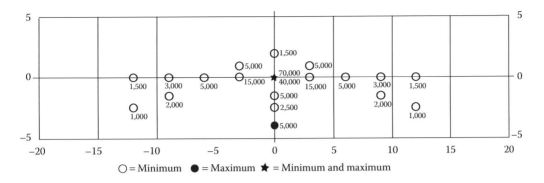

O = Minimum ● = Maximum ★ = Minimum and maximum

FIGURE 7.3 High beam photometry test point locations with respect to the lamp axis. (The numbers next to each test point symbol present the minimum, maximum, or both intensity requirement values in candela at the test point.)

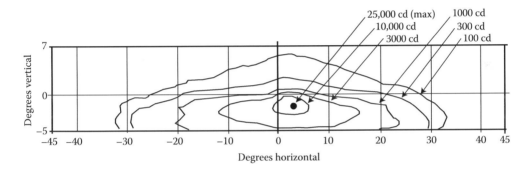

FIGURE 7.4 Distribution of a typical low beam pattern. (Redrawn from Schoettle, B., M. Sivak, and M. Flannagan. High-Beam and Low-Beam Headlighting Patterns in the U.S. and Europe at the Turn of the Millennium, Report UMTRI-2001-19, The University of Michigan Transportation Research Institute, Ann Arbor, MI, 2001.)

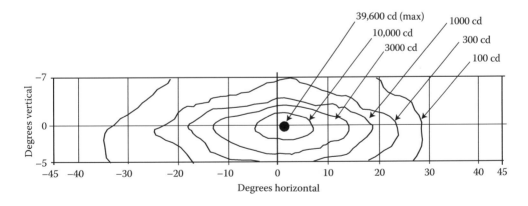

FIGURE 7.5 Distribution of a typical high beam pattern. (Redrawn from Schoettle, B., M. Sivak, and M. Flannagan, High-Beam and Low-Beam Headlighting Patterns in the U.S. and Europe at the Turn of the Millennium, Report UMTRI-2001-19, The University of Michigan Transportation Research Institute, Ann Arbor, MI, 2001.)

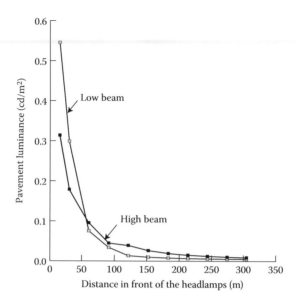

FIGURE 7.6 Pavement luminance in the middle of the driving lane with low and high beam.

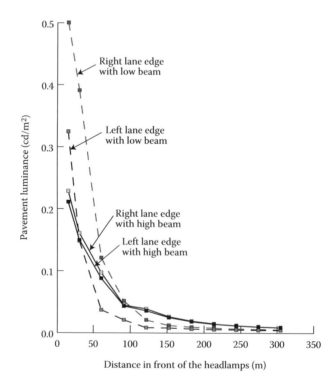

FIGURE 7.7 Pavement luminance at the lane edges of the driving lane with low and high beam.

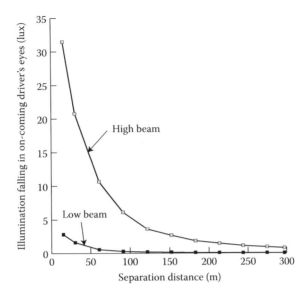

FIGURE 7.8   Illumination falling into oncoming driver's eyes from properly aimed high and low beams on a two-lane straight-level roadway.

between the observer and the oncoming passenger vehicles. The figure shows that the oncoming driver will experience substantially less glare illumination from the low beam as compared with the high beam.

PHOTOMETRIC REQUIREMENTS FOR SIGNAL LAMPS

FMVSS 108 (NHTSA, 2010) and SAE standards (SAE, 2009) provide photometric requirements in terms of minimum and maximum luminous intensity requirements for different lamps at different test points (angular locations with respect to the lamp axis). Table 7.3 provides the minimum and maximum intensity requirements for tail, stop, turn, parking, and side marker lamps. The maximum intensity requirements apply to any test point within the beam patterns of a given lamp, whereas the minimum intensity requirements, shown in Table 7.3, apply to the test point located at (0,0) (i.e., at the lamp axis).

Figure 7.9 illustrates minimum luminous intensity requirements in terms of the percentage of minimum values shown in Table 7.3 for different test points for stop and rear turn signal lamps. The requirements for other signal lamps are specified similarly using different percentage values. In addition, there are requirements on the sum of intensities for test points within different groups (zones). For more details, the reader should refer to FMVSS 108 and SAE standards (refer to SAE standards J222, J585, J586, J588, and J592 in the *SAE Handbook* [SAE, 2009]). To reduce confusion between tail and stop signals incorporated in the same lamp, there are also requirements on the minimum ratio of stop to tail lamp intensities at different test points.

HEADLAMP EVALUATION METHODS

A number of objective and subjective methods are used in the industry to evaluate headlighting systems. Five of these methods are described below.

1. Photometric measurements and compliance evaluations
    This headlamp evaluation method involves the measurement of headlamp intensity (candlepower) output at angular locations (test points) that define the photometric requirements

TABLE 7.3
Current Signal Lighting Requirements

Lamp	Lighted Sections	Luminous Intensity (cd)	
		Minimum	Maximum
Stop (rear red)	1	80	300
	2	95	360
	3	110	420
Tail (rear red)	1	2	18
	2	3.5	20
	3	5	25
Red rear turn signal	1	80	300
	2	95	360
	3	110	420
Yellow rear turn signal	1	130	750
	2	150	900
	3	175	1050
Yellow front turn signal	1	200	–
	2	240	–
	3	275	–
Yellow front turn signal (when signal lamp spaced	1	500	–
less than 100 mm from lighted edge of headlamp)	2	600	–
	3	685	–
Front parking	1	4	–
Center high-mounted stop	1	25	160
Front yellow side marker	1	0.62	–
Rear red side marker	1	0.25	–

Source: FMVSS 108, National Highway Traffic Safety Administration, Federal Motor Vehicle Safety Standards, Federal Register, Code of Federal Regulations, Title 49, Part 571, U.S. Department of Transportation, 2010.

(e.g., Table 7.1 for low beam and Table 7.2 for high beam). To conduct the measurements, a headlamp is mounted on a computer-controlled goniometer and turned on using a regulated power supply at a specified voltage (12.0 or 12.8 V) and the lamp intensity (in candela) is measured by aiming the lamp in front of a photocell mounted at a distance at least 18.3 m (60 ft) from the lamp. The lamp axis is reaimed after each measurement to cover all the required test point locations. This method is primarily used by the headlamp and the vehicle manufacturers to check if a given headlamp meets the lamp output requirement at each test point.

2. Static field tests

The static field test involves parking a vehicle equipped with a test headlight system on a roadway. A number of targets are placed at different longitudinal and lateral locations, and selected drivers are asked to sit in the driver's seat. The field tests are generally conducted in dark areas at night away from any external light sources and populated areas to reduce the effect of ambient illumination caused by the scattered light from city lights (city glow). The driver is typically asked to identify the targets (e.g., full-size pedestrian silhouettes or rectangular targets of different dimensions) that he or she can see and is also asked to provide ratings on the glare produced by the headlamps of a parked vehicle (if present) facing the evaluator. In this type of static test, the drivers get more time to make

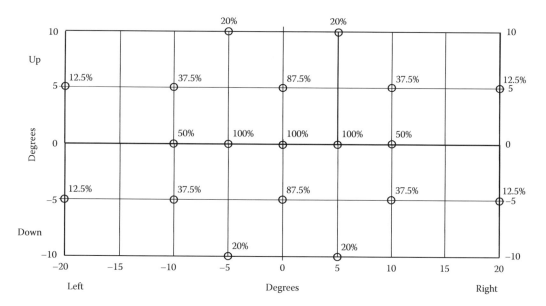

FIGURE 7.9 Rear signal test points grid showing percentages of minimum luminous intensity specified in Table 7.3 for stop and turn signal lamps.

their responses (i.e., tell the experimenter if he or she could see each target) as compared with the time they would have in actual driving situations. Further, such static tests are generally conducted under somewhat unrealistic test conditions as they involve straight, level, and dry roadways; alerted and younger drivers; clean headlamps; and windshields, properly aimed headlamps, etc. If such static tests involve the presence of a glare vehicle, the glare exposure that the subjects get is considerably different than the changing glare that the drivers experience from the motion of both the oncoming and subject vehicles in meeting situations. Thus, the static headlighting tests have limited usefulness, but they are conducted because they are less time consuming and convenient to set up and they allow visualization of beam patterns on a roadway.

3. Dynamic field tests (target detection, glare judgments, and appearance of beam pattern)

Dynamic tests are conducted to evaluate headlamps under more realistic driving conditions on public roads, proving grounds, or airport runways where motion cues such as the pitching of the car, relative motion between the target and the driver, relative motion between the subject car and the glare car, moving shadows of the stand-up targets can be provided. For seeing distance tests, targets are usually located at different points on the roadway, and the drivers are asked to indicate as soon as they can see a target and also identify the target. On-board recorders are generally used to record timings of events (e.g., subject response), vehicle velocity, and distance traveled on the roadway. The driver's attentional demands can be controlled by providing needed instructions, but generally, such tests are conducted with "alerted" subjects (whereas in the real world, the subjects generally are "unalert", that is, they are not aware of approaching targets). Use of regulated power supply for the headlamps is desirable to remove effects of fluctuations in headlamp voltages. The drivers can be also asked to provide subjective ratings on discomfort glare and appearance of beam pattern on the roadway (Jack et al., 1994 and 1995).

4. Computer visualization

Because of advances in computer simulation and visualization graphics, the driver's view from the vehicle with a given headlamp beam pattern can be simulated under different driving conditions. Farber and Bhise (1984) have described a method used to develop computer-generated driver's views under headlamp illumination and veiling glare experienced by the oncoming drivers.

5. Simulations and prediction of driver's performance

A number of computer models have been developed to predict target detection distances with and without the presence of glare sources (e.g., glare illumination from oncoming vehicle headlamps) and discomfort glare levels experienced by the drivers (Bhise et al., 1977a). As an example, the outputs of the seeing distance model developed by Bhise, Farber, and McMahan (1977a) are presented in Figure 7.10. The figure presents predicted visibility distances for 1.83-m-high (6-ft-high) 7% reflectance pedestrian targets located on the right-hand shoulder of a two-lane roadway under low beam illumination (Bhise and Matle, 1989). The seeing distances were predicted for 5th, 50th, and 95th percentile visual capabilities (based on distributions of contrast thresholds) for drivers ranging from 20 to 80 years of age. The figure shows that visibility distances of the pedestrian targets decreased as the driver's age increased. The visibility distances to pedestrian targets were lower under opposed situations (data shown in dashed lines, when a glare vehicle using the low beam was placed at 122-m (400-ft) separation distance) as compared with under the unopposed situation (data shown in solid lines). Further, it should be noted that the largest seeing distance with 95th percentile (best) visual capability of a 20-year-old driver under unopposed situation was about 120 m (394 ft), whereas the shortest seeing distance of an 80-year-old driver with 5th percentile (worst) visual capability under opposed driving situation was about 30 m (98 ft). Thus, the ratio of best-case to worse-case seeing distances was 4 (120 m/30 m).

More details of the seeing distance prediction and discomfort glare models are presented in Chapter 12.

Bhise et al. (1977b) developed a comprehensive headlamp environment systems simulation (CHESS) model to evaluate headlighting systems. The model simulates night-driving situations in a representative U.S. night-driving environment (called the standardized test route with different types of roadways, traffic speeds, traffic density, road geometry, and photometric variables to represent different reflectances [e.g., pavement reflectance],

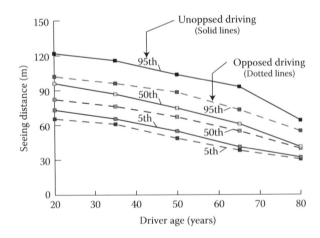

FIGURE 7.10 Comparison of seeing distances for 5th, 50th, and 95th percentile visual capabilities of drivers as a function of driver's age and unopposed and opposed driving with low beams. (A glare vehicle with low beam was placed at 122-m (400-ft) separation distances in the opposing lane.) (From Bhise, V. D., and C. C. Matle, Effects of Headlamp Aim and Aiming Variability on Visual Performance in Night Driving, *Transportation Research Record*, 1247, Transportation Research Board, Washington, DC. Reprinted from SAE paper 890873, Copyright 1989 SAE International. With permission.)

ambient lighting conditions, and external light sources [e.g., street lights]) and creates thousands of night-driving encounters involving drivers of different characteristics, pedestrian targets, lane marking targets, and oncoming vehicles. The model evaluates input beam patterns by determining if the drivers can see lane lines far enough ahead to maintain lane position (at least 2 s ahead at the vehicle speed), detect pedestrian targets (from perception and stopping distances) to avoid accidents, and not discomfort drivers in the oncoming vehicles (discomfort not more than properly aimed U.S. low beams). The output of the model is a figure of merit that represents the percentage of driving in which all the three visibility criteria simultaneously are met. Figure 7.11 shows a simple diagram illustrating the basic working of the model with inputs and outputs. Table 7.4 presents a list of variables simulated in the model. Detailed description of the model and its applications are available in Bhise et al. (1977b, 1988).

The basic outputs of a low beam evaluation presented in Bhise et al. (1988) are summarized below:

a. Figure of merit of a properly aimed U.S. low beam was 69.1. This means that, out of the over 6400 random night-driving encounters simulated by the CHESS model on different roadways, 69.1% of the encounters met all the three visibility criteria described in Figure 7.11.

b. Percentages of pedestrians detected by unalerted drivers in time to avoid accidents with the vehicles were 43.8% and 32.8% under unopposed and opposed situations, respectively.

c. 88.9% and 88.3% of delineation lines (102-mm-wide [4-in.-wide] painted lane lines) were visible (over 2 s of driving distance) in unopposed (no oncoming vehicle present) and opposed driving encounters (with oncoming vehicles), respectively.

d. Percentage of drivers discomforted under opposed encounters were 2.0%.

The above results thus demonstrate that improving low beam performance still remains to be a challenge due to the basic trade-off between visibility and glare.

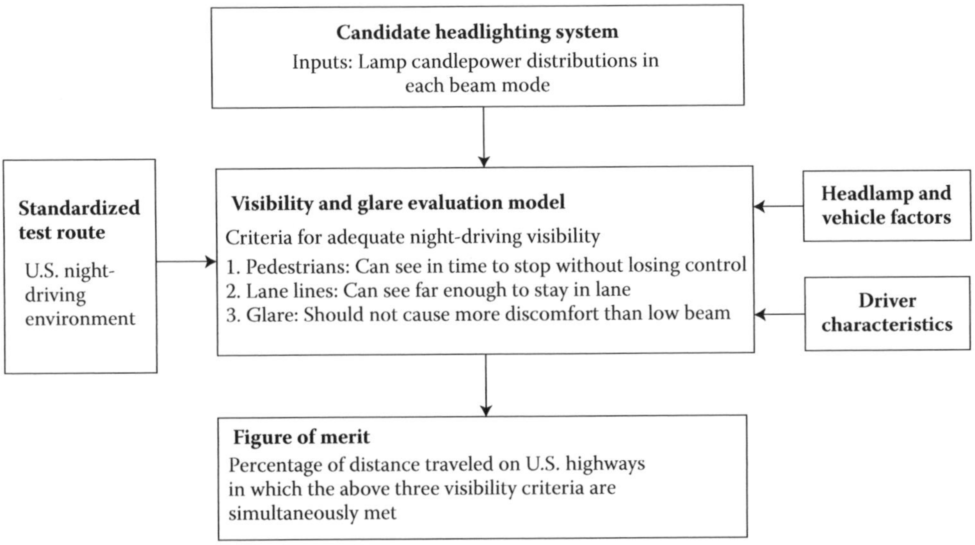

FIGURE 7.11 CHESS model inputs and outputs. (From Bhise, V. D., C. C. Matle, and D. H. Hoffmeister, CHESS Model Applications in Headlamp Systems Evaluations, SAE Paper 840046, presented at the SAE International Congress, Detroit, MI, 1984, reprinted from SAE paper 840046, Copyright 1984 SAE International. With permission.)

TABLE 7.4
List of Variables Simulated in the CHESS Model

Road–Environment Characteristics	Vehicle Characteristics: Observer and Oncoming
Area: urban/rural	Velocity
Illuminated/nonilluminated	Headlamp height and separation
Lane configuration	Headlamp aim
Lane delineation	Beam patterns
Road topography/curvatures	Intensity multipliers for dirt and
Ambient luminances	voltage
Ambient veiling luminances	Lamp manufacturing variations
Road surface type	**Driver–vehicle variables**
Road wetness	Driver's eye location
Pavement friction	Beam mode selection
Traffic density	**Driver variables**
Pedestrian exposure	Driver's age
Pedestrian characteristics	Visual contrast thresholds
Pedestrian action	Percentile visual capability
Walking speed	Glare sensitivity
Clothing reflectance: top and bottom	Alertness
Dimensions: height and width	Reaction time

Source: Bhise, V. D., C. C. Matle, and D. H. Hoffmeister, CHESS Model Applications in Headlamp Systems Evaluations, SAE paper 840046, presented at the SAE International Congress, Detroit, MI, 1984. With permission.

SIGNAL LIGHTING EVALUATION METHODS

1. Photometric measurements and compliance evaluation

   This signal lamp evaluation method involves the measurement of lamp candlepower output at different angles that define their photometric requirements (e.g., Figure 7.9). To conduct the measurements, a lamp is mounted on a computer-controlled goniometer and powered on using a regulated power supply at a specified voltage (12.0 or 12.8 V) and the lamp intensity (in candela) is measured by aiming the lamp in front of a photocell mounted at a distance at least 3 m or at least 10 times the maximum linear extent of the effective projected luminous area of the lamp, whichever is greater. The lamp axis is reaimed after each measurement to cover all the required test point locations. This method is primarily used by the lamp and vehicle manufacturers to check if a given signal lamp meets the lamp output (luminous intensity) requirements at each test point.

2. Field observations and evaluations

   This signal lighting evaluation method involves asking a group of subjects (either individually or in a group) to view test signal lamps of different signal characteristics (e.g., intensity, illuminated area, shape of lamp, lamp source type/technology) in a series of trials. The trials are set up from selected viewing locations under various ambient lighting conditions, and the subjects are asked to provide subjective ratings on visibility, conspicuity, or effectiveness of the signals. In some field observations, the subjects are presented with a pair of signals (a reference signal along with a test signal) and asked to compare the test signal with the reference signal. The method of paired comparisons is generally superior as humans are very good at discriminating between two signals presented at the same time. The method provides more reliable ratings on equivalency or relative visibility, conspicuity, or signal effectiveness than when a lamp is viewed individually without a reference signal. (See Chapter 15 for more information evaluation methods using paired comparisons.)

a. Identification of tail and stop lamps

Figures 7.12 and 7.13 illustrate the results of some of the tail versus stop lamp discrimination studies based on lamp intensity conducted by researchers at the Ford Motor Company (Troell et al., 1978). Figure 7.12 shows the relationship between percent confusion in recognizing a lamp (such a tail lamp or a stop lamp) between a pair of lamps as a function of ratio of intensities of the brighter to dimmer lamp. The curve suggests that at least a 10:1 ratio of intensities between the two lamps should be maintained for less than 10% confusion error. It should be noted that current signal lighting standards for rear signal lamps with combined tail and stop signals require that the stop signal intensities to be at least five times more than the tail signal intensities in the middle photometric zone closer to the lamp axis (FMVSS 108, NHTSA, 2010; J586, SAE 2009).

Figure 7.13 presents results of a field test in which a vehicle driven with a variable intensity red rear signal lamps was viewed by the subjects from over 100 m. In each trial, the lamp intensity was varied, and the subjects were asked if the red signals were tail or stop lamps. The figure shows that lamps over 60 cd were judged as stop lamps in over 90% of the trials. Conversely, lamps below 20 cd were judged as tail lamps in about 80% or more trials. The recognition performance was also affected by the illuminated area of the lamps. The large 51.6 cm² (8 in²) lamps were recognized as stop lamps in about 10 more percent of the trials as compared with the small 25.8 cm² (4 in.²) lamps of the same intensity as that of the larger lamps.

b. Measurement of reaction times and signal recognition errors

Driver reaction times to recognize various rear signals and signal recognition errors have been measured in actual car-following situations as well as in studies using driving simulators. Figures 7.14 through 7.17 illustrate some examples from the

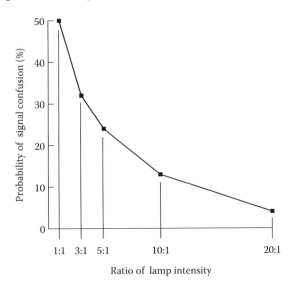

FIGURE 7.12 Probability of signal confusion as a function of stop to tail lamp intensity ratio. (From Troell, G., D. H. Hoffmeister, and V. D. Bhise, Identification of Steady Burning Red Lamps as Tail or Stop Signal in Night Driving, *Proceedings of the 1978 Annual Meeting of the Human Factors Society,* Detroit, MI, 1978. Reproduced from *Proceedings of the Human Factors and Ergonomics Society 22nd Annual Meeting.* With permission. Copyright 1978 by the Human Factors and Ergonomics Society. All rights reserved.)

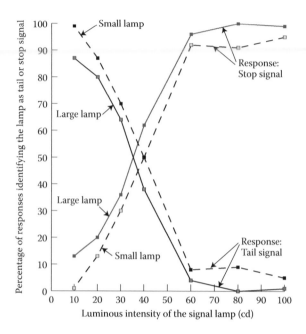

FIGURE 7.13 Recognition of tail or stop lamp as a function of lamp intensity (in candela) and lamp area (small lamp of 25.8 cm² [4 in²] and large lamp of 51.6 cm² [8 in²]) at nighttime. (Redrawn from Troell, G., D. H. Hoffmeister, and V. D. Bhise, Identification of Steady Burning Red Lamps as Tail or Stop Signal in Night Driving, *Proceedings of the 1978 Annual Meeting of the Human Factors Society,* Detroit, MI, 1978. Reproduced from *Proceedings of the Human Factors and Ergonomics Society 22nd Annual Meeting.* With permission. Copyright 1978 by the Human Factors and Ergonomics Society. All rights reserved.)

studies conducted on driver reaction times to various rear signals (Luoma et al., 1995; Mortimer, 1970; Rockwell and Safford, 1969; Rockwell and Banasik, 1968). Rockwell and Safford (1969) conducted car-following studies and measured the reaction time of following drivers to brake signal by varying the ratio of red stop lamp to red tail lamp intensity ratio. They found that as the stop lamp to tail intensity ratio increased from 1:1 to 5:1 the following driver's reaction to brake signal decreased from about 2.2 s to 1.0 s. Increases beyond 5:1 ratio did not yield significant decrease in the reaction time. Mortimer (1970) measured the reaction times of the following drivers to four different signal modes (brake, turn, brake and turn, and turn and brake) presented for three rear signal systems shown in Figure 7.14. The data presented in Figure 7.15 show that the reaction time of following drivers was reduced when the tail, stop, and turn signals were coded by green, red, and yellow colors, respectively (in system B), as compared with other two rear signal systems that used all red-colored signals.

Figure 7.16 shows that the reaction time to a rear red brake (stop) signal in the presence of the turn signal is more affected by the color of the turn signal than its intensity. The brake signal reaction time was shortest in the presence of the 130-cd yellow turn signal. The turn signal color had a greater effect on the stop signal reaction time than the intensity because the yellow turn signal could be more quickly distinguished from the red stop lamps.

Figure 7.17 shows the reaction time of the following driver to stop signal and braking deceleration level while following a lead vehicle equipped with four different rear signaling systems (Rockwell and Safford, 1969; Rockwell and Banasik, 1968). The Tri-light system, which provided color-coded information on the state of the lead vehicle to the following driver, was found to be most effective in reducing the following driver's response time. The NIL systems (having no rear signals) had reaction times of 2 s or more.

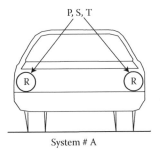

System # A

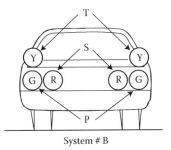

System # B

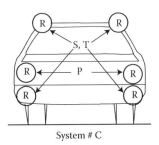

System # C

FIGURE 7.14 Three rear signal systems evaluated by Mortimer. (Note: P = parking [tail] lamps, S = stop lamps, and T = turn signal lamps; R = red signal color, Y = yellow signal color and R = red signal color.)

It should be noted that despite of low reaction times for the Tri-light system, it could not be considered a practical alternative because the tricolor-equipped vehicles would create confusions in traffic situations involving presence of many lead vehicles. The problem was somewhat solved by incorporation of the center high-mounted stop lamp (CHMSL), and it is covered in the next section.

c. Analysis of accident data

Rear-end accident data have been analyzed to determine if the rear-end accident rates are affected by rear signal lamp characteristics such as the separation of tail lamps from stop lamps, incorporation of a CHMSL, rear turn signal lamp color (yellow vs. red), etc. A fleet study conducted by Malone et al. (1978) is described in the next section.

CHMSL FLEET STUDY

Under a contract sponsored by the U.S. Department of Transportation, the Essex Corporation conducted a 12-month fleet study of four rear signaling systems installed on 2101 taxicabs in the Washington D.C. area (Malone et al., 1978). The four signaling systems are shown in Figure 7.18. Configuration 1 involved an addition of a CHMSL on the top of the trunk to taxicabs equipped with production red rear lamps, which displayed three signals, namely, tail (P), stop (S), and red

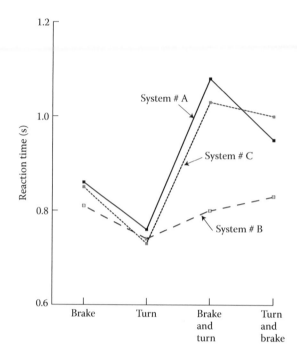

FIGURE 7.15 Reaction time of the following driver to different signals in three rear signal systems in four signal modes. (Note: Brake = brake (stop) signal only, Turn = turn signal only, Brake and Turn = brake signal changing to brake plus a turn signal, and Turn and Brake = turn signal changing to turn plus brake signal.) (Redrawn from Mortimer, R. G., Automotive Rear Lighting and Signaling Research, Final Report, Contract FH-11-6936, U.S. Department of Transportation, Highway Safety Research Institute, University of Michigan, Report HuF-5, 1970.)

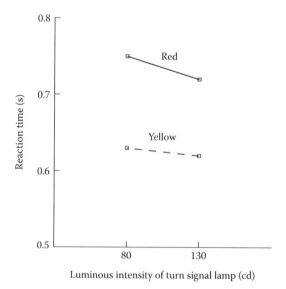

FIGURE 7.16 Mean reaction to brake lamps in the presence of red or yellow turn signal activations. (Redrawn from Luoma, J., M. J. Flannagan, M. Sivak, M. Aoki, and E. Traube, Effects of Turn-Signal Color on Reaction Time to Brake Signals, Report UMTRI 95-5, The University of Michigan Transportation Research Institute, Ann Arbor, MI, 1995.)

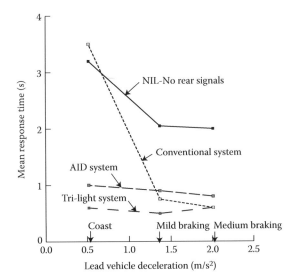

FIGURE 7.17 Following driver's response time to different rear signal systems as a function of lead vehicle deceleration under four rear signal systems. (Note: NIL = no rear signals presented; Conventional = red tail lamp and red stop lamp system; Tri-light = rear signal was green when the lead car driver's foot was on the gas pedal, the rear signal turned yellow when the lead car driver's foot released the gas pedal, and the signal turned red when the brake pedal was pressed by the lead car driver; AID = the rear signal provided the level of acceleration and deceleration of the lead vehicle by lighting up a number of horizontal rows of green or red lamps, respectively, on the back of the lead vehicle.) (Redrawn from Rockwell, T. H. and R. R. Safford, An Evaluation of Rear-end Signal System Characteristics in Night Driving. Paper presented at the 46th Annual Meeting of the Highway Research Board, Washington, DC, 1969.)

turn signals (T) in red rear lamps mounted on each side of the car. Configuration 2 was similar to configuration 1 except that it had two high-mounted stop lamps mounted near the outboard sides of the trunk. Configuration 3 did not have any high-mounted stop lamps. However, its tail lamps (P) were in separate outboard compartments, and the stop and turn signals were combined in inboard compartments, as shown in Figure 7.18. Configuration 4 had the conventional rear signal system with all the three signals, that is, the tail (P), stop (S), and turn (T) combined into the red rear lamps mounted on each side of the car. The four signal systems were installed in 2101 taxicabs with about equal installation of each configuration (about 525 taxicabs with each configuration). The four signal systems were also distributed evenly among the taxicabs available from eight different taxicab companies that provided their vehicles for the study.

The research team monitored each taxicab in terms of the mileage and accidents experienced over the 12-month study period. The taxicabs accumulated a total of 60 million vehicle miles and 1470 accidents over the 12-month study period. Out of the 1470 accidents, 217 accidents involved the taxicabs being struck in the rear. The data on the rate of the rear-end accidents (rear-end accidents per million vehicle miles traveled) showed that the vehicles equipped with the CHMSLs had about a 50% reduction in their rear-end accident rate as compared with the rates for the other three rear-end configurations (see Figure 7.19).

Several other fleet studies conducted subsequently also provided similar reductions in rear collisions by incorporation of the CHMSLs. Based on these studies, the FMVSS 108 was amended to require the center mounted stop lamps on all passenger cars from the 1986 model year and all light truck products from the 1994 model year. Kahane and Hertz (1998) evaluated long-term effectiveness of the CHMSL and found that the CHMSLs have continued to show reduction in rear-end accidents. However, the reductions in rear accidents during 1989–1995 was far less—about 4.3%, as compared with the near 50% reductions observed in several fleet studies conducted in late 1970s and early 1980s.

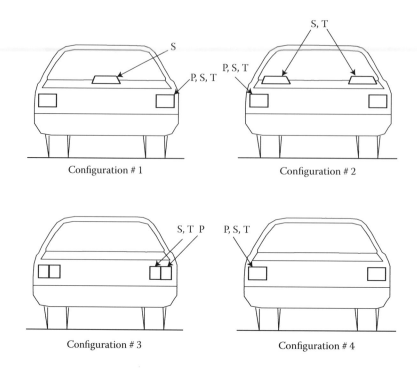

FIGURE 7.18 Four rear signal lamp configurations evaluated in the fleet study. (Note: P = red tail lamp, S = red stop lamp and T = red turn signal lamp.) (Redrawn from Malone, T. B., M. Kirkpatrick, J. S. Kohl, and C. Baler, Field Test Evaluation of Rear Lighting Systems, NHTSA, DOT Report DOT-HS-5-01228, 1978.)

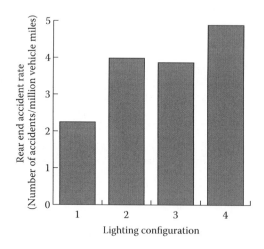

FIGURE 7.19 Rear-end accident rates by configuration of rear signal system. (Redrawn from Malone, T. B., M. Kirkpatrick, J. S. Kohl, and C. Baler, Field Test Evaluation of Rear Lighting Systems, NHTSA, DOT Report DOT-HS-5-01228, 1978.)

OTHER SIGNAL LIGHTING STUDIES

The SAE Vehicle Lighting Committee and Motor Vehicle Manufacturers Association (until about mid-1990s) Lighting Task Force have conducted a number of studies (generally conducted as lighting demonstrations to show proposed lighting configurations, intensities, etc., to familiarize the committee members prior to balloting for approval or disapproval of proposed SAE practices) (Bhise 1981,

1983; Moore and Rumar 1999; Mortimer 1970; Mortimer et al. 1973). The above-mentioned and other research studies have included evaluations of signals from different viewing locations on issues such as (a) signal effectiveness (through measurements of driver ratings and responses, e.g., response time and correct identification of signal modes); (b) minimum perceptible difference in signal intensities; (c) equivalency of signal value; (d) effect of lamp size, shape, and length-to-width ratio; (e) international harmonization between SAE, ECE (United Nations Economic Commission for Europe), and Japanese lighting requirements; and (f) effectiveness of the CHMSL.

CONCLUDING REMARKS

With the advances in new lighting technologies, the lighting systems in vehicles have been improving continuously. The progress in the implementation of new developments has been somewhat limited because of the cost of the new equipment and the lengthy approval process of government regulations. Since the lamps are directly related to safety and are regulated by the government agencies, new technologies have been able to replace old technologies only after substantiated supporting evidence for improving safety, economy, and durability.

Some issues in the technological changes are as follows:

1. Costs: The new technology lighting sources (e.g., LEDs, neon, and HID) are in general more costly than the traditional incandescent bulbs. However, their costs have been steadily decreasing after introduction in production vehicles.
2. LEDs: LEDs are now being used at increasing rate in signal lamps. The use of LEDs as a source for headlighting is just emerging with the introduction of the LED headlamps in a few luxury vehicles and in aftermarket applications. The design considerations with LEDs are
   a. Shorter rise time (by 170–200 ms) as compared with the incandescent lamps.
   b. Small packaging space: The space required (especially depth of lamps) to incorporate the LED sources is smaller than the traditional bulb type sources.
   c. Up to 60%–80% less power required than the comparable tungsten or halogen lamps.
   d. Vibration and shock resistant (no filament in the LED and HID light sources to break).
   e. Ultralong service life (about 10,000 h).
   f. Many different coded signals can be displayed (based on combinations of lighted patterns, colors, and words/messages/symbols, etc.).
   g. The LED lamp will not completely burnout as the percentage of inoperative LEDs during the useful vehicle life will be very small.
   h. The lamp output is affected by temperature.
3. Neon: Neon lamps are used on a limited basis, primarily in rear signal lamps (e.g., high-mounted stop lamps). The design considerations with neon technology are
   a. Shorter rise time as compared with traditional filament light sources.
   b. Longer life (2500 hours) with 1 million on/off cycles.
   c. Design flexibility and continuous smooth uniform appearance and seamless lamp.
4. Fiber optics and light piping: Fiber optics systems (and other light piping materials) allow for the piping of light flux from a larger, more efficient light source placed at a less proximate location. The design considerations with this technology are
   a. Packaging flexibility.
   b. One energy-efficient light source can distribute light flux to different locations (called the light engine concept).
   c. Total electrical isolation—immune to electrical interferences (electromagnetic and radio frequency interferences).

5. New signal functions: Besides the presence, parking, stop, and turn information, signal lamps based on new technologies have the potential to provide additional information to other drivers. Thus, some future application issues are

    a. A new "stopped vehicle signal" may have the potential to reduce rear-end collisions.

    b. Improving effectiveness of stop lamps by incorporating a "stopping" vehicle signal that provides information on the rate of closure during stopping (i.e., providing information about the level of lead vehicle deceleration).

    c. Displaying state of other decelerating devices (e.g., retarders on heavy trucks, vehicle stability control systems).

    d. Displaying other information (e.g., disabled vehicle).

    e. Car-following aids (e.g., safe headway/following distance signal).

    f. Variable-intensity signal lamps (that adjust according to the changing ambient lighting conditions, e.g., bright sun sunlight [100,000 lx of incident illumination] to dusk to night).

    g. If the future lighting system concepts do not comply with the existing FMVSS 108 requirements (NHTSA, 2010), then the developer must make a petition to the NHTSA to allow the use of such systems. The petition evaluation and approval process has been lengthy and can take many years.

REFERENCES

Bhise, V. D. 1981. Effects of area and intensity on the performance of red signal lamps. A report on the SAE Lighting Committee Demonstrations conducted at the Ford Dearborn Proving Ground on September 9–10, 1981. Society of Automotive Engineers, Inc., Warrendale, PA.

Bhise, V. D. 1983. Effects of luminance distribution, luminance intensity and lens area on the conspicuity of red signal lamps. A report on the SAE Lighting Committee Demonstrations conducted at the Rockcliffe AFB, Ottawa, October 3, 1982. Society of Automotive Engineers, Inc., Warrendale, PA.

Bhise, V. D., E. I. Farber, and P. B. McMahan. 1977a. Predicting target detection distance with headlights. *Transportation Research Record*, 611, Transportation Research Board, Washington, DC.

Bhise, V. D., E. I. Farber, C. S. Saunby, J. B. Walnus, and G. M. Troell. 1977b. Modeling vision with headlights in a systems context. SAE Paper 770238. Presented at the 1977 SAE International Automotive Engineering Congress, Detroit, MI, pp. 54.

Bhise, V. D., and C. C. Matle. 1989. Effects of headlamp aim and aiming variability on visual performance in night driving. *Transportation Research Record*, 1247, Transportation Research Board, Washington, DC.

Bhise, V. D., C. C. Matle, and E. I. Farber. 1988. Predicting effects of driver age on visual performance in night driving. SAE Paper 881755 (also 890873). Presented at the 1988 SAE Passenger Car Meeting, Dearborn, MI.

Bhise, V. D., C.C. Matle, and D. H. Hoffmeister. 1984. CHESS model applications in headlamp systems evaluations. SAE Paper 840046. Presented at the 1984 SAE International Congress, Detroit, MI.

Blackwell, H. R. 1952. Brightness discrimination data for the specification of quantity of illumination. *Illuminating Engineering*, 47(11), 602–609.

Farber, E. I., and V. D. Bhise. 1984. Using computer-generated pictures to evaluate headlamp beam patterns. *Transportation Research Record*, 996, 47–53. Transportation Research Board, Washington, DC.

Jack, D. D., S. M. O'Day, and V. D. Bhise. 1994. Headlight beam pattern evaluation: Customer to engineer to customer. SAE Paper 940639. Presented at the 1994 SAE International Congress, Detroit, MI.

Jack, D. D., S. M. O'Day, and V. D. Bhise. 1995. Headlight beam pattern evaluation: Customer to engineer to customer—A continuation. SAE Paper 950592. Presented at the 1995 SAE International Congress, Detroit, MI.

Kahane, C. J., and E. Hertz. 1998. The long-term effectiveness of center high mounted stop lamps in passenger cars and light trucks. NHTSA Technical Report DOT HS 808 696. Available at http://www.nhtsa.gov/cars/problems/Equipment/CHMSL.html.

Luoma, J., M. J. Flannagan, M. Sivak, M. Aoki, and E. Traube. 1995. Effects of turn-signal color on reaction time to brake signals. Report UMTRI 95-5. The University of Michigan Transportation Research Institute, Ann Arbor, MI.

Malone, T. B., M. Kirkpatrick, J. S. Kohl, and C. Baler. 1978. Field test evaluation of rear lighting systems. NHTSA, DOT Report DOT-HS-5-01228.

Moore, D. W., and K. Rumar. 1999. Historical development and current effectiveness of rear lighting systems. Report UMTRI-99-31. The University of Michigan Transportation Research Institute, Ann Arbor, MI.

Mortimer, R. G. 1970. Automotive rear lighting and signaling research. Final Report, Contract FH-11-6936. U.S. Department of Transportation, Highway Safety Research Institute, University of Michigan, Report HuF-5.

Mortimer, R. G., C. D. Moore Jr., C. M. Jorgeson, and J. K. Thomas. 1973. Passenger car and truck signaling and marking research: I. Regulations, intensity requirements and color filter characteristics. Report UM-HSRI-HF-73-18. Highway Safety Research Institute, University of Michigan, Ann Arbor, MI.

National Highway Traffic Safety Administration. 2010. Federal Motor Vehicle Safety Standards. *Federal Register*, Code of Federal Regulations, Title 49, Part 571, U.S. Department of Transportation, Washington, DC. Available at http://www.gpo.gov/nara/cfr/waisidx_04/49cfr571_04.html.

Rockwell, T. H., and R. C. Banasik. 1968. Experimental highway testing of alternate vehicle rear lighting systems. RF Project 2475. Systems Research Group, Department of Industrial and Systems Engineering, Ohio State University, Columbus, OH.

Rockwell, T. H., and R. R. Safford. 1969. An evaluation of rear-end signal system characteristics. Paper presented at the 46th Annual Meeting of the Highway Research Board, Washington, DC.

Schoettle, B., M. Sivak, and M. Flannagan. 2001. High-beam and low-beam headlighting patterns in the U.S. and Europe at the turn of the millennium. Report UMTRI-2001-19. The University of Michigan Transportation Research Institute, Ann Arbor, MI.

Society of Automotive Engineer Inc., 2009. *SAE Handbook*. Warrendale, PA: SAE.

Troell, G., D. H. Hoffmeister, and V. D. Bhise. 1978. Identification of steady burning red lamps as tail or stop signal in night driving. *Proceedings of the 1978 Annual Meeting of the Human Factors Society*, Detroit, MI.

8 Entry and Exit from Automotive Vehicles

INTRODUCTION TO ENTRY AND EXIT

The driver and occupants should be able to enter and exit from the vehicle quickly and comfortably, without any awkward postures or high physical efforts that may involve excessive bending, turning, twisting, stretching, leaning, and hitting of body parts on the vehicle components. In this chapter, we will cover many problems that the drivers and passengers experience during entry and exit from vehicles and relate these to different vehicle package dimensions.

Drivers experience different problems while entering and exiting vehicles with different body styles, from low sports cars to sedans to SUVs to pickups and heavy trucks. Assuming that a vehicle or a physical buck is available, the best method to uncover these problems would be to ask a number of male and female drivers with different anthropometric characteristics (e.g., tall, short, slim, obese) to get in and get out of the vehicles, preferably after they have adjusted the seat and steering wheel to their preferred driving positions, and observe (or video record and replay in slow motion) and ask them to also describe the difficulties that they encountered. Such exercises are usually performed by package engineers during the evaluation of a physical buck of a new vehicle. Comparisons are also made with different benchmarked vehicles to understand the differences in vehicle dimensional characteristics and assist features (e.g., door handles, grab handles, steps, hidden rockers) used during entry/exit performance and difficulty/ease ratings to determine if a given vehicle package will be acceptable or will need improvements.

PROBLEMS DURING ENTRY AND EXIT

The problems that drivers experience while entering or exiting from a passenger car depend on their gender and anthropometric characteristics as follows:

1. Drivers with short legs (predominantly women) will complain, saying the following:
    a. The seat and step-up (top of the rocker panel) is too high. The rocker panel is the lower part on the side of the vehicle body under the doors (which in effect creates the lower part of the door frame) over which the occupant's feet move over during entry and exit (see Figure 8.1 for picture of the rocker panel and Figure 8.2 for its lateral cross section at the seating reference point [SgRP]).
    b. The step-over is too wide (rocker panel is too far out). The lateral distance between the outer edge of the vehicle (i.e., the rocker panel) from the driver centerline is too far out to move the lagging leg during entry from the ground to inside the vehicle (see dimension W in Figure 8.2).
    c. The clearance between the driver's knees and the instrument panel and/or the steering column (due to seat moved forward to accommodate the driver's short legs to reach the pedals) is insufficient.
2. Older, obese, mobility-challenged drivers will complain, saying the following:
    a. The seat is either too high or too low (see dimension H5 in Figure 8.2). This indicates that the driver had difficulty climbing up into the seat (e.g., strain in the knees) or

sitting down into the seat due to larger muscular forces needed in the leg and back muscles to move the driver's body on to the seat during entry.

b. The upper part of the body door opening is too low (entrance height defined as H11; see Figure 3.9). The person will experience difficulty in moving his or her head under the lower edge of the upper body opening (see Figure 8.1).

c. The step-over is too wide.

d. The thigh clearance is insufficient (between the bottom edge of the steering wheel and top surface of the seat; SAE dimension H74, SAE J1100, SAE, 2009).

e. The steering-wheel-to-stomach clearance (between the bottom edge of the steering wheel and the driver's stomach) is insufficient.

f. The door does not open wide enough (i.e., the space between the inner door trim of the opened door and the vehicle body side is insufficient).

3. Drivers with a tall torso will complain, saying the following:

a. The upper body opening (entrance height, H11) is too low.

b. The A-pillar (front-roof pillar) is too close to their head while bending the torso forward (point H in Figure 8.6).

c. The seat bolsters (i.e., the raised sides of the seat cushion) are too high.

d. The head clearance is insufficient.

4. Drivers with long legs will complain, saying the following:

a. The seat track does not extend sufficiently rearward (seat track is too short and placed more forward in the vehicle).

b. The front edge of the B-pillar (roof pillar between the front side window and the rear side window) is too far forward. The seat back in this case is moved rearward of the front edge of the B-pillar, requiring the driver to brush past the B-pillar to get into the seat (see Figure 8.1).

c. The lower forward edge (cowl side) of the door opening is too far rearward (points F in Figure 8.6). (Note: Cowl panel is the body panel between the rear edge of the engine hood and lower edge of the windshield.) The legroom for the driver is not sufficient to move his or her legs from the ground to inside the vehicle. This problem usually results into the drivers' shoes hitting the door opening edge under the cowl side (look for shoe scuff marks on the trim parts on the forward and lower part of the door opening).

d. The door does not open wide enough (i.e., the space between the opened door to vehicle body is insufficient).

Thus, the design of the door opening size and shape (created by rocker panel, B-pillar, roof rail [vehicle body component above the doors and mounted on the sides of the vehicle roof], A-pillar [front-roof pillar], door frame under the A-pillar on the cowl side, and the knee bolster), the door opening angles, positioning of the seat, the steering wheel, and door grab and opening handles—all affect the driver's ease during entry and exit.

VEHICLE FEATURES AND DIMENSIONS RELATED TO ENTRY AND EXIT

The vehicle features and vehicle dimensions that the vehicle designers and engineers should pay attention to facilitate ease during entry/egress are as follows:

DOOR HANDLES

1. Height of the outside door handle: the short 5th percentile woman should be able to grasp the door handle without raising her hand over her standing shoulder height, and the tall

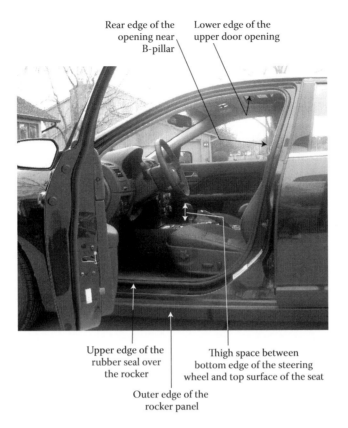

Rear edge of the opening near B-pillar

Lower edge of the upper door opening

Upper edge of the rubber seal over the rocker

Thigh space between bottom edge of the steering wheel and top surface of the seat

Outer edge of the rocker panel

FIGURE 8.1 Space available for the driver to slide into the seat is limited by the thigh space between the steering wheel and the top of seat, head, and torso space between the top of seat to the lower edge of the upper body opening and the B-pillar location.

95th percentile male should be able to grasp the handle without bending down (i.e., the handle should not be below the standing wrist height of 95th percentile male).

2. Longitudinal location of the outside door handle: The handle should be placed as close to the rear edge of the door as possible to avoid the lower right corner of the driver's door hitting on the driver's shin during door opening.

3. Insider door handle location: While closing the door (after the driver has entered the vehicle and is seated in the driver's seat), the inside door grasp (or grab) handle should not require the driver to get into the "chicken winging" type wrist posture. This means that the inside door handle should be placed (a) forward of the minimum reach zone (see Chapter 3), (b) rearward of the maximum reach zone (see Chapter 3), (c) placed not below the door armrest height, and (d) placed not above the seated shoulder height. While exiting, the inside door opening handle location should also meet the above location requirements.

4. Handle grasps: The grasp area clearances should be checked to assure that the outside door handle and the inside grasp handle (or pull cup) can allow the insertion of four fingers of the 95th percentile male's palm (by considering palm width, finger widths, and finger thickness). Further, to facilitate gloved hand operation, additional clearances would be needed depending on the winter usage and type of gloves used by the population where the vehicle is to be marketed. Further, additional clearances to avoid scratches on nearby surfaces due to finger rings and long finger nails should be considered.

LATERAL SECTIONS AT THE SGRP AND FOOT MOVEMENT AREAS

Figure 8.2 shows a cross section of a passenger car on a vertical plane passing through the driver's SgRP and perpendicular to the vehicle X-axis. The following vehicle dimensions shown in Figure 8.2 are important for ease during entry and exit.

1. Vertical height of the SgRP from the ground (H5)
2. Lateral distance of the SgRP from outside edge of the rocker (W)
3. Lateral distance of the outside of seat cushion to outside of rocker (S)
4. Lateral overlap thickness of lower door (T)
5. Vertical top of rocker to the ground (G)
6. Vertical top of the vehicle floor to the top of the rocker (D)
7. Curb clearance of doors at design weight (C)

To improve ease of the driver's entry and exit, the magnitudes of the above dimensions (separately and in combinations) need to be considered during early stages of the vehicle design.

The H5 dimension should allow drivers to easily slide in and out of their seat without climbing up into the seat or sitting down into the seat. Thus, H5 should be about 50 mm below the buttock height of most of the users (by considering the up/down adjustment of the seat cushion). The top of the seat to ground distance of about 600–750 mm is generally considered to facilitate easy ingress and egress for the U.S. population.

The dimension W should be as short as possible (see Figure 8.2). This means that dimension S, the lateral distance from the outer edge of the seat and the outer edge of the rocker, should be short enough to allow the driver's foot to be placed on the ground during entry/exit. The lateral distance of the outer edge of the rocker panel to the SgRP (i.e., distance W) should be about 420–480 mm to accommodate most drivers. Figure 8.3 shows a picture of a vehicle with a wide rocker and a large W dimension. A smaller width rocker section or a hidden rocker design (where the door extends down and overlaps the rocker and the rocker is tucked in toward the vehicle center; see Figure 8.4) will help in lowering dimension W. The height of the lower door edge C (at maximum vehicle weight) should be sufficient to clear the curb height so that the door swings over the curb and does not hit most curbs (see Figure 8.2).

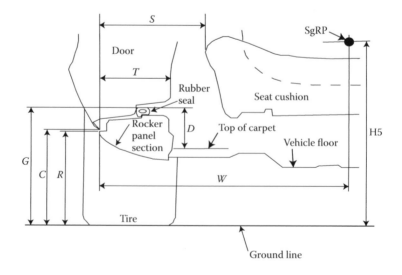

FIGURE 8.2 Cross section of the vehicle at SgRP (in the rearview) showing dimensions relevant for entry and exit.

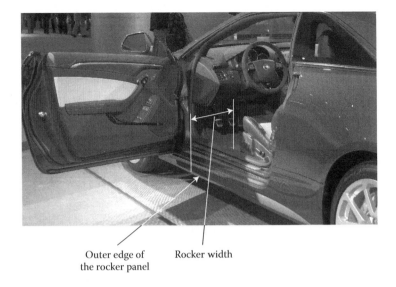

Outer edge of   Rocker width
the rocker panel

FIGURE 8.3 Wide rocker panels make the entry/exit more difficult as the driver needs to take a wider lateral step from the outer edge of the rocker to move into the seat.

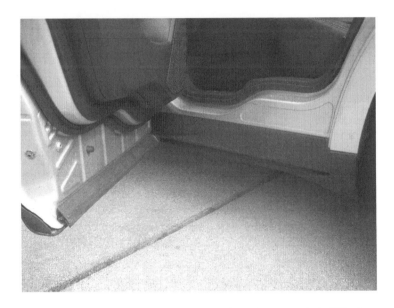

FIGURE 8.4 Hidden rocker and foot access space provided to reduce entry/exit difficulty.

The top of the rocker from the ground (dimension G) and rocker top from the vehicle floor (dimension D) should be as small as possible to reduce foot lifting during entry and exit. These dimensions are dependent on ground and curb clearances (see Figure 8.2).

The width dimension T, which is the lateral dimension from the outer edge of the rocker to the lower inward edge of the inner door trim panel (which is generally the lower inward-protruding edge of the map pocket), should be as small as possible. The smaller the dimension, the more foot passage space will be available during entry and exit (see Figure 8.5). This is especially important when the door cannot be opened wide due to restricted space on the side, such as in garages and when parking close to another vehicle (side by side).

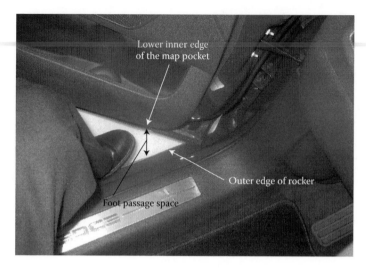

FIGURE 8.5   Foot movement space between the outer edge of the rocker and the inner edge of the map pocket.

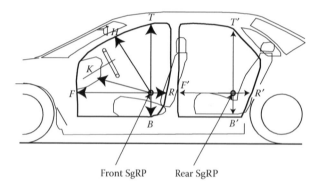

FIGURE 8.6   Side view showing body opening points with respect to the SgRPs of front and rear seats.

BODY OPENING CLEARANCES FROM SGRP LOCATIONS

Figure 8.6 shows a number of points on the door openings and the instrument panel measured from the SgRPs of the front and rear occupant positions. Larger distances of these points from the respective SgRPs will allow more room during entry and exit. For the driver's door opening, point T defines the entrance height for the head clearance, point H defines the head swing clearance to the A-pillar during leaning and torso bend, point K defines the knee clearance to the lower portion of the instrument panel, point F defines the foot clearance to the lower front area of the door opening, point B defines the top of the rocker, and point R defines the forward edge of the B-pillar; similarly, points T', F', B', and R' define clearances for the rear passenger entry and egress. The package engineers generally compare the dimensions with the above points and the SgRP locations of the vehicle being designed to other benchmarked vehicles.

DOOR AND HINGE ANGLES

1. Door fully opened angles should be designed as follows:
    a. Normally about 65–70 degrees for front doors and about 70–80 degrees for rear doors.
    b. Too large of an opened angle will increase the difficulty of reaching the door grab handle/area to close it.

2. Door hinge centerline
   a. The amount of space available between the open door and the vehicle body is affected by the position and orientation of the hinge centerline of the door. The hinge centerline is the axis passing through the upper and lower door hinges.
   b. The top of the door hinge centerline should be slightly inward and forward to improve head and shoulder clearance to the upper front corner of opened door. This will also reduce door closing effort (due to door weight assisting in door closure).

SEAT BOLSTERS, LOCATION, AND MATERIALS

1. Very high (raised) side bolsters on the seat cushion and the seat back make entry/exit more difficult as the occupant needs to slide over the raised bolsters.
2. The seat should be located as outboard as possible for easier entry/exit by reducing the lateral foot movement distance between the occupant centerline and the outer edge of the rocker.
3. The surface friction characteristics of the seat material and clothing worn by the occupants affect how easily the occupants can slide in and out of the seat. Higher friction upholstery fabrics make entry/exit more difficult. Leather seating surfaces, thus, aid in sliding and repositioning during entry/exit.

SEAT HARDWARE

1. The recliner handles, side seat shields (trim parts on the side of the seat cushion and lower part of the seat), and other seat controls should be located below the point where the seat compresses during entry/exit.
2. The seat track should be located out of way of the rear occupant's foot passage area to reduce intrusions during entry/exit.
3. The automatic rearward movement of the driver's seat and upward and forward movement of the steering wheel (after the engine is turned off and when the driver's door is opened) can reduce entry/exit difficulties.

TIRES AND ROCKER PANELS

The current design trend is to increase the tread width of the front and rear wheels as much as possible to improve vehicle stability and external appearance. Further, locating the outer side surfaces of the wheels flush with the outer sides of the fenders will also help in reducing aerodynamic drag associated with the fenders and wheels. However, more outward positioning of the wheels causes another interesting problem—called "stone pecking." The term "stone pecking" refers to the action of the spinning tires to pick up road dirt and small stones and spray them against the lower sides of the vehicle. Stone pecking can cause erosion (or sanding due to the spraying of stones and dirt) of the paint on the lower body side of the vehicle and also make the rocker dirty. One solution to the paint erosion problem is to increase the width of the rocker panels. However, the increase in the width of the rockers will increase the step-over width and make the entry/exit tasks more difficult. Further, leg contact with the wider and slushy/muddy/wet rockers can stain pant legs, and rough rocker surfaces or debris on the rocker can also snag stockings. Thus, this design trend creates conflicts in improving the occupant's ease and reducing customer complaints related to entry and egress. It is thus a trade-off issue that ergonomics engineer has to resolve with the exterior styling and body designers.

RUNNING BOARDS

The entry/exit ease from taller vehicles such as SUVs and pickup trucks can be improved by providing running boards. Shorter drivers can benefit more from the running boards as compared with some tall drivers who may find the running boards to be an intrusion in the foot placement space on the ground next to the vehicle.

Figure 8.7 shows a lateral cross section through a running board with important dimensions related to entry and exit. Dimension L should be wide enough (at least 50–55 mm) to allow foot placement. Dimension T should be large enough to accommodate the toe height of most shoes (about 50 mm or more), and dimension W should allow for a shoe support width of at least 125–150 mm. The height of the step (H) will depend on the state of suspension (i.e., loading) and tires. However, from the viewpoint of ease in climbing the step, dimension H should not be greater than about 450 mm from the ground. Dimension X (not shown in Figure 8.7) represents the longitudinal length of the running board. It should be designed to assure that sufficient longitudinal running board length for foot support is provided forward of each occupant's SgRP.

THIRD ROW AND REAR SEAT ENTRY FROM TWO-DOOR VEHICLES

1. Entry/exit to the third row of minivans (or large SUVs) or the back seat of two-door coupes is challenging for many customers because of (a) lack of direct door access, (b) the need to climb up over and through a small opening to reach the rear seat, and (c) the need to reposition one's body before sitting on the seat.
2. Provision of grab handles and their locations are important in improving ease during entry and exit. The grab handles can be located as follows: (a) locate high on the B-pillar for the second row and the C-pillar for third row entry/exit assistance, (b) particularly helpful for third row exit is to pull an occupant up and out of the seat, and (c) also a grab handle will aid during repositioning and transferring weight between sitting, standing, and climbing.

HEAVY-TRUCK CAB ENTRY AND EXIT

The heavy-truck cab floor is generally high above the ground, so accessing the cab usually requires climbing up about three steps (see Figure 8.8). The basic rule in designing steps and grab handles is that the driver or the passenger must be able to always be in "three points of contact" with the truck during the entire entry or exit process. The contact points are defined as the user's foot contacts and hand contacts with the truck. Thus, while entering the cab from the ground, two hands on the grab bars (driver's left hand on the door grab bar or the steering wheel and the right hand on the inside or outside rear grab bar) and one foot on a step, or two feet on one or more steps and one hand on a grab bar must be

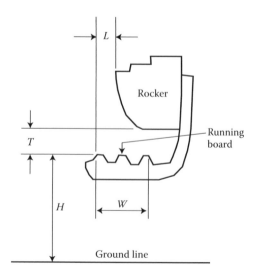

FIGURE 8.7   Running board dimensions.

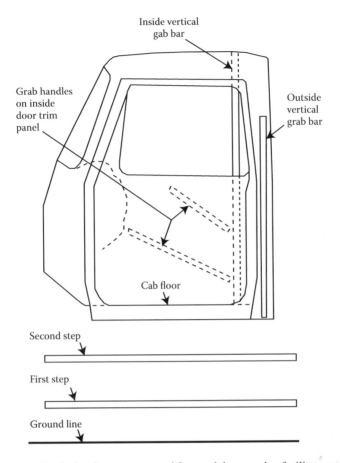

FIGURE 8.8   Heavy-truck cab showing two steps and four grab bars used to facilitate entry and exit.

minimally accommodated. The cabs should have two long vertical grab bars on the rear side of the door opening (one mounted outside and one inside the cab; see Figure 8.8). Further, at least two grab handles or grasp areas must be incorporated into the interior door trim panel. One grab handle should be located near the lower rear corner of the interior door trim panel so that the user can grab the door handle while on the ground or the first step. The second grab handle should be high on the door trim panel, which can be easily reached (must be placed rearward of the maximum reach envelope; see Chapter 3) to close the door while the occupant is in the seat. The upper grab handle can be also used by the occupant while exiting from the seat and repositioning during his or her descend (see Figure 8.8). The surface of the steps should be designed to minimize foot slippage and accumulation of snow and dirt.

METHODS TO EVALUATE ENTRY AND EXIT

Several methods are generally used in the industry to assure that entry and exit tasks are convenient for most users. The methods are described below.

1. Application of available guidelines and requirements: Auto manufacturers generally create corporate design guidelines (mostly based on research and customer feedback) to address many issues covered in earlier sections of this chapter.
2. Usage trials involving representative subjects: A large number of representative subjects are invited to a product evaluation clinic or a study, and the subjects are asked to perform a number of tasks (e.g., open the door, sit in the seat, adjust the seat for driving). The subjects

are observed in performing different tasks to uncover problems experienced (e.g., hitting heads or knees while entering the vehicle) in performing each task and the types of body motions used during the tasks. Further, the subjects can be asked to provide ratings on the ease or difficulty in performing each task. In addition, measurements can be made on the time taken to perform the tasks, number and types of body motions, and muscle activity (e.g., electromyographic measurements [EMG]).

3. Experimentation: The design of the product to be evaluated can be configured by systematically changing a number of independent variables that define the product, for example, for entry/exit evaluation, the seat height and/or the rocker width can be changed to determine the effect of the independent variables and their levels on the user's entry/exit performance and their perceived ease in performing the tasks. The changes can be made by using adjustable bucks (e.g., programmable vehicle models) or by incorporating parts that can be quickly changed between subject trials.

In both methods (2) and (3), different methods of observation and communication can be incorporated to obtain more information on the subject's experience in performing the task. Some user's observations can include the following: (a) observe what subjects do, (b) observe how they do it, (c) record hand/foot contact points and body postures, (d) track errors committed, and (e) measure time taken to complete the task. Some user's communication methods can include the following: (a) ask the subjects about problems encountered/complaints, (b) ask them to provide feedback on dimensions of certain details (examples of questions: is the grab bar diameter too large, too small, or just right to hold? Is the step too wide, too narrow, or just right?), (c) ask them to rate the ease or difficulty in performing the task, and (d) ask them about product features or items that they liked or disliked very much, etc. (see Chapter 15 for more information on methods used in vehicle evaluation).

4. Use of mathematical and manikin models: Use of motion analysis, biomechanical, and manikin models to predict entry and exit performance has had limited reliability because of complex changes in body postures performed by subjects with different anthropometric characteristics and the complex three-dimensional space in the vehicle body. Some design guidelines based on such applications of biomechanical principles are generally more useful (see item (1) above) than the use of complex modeling required to predict and analyze the body motions.

5. Task analysis: Task analysis is a simple but powerful method to determine user problems in product designs. The next section provides additional information on this technique.

TASK ANALYSIS

Task analysis is one of the basic tools used by ergonomists in investigating and designing tasks. It provides a formal comparison between the demands that each task places on the human operator and the capabilities the human operator possesses to deal with the demands. Task analysis can be conducted with or without a real product or a process. However, it is easier if the real product or equipment is available and the task can be performed by actual (representative) users under real usage situations to understand the subtasks.

The analysis involves breaking the task or an operation into smaller units (called the subtasks) and analyzing subtask demands with respect to the user capabilities. The subtasks are the smallest units of behavior, which needs to be differentiated to solve the problem at hand, for example, grasp a handle, read a display, and set speed. The user capabilities that are considered here are generally as follows: sensing, use of memory, information processing, and response execution (movements, reaches, accuracy, postures, forces, time constraints, etc.).

Task analysis can be conducted by using different formats. Table 8.1 presents a format presented by Drury (1983) that the author found to be useful in analyzing products during their design process.

TABLE 8.1
Tabular Format of Task Analysis

		Task Description			Task Analysis					
Task	Subtask No.	Subtask/Step Description	Subtask Purpose	Trigger	Scanning and Seeing	Memory	Interpolating	Manipulating	Possible Errors
Egress from driver's seat of a heavy truck	1	Pull door latch handle	Disengage the door latch mechanism	Need to open the door	Locate the inside door opening handle	Remember the location of the door handle	Determine if the door latch is already disengaged	Pull handle until the latch disengages	Door latch handle not pulled far enough Latch opening force too high
	2	Push door open	Position door to facilitate exit from the cab	Need to exit the cab	See when the door is open enough to egress from the cab	Remember door opening angle	Determine the position of the opened door	Push door until opened a sufficient angle/distance to exit the cab	Door is pushed hard and damaged and the door is not opened far enough
	3	Rotate torso to face inward	To leave the driver's seat and be properly positioned for descending down the steps	Need to leave the cab	See that the cab floor and steps are clear	Remember to face the proper direction and hold grab handle or steering wheel	Determine if you are ready to egress	Look to see if you are clear of obstructions	Operator faces wrong direction and slips while exiting
	4	Grab handles	To anchor body while climbing down and avoid fall	Need to prevent injury from falling	Look for the grab handles and foot placement areas on the steps	Remember the grab handles and steps	Determine the positions of the grab handle and steps	Grab handle and position feet	Operator slips due to not being able to grasp a handle
	5	Climb down the cab step	To reach ground	Need to leave the vehicle	Look for the cab steps	Remember to climb down by holding on both hands. Remember to use three-point contact with the cab all the time during egress	Determine the positions of the grab handles and steps with respect to your body	Transfer weight to cab step	Operator misses cab step and falls off
	6	Climb down to ground level	To reach ground	Need to leave the vehicle	Look for the ground	Remember to get feedback from the leading foot when it touches the ground	Determine the position and status of the ground	Transfer weight to ground	Operator steps in incorrect place and injury occurs

The left-hand side of Table 8.1 describes the subtasks involved in the task along with the purpose of each subtask and the event that triggers each subtask. Thus, the description of each subtask makes the analyst think about the need for each subtask and perhaps to even suggest a better way to do the task. The right-hand columns of the table force the analyst to consider different human functional capabilities such as searching and scanning, retrieving information from the memory, interpolating (information processing), and manipulating (e.g., hand finger movements) required in performing each subtask. The last column requires the analyst to think about possible errors that can occur in each subtask. This last column is the most important output that can be used to improve the task and/or improve the product to reduce errors, problems, and difficulties involved in performing the task.

If the task analysis is performed on a product that already exists (or its mock-up, prototype, or a simulation), then a number of user trials can be performed and information can be gathered on how different users perform each subtask and the problems, difficulties, and errors experienced by the users. The information then can be used in creating the task analysis table (Table 8.1). Even if a product is not available, the task analysis can be preformed on an early product concept by predicting the possible sequence of subtasks needed to use the product in performing each task (or product usage).

The task analyzed in Table 8.1 is for egressing from the driver's seat of a heavy truck. The last column of possible errors can provide insights that will assist in designing and locating components, such as door opening handles, seats, grab handles, steps.

EFFECT OF VEHICLE BODY STYLE ON VEHICLE ENTRY AND EXIT

Bodenmiller et al. (2002) conducted a study to determine differences in older (over the age of 55 years) and younger (under the age of 35 years) male and female drivers while entering and exiting vehicles with three different body styles—namely, a large sedan, a minivan, and a full-size pickup truck. The test vehicles were (a) 2001 Pontiac Bonneville 4-Door Sedan (rocker height: 381 mm [15″], seat height: 521 mm [20.5″]), (b) 2001 Oldsmobile Silhouette Minivan (rocker height: 419 mm [16.5″], seat height: 686 mm [27″]), and (c) 2000 Chevrolet Full-Size Silverado (4WD extended cab, long bed, pickup [1500 series], rocker height: 533 mm [21″], seat height: 889 mm [35″]).

Thirty-six drivers (males and females, aged 25–89 years) who participated in this study were first measured for their anthropometric, strength, and body flexibility measures relevant to the entry/exit tasks. They were asked to first get in each vehicle and adjust the seat to their preferred seating positions. Then, they were asked to get in the vehicle, and their entry time was measured. Their entry maneuver was also recorded on video, and they were asked to rate the level of ease/difficulty (using a 5-point scale) in entering. A similar procedure was conducted during their exit from each vehicle.

In addition to the expected effects of differences due to gender (males stronger than females and decrements in strength and flexibility with increase in age), the following results were observed on the entry/exit tasks: (1) Overall, the minivan was rated as the easiest vehicle to get in and out of, and the full-size pickup was the most difficult vehicle to get in and out of. (2) The shortest average entry time was 6.4 s for the large sedan, and the shortest average exit time was 5.9 s for the minivan. (3) Both entry and exit performance (as measured by total time to enter) and ratings of older women were significantly lower (worse) than those of males and younger females.

The minivan was found to have the highest ratings for both entry and exit. The 5-point rating system was based on ease in entry or exit, with a rating of 1 representing "very difficult" and a rating of 5 signifying "very easy." The mean entry ratings were as follows: 4.2 for the minivan, 4.0 for the sedan, and 3.1 for the pickup truck. The corresponding mean exit ratings were 4.2 for the minivan, 4.0 for the sedan, and 3.3 for the pickup truck.

The mean times of entry and exit, however, did not exactly mirror the mean ratings. The mean entry times were as follows: 6.6 s for the minivan, 6.4 s for the sedan, and 6.5 s for the pickup truck.

The corresponding mean exit times were 5.9 s for the minivan, 6.2 s for the sedan, and 5.9 s for the pickup truck.

On average, the subjects took slightly longer (a couple tenths of a second) to get into the minivan than into the sedan. The researchers attributed this minor difference to the fact that many of the test subjects were observed to simply drop right into the lower seat of the sedan. The exit from the minivan was the quickest. The average time differences between the three vehicles were not very large. The sedan took the longest average exit time (6.2 s). The researchers attributed this to the low rocker and the low seat height of the sedan. The sedan package appeared to require the test subjects to utilize their legs to push themselves up and out of the vehicle as compared with the other two vehicles with higher seat heights.

Analyzing the mean entry ratings and times for each of the three test vehicles by four subject groups (younger males, older males, younger females, and older females), the following conclusions were made:

1. The minivan received the highest ratings from three of the four test subject groups, with the exception being the older females. The older females rated entry into the sedan easier than that of the minivan. However, it actually took longer for the older females, on average, to enter the sedan than the minivan. The researchers theorized that the senior female subjects were more familiar with sedans as they all own and drive them. None of the senior female subjects in this field study owned a minivan, so it appears that they were reluctant to give it a higher rating than the sedan, despite the quicker entry times associated with the minivan.
2. The older female test subjects gave the lowest ratings to the pickup truck, which also had the highest entry times for them. This was consistent with the observations made throughout the field testing on the pickup truck that many of the female seniors had extreme difficulty getting both into and out of the truck due to higher sill and seat heights of the pickup.

For additional data, plots, and analyses, interested readers should refer to the article by Bodenmiller et al. (2002).

CONCLUDING COMMENTS

The entry and exit performance and preferences of drivers and occupants are affected by combinations of a number of vehicle dimensions as well as combinations of the characteristics of subjects. Therefore, ergonomics engineers should understand the basic issues and guidelines presented in this chapter and develop a physical mock-up of the proposed vehicle for further verification. The only reliable method to verify entry and exit performance is to conduct a study where the subjects would be asked to rate overall ease in entry and exit as well as collect additional data by measurements such as time taken, EMG, and observational measures such as number of problems observed and problems reported by the subjects. In addition, the inclusion of other existing vehicles (prior models of the vehicle being designed and its competitors) in the test would provide useful information for reference and comparisons.

REFERENCES

Bodenmiller, F., J. Hart, and V. Bhise. 2002. Effect of vehicle body style on vehicle entry/exit performance and preferences of older and younger drivers. SAE Paper 2002-01-00911. Paper presented at the 2002 SAE International Congress in Detroit, MI.
Drury, C. 1983. Task analysis methods in industry. *Applied Ergonomics*, 14, 19–28.
Society of Automotive Engineers Inc. 2009. *SAE Handbook*. Warrendale, PA: Society of Automotive Engineers Inc.

9 Automotive Exterior Interfaces: Service and Loading/Unloading Tasks

INTRODUCTION TO EXTERIOR INTERFACES

This chapter covers the issues that the vehicle users and service personnel experience with the vehicle exterior items during situations or operations such as opening the hood, servicing the engine, loading and unloading items in the trunk and cargo areas, refueling the vehicle, changing a flat tire, cleaning the vehicle exterior. The vehicle should be designed to assure that customers can do all such tasks easily and without any problems.

The objectives of this chapter are (a) to assure that the exterior design will allow the users to perform all interfacing tasks easily, quickly, and without errors and (b) to suggest improvements to the vehicle during early design stages.

EXTERIOR INTERFACING ISSUES

Some of the tasks and subtasks associated with servicing items located in different vehicle areas are listed below:

1. Engine compartment service
   a. Opening the hood involves the following steps:
      i. Finding and operating the inside hood release control (located at or near the instrument panel)
      ii. Finding and operating the secondary hood opening lever (located at the hood opening; see Figure 9.1 for the visibility of the secondary hood release lever)
      iii. Raising the hood (lifting the hood weight and reaching to its raised location)
      iv. Finding, positioning, and inserting the hood prop-up rod (see Figure 9.2 for the use of a rubberized and color-coded sleeve to assist in finding, grasping and inserting the prop rod)
      v. Turning on a hood lamp (for roadside engine service at night)
   b. Checking engine oil (involves the following steps):
      i. Finding the engine oil dipstick (see Figure 9.3 for color-coded grasp handle of the dipstick)
      ii. Reaching, grasping, and pulling out the dipstick
      iii. Cleaning the oil on the dipstick
      iv. Inserting the dipstick in the engine
      v. Pulling out the dipstick
      vi. Reading the oil level (note: easy-to-find holes or notches on the dipstick will facilitate reading the oil level)
      vii. Inserting the dipstick in the engine
      viii. Removing the engine oil filler cap
      ix. Filling engine oil

FIGURE 9.1 Illustration of difficulty in finding the secondary hood release lever. (The top picture shows the partially open hood, which does not allow good visibility of the hood release handle. The bottom figure shows the inside hood release lever that is located over 100 mm deep from the edge of the hood and thus requires a palm insertion of about 125 mm to activate the lever.)

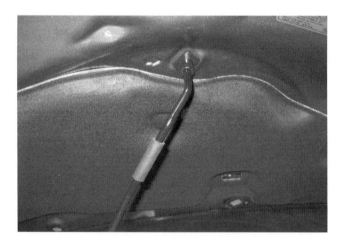

FIGURE 9.2 Illustration of rubberized and color-coded grasp sleeve around the prop-up rod and the rectangular insertion bracket to improve the usability of the prop-up rod.

Other engine compartment tasks include checking washer fluid, filling washer fluid, checking coolant level, filling coolant, inspecting the fan belt, inspecting the brake fluid level, inspecting power steering fluid level, inspecting air filter, inspecting battery terminals, finding the fuse box, opening the fuse box, identifying problem fuse location and replacing a new fuse, checking the transmission fluid level, replacing a headlamp bulb, etc.

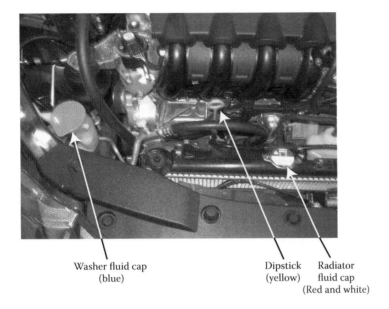

Washer fluid cap   Dipstick   Radiator
(blue)   (yellow)   fluid cap
                                   (Red and white)

FIGURE 9.3   Illustration of easy-to-find tall yellow dipstick handle and color-coded washer fluid and radiator fluid caps.

2. Trunk-compartment-related tasks
   a. Unlocking the trunk
      i. Using the remote key fob
      ii. Using the inside trunk release lever/button
      iii. Using the key
   b. Loading or unloading items
   c. Finding the spare tire
   d. Finding the jack
3. Changing a tire
   a. Removing the spare tire, the jack, and the wheel-nut wrench
   b. Positioning the jack and jacking up the vehicle
   c. Locating special wrench used for wheel nuts removal and tightening
   d. Loosening the wheel nuts
   e. Removing the flat tire
   f. Installing the spare tire
   g. Replacing the jack, wrench, and the flat tire in the trunk compartment
4. Refueling the vehicle
   a. Finding and operating the fuel door release lever/button
   b. Opening the fuel door
   c. Removing the cap (if not capless filler)
   d. Inserting the fuel pump nozzle (height of the filler above the ground and its orientation angle are important variables influencing comfort while filling gas)
   e. Fueling
   f. Removing the nozzle
   g. Replacing the cap and closing the fuel door

With the incorporation of the capless fuel fillers in newer vehicles, many cap-turning motions, cap removal and installation times, and hand injuries (from sharp edges on the surrounding sheet metal) are eliminated (see Figure 9.4).

FIGURE 9.4   Capless fuel filler.

5. Other interactions with the vehicle
   a. Cleaning the windshield, backlite, headlamps, and tail lamps
   b. Aiming the headlamps
   c. Checking tire pressure and filling air
   d. Removing the snow
   e. Installing wiper blades

The above list of tasks and the steps within each task were provided purposely for the reader to realize that many of the steps can be simplified or even eliminated by improving the design of the interfacing equipment.

METHODS AND ISSUES TO STUDY

To assure that all the tasks such as those listed above can be performed easily by the users, the ergonomics engineer must analyze each of the tasks, conduct necessary evaluations, and provide necessary guidelines, design suggestions, etc., to other designers and engineers associated with the design of the components and systems involved with the tasks.

The following methods can be used to study and improve the tasks:

STANDARDS, DESIGN GUIDELINES, AND REQUIREMENTS

The automotive companies generally have internal design standards that include design guidelines and requirements to provide the designers and engineers with information on how different systems are to be designed to assure that customers will be satisfied. The requirements in the standards are provided for the following purposes: (a) to provide some uniformity in design so that the customers of a given brand can expect design consistency across different vehicle models, (b) to prevent repetition of problems in previous designs (by incorporating lessons learned), and (c) to reduce time involved during design stages (i.e., the engineers do not spend time researching issues that were solved in previous design exercises).

For example, the important dimensions or variables in designing the trunk of a passenger vehicle include (see Figure 9.5): (a) liftover (sill) height from the ground; (b) trunk opening width and height; (c) trunk lid design: weight, hinge design (e.g., gooseneck hinges occupy trunk space as compared to four-bar linkages), reach/grasp height to close the opened lid, head injury protection against sharp protruding edges, corners, latches, etc.; (d) sill design (e.g., sill width to rest and reposition lifted items

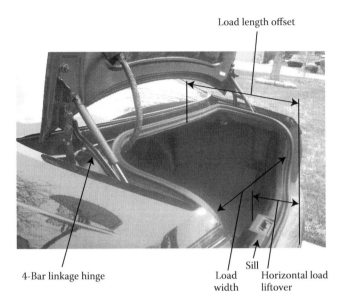

FIGURE 9.5 Illustration of trunk opening space for loading and unloading. (Note: Wider opening from the front and top, lower sill height, smaller horizontal liftover distance, and space-saving four-bar linkage hinges.)

during moves, durability of the sill and seal material, interference and protection of protruding seals, etc.); (e) sill-to-load floor depth (which may require excessive leaning and torso bending to reach inside the trunk); (f) trunk depth (i.e., longitudinal load length and ability of person to reach all corners of the trunk space and to slide and move loads); (g) spare wheel location and access; (h) stowage space (space for jack and wheel-nut wrench, tool boxes, first-aid kit, tie down latches, etc.); (i) release and locking controls and mechanisms for spare wheel cover, storage compartments, rear seat folding, fuel door, child lock-out trunk opening release, etc.; and (j) access and clearance space to change rear lamp bulbs.

CHECKLISTS

The checklists included in Chapter 5 for controls and displays (see Tables 5.1 and 5.2) will be useful to evaluate various service-related items because to service, that is, to use an item, the following sequence of ergonomic issues need to be considered: (a) findability—visibility, identification, and expected locations; (b) reading and understanding labels to determine what needs to be done; (c) accessing and reaching a needed item (e.g., 5th percentile female should be able to reach and 95th male should be able to fit within the space available under the trunk lid or liftgate [provide clearances for his hands, head, torso, etc.]); (d) grasping and operating (i.e., turning, pulling, pushing, etc.); and (e) obtaining feedback on completed movement or operation.

BIOMECHANICAL GUIDELINES FOR LOADING AND UNLOADING TASKS

For analyzing the tasks involving loading and unloading items into/from the trunk and cargo compartments, biomechanical guidelines related to lifting and carrying will be useful. Some guidelines available in this area (Chaffin et al., 1999; Konz and Johnson, 2004) are provided below:

1. Keep load close to body: This guideline will reduce the moment created by the load being lifted and stored in the vehicle. Leaning forward to reach or place a load deeper (further forward in a trunk or cargo area) will increase the moment and load in the back muscles

and the spinal column (disc between the L5/S1 joint; L5 = 5th lumbar vertebrae and S1 = 1st sacral vertebrae). The guideline also suggests that reducing longitudinal distance from the bumper edge to load area or engine compartment will reduce the bending involved in placing loads and lifting hoods.

2. Bend at knees rather than bend at back: This guideline will help reduce stresses in the lower back (L5/S1 region) and place more load in the leg muscles during lifting.

3. Work at knuckle height: Lifting load from a lower height (lower than standing knuckle height) will increase load in the L5/S1 disc. Lifting load above the standing chest and shoulder levels will increase load in the upper extremity muscles (shoulder and neck regions). Therefore, working at about the knuckle height (or about 50 mm below the elbow height) is generally less stressful on the human body. (Note: The load floor heights in minivans, SUVs, and pickups are higher and more convenient for loading/unloading than the lower trunk floor height in passenger cars.)

4. Do not twist body during the moves: Turning and twisting of the body will place more stresses in one side of the body (i.e., it will create asymmetrical loading, which will be greater at some joints than if the load is held in a symmetric position with respect to the body). Provision of wider load-access openings in trunks and cargo areas will reduce turning and twisting during loading and unloading.

5. Do not slip or jerk: Any slipping or jerking motion while carrying or lifting a load will increase dynamic loading on the body (remember that a jerk involves acceleration, and the stress in the body will be proportional to the mass moved times its acceleration).

6. Get a good grip: A good gripping of the load and gripping on the vehicle (body or grab handles) during loading will reduce the load in the spinal column. A good grip of the load will also reduce sliding and slipping of the load, which can create larger dynamic loading in the body. Supporting the body during bending (by a hand gripping on a firm support on the vehicle body) will also reduce stress in the spinal column.

APPLICATIONS OF MANUAL LIFTING MODELS

Many biomechanical models are available (e.g., University of Michigan two- and three-dimensional static strength models [Chaffin et al., 1999] and the NIOSH Lifting formula [Konz and Johnson, 2004]), which can be used to analyze lifting situations involved in lifting/opening vehicle body closures (hoods, trunk lids, and liftgates in hatchbacks, wagons, minivans, and SUVs) and loading/unloading items (e.g., boxes, suitcases, golf bags, grocery bags) in the vehicles. The vehicle dimensional parameters related to load floor heights, forward offset (leaning) distances (e.g., rear bumper to trunk storage area, front bumper to a service point in the engine compartment), and weights of hood, trunk lid, etc., along with the lift support provided by the struts can be evaluated by use of the biomechanical lifting models.

TASK ANALYSIS

Task analysis is a simple but powerful method to determine user problems in product designs. It involves an ergonomics engineer to break down a tasks into a series of subtasks or steps and analyze the demands placed on the user in performing each step and compare against the capabilities and limitations of the users. The analysis reveals a number of possible user problems and errors that the users can make in using the product. Table 9.1 provides an example of a task analysis for checking engine oil level. The description of the task analysis technique is provided in Chapter 8 (Drury, 1983).

METHODS OF OBSERVATION, COMMUNICATION, AND EXPERIMENTATION

In order to understand the problems encountered by people in interfacing with a vehicle from the exterior, the basic methods of data collection involving observation and communication would be

TABLE 9.1
Illustration of a Task Analysis for Checking Engine Oil Level

Task	Subtask No.	Task Description			Task Analysis				
		Subtask/Step Description	Subtask Purpose	Trigger	Scanning and Seeing	Memory	Interpolating	Manipulating	Possible Errors
Opening the hood and checking engine oil	1	Pull hood release lever from inside the vehicle	To open the hood	To check fluid levels	Look for the inside release lever	Remember the location	Determine how to operate	Use fingers to pull the inside lever	Released the parking brake instead
	2	Lift the hood release lever under the hood	To open the hood	To check fluid levels	Look for the hood release lever	Remember the location of the hood release	Move fingers in the gap between the hood and the grill to find the lever	Use fingers to lift the hood release lever	Difficulty in seeing and finding the lever
	3	Prop up the hood	To keep hood open	To check fluid levels	Look for the prop rod and the hole in the hood	Remember the locations of the prop rod and hole in the hood	Guess which hole in the hood is closest to the top end of the prop rod	Keep holding the hood with the left hand and insert the prop rod in the hole with the right hand	Placed prop rod in the wrong hole
	4	Locate oil dipstick	To check oil level	Need to check oil	Look for the dipstick	Remember how the dipstick looks and its location	Guess which side of the engine to look for	Use fingers to grasp the dipstick handle and pull it out	Selected the transmission oil dipstick instead
	5	Pull out, clean, and reinsert the dipstick and pull again and check the oil level	To check oil level	Need to check oil	Look for minimum and maximum marks on the dipstick and oil line	Remember to clean the dipstick and reinsert	Reinsert clean dipstick, hold for 10 s, and pull it out	Reorient the dipstick to get a good look at the oil level	Oil dripped on the floor
	6	Reinsert the dipstick	To place the dipstick back	Need to put back the dipstick	Look for the hole to insert the dipstick	Remember the hold location	Orient the dipstick over the hole	Slide the dipstick into the hole	Missed the hole two times before correctly inserting the dipstick in the engine

very useful. In the methods of observation, the user is observed by a trained human observer, and the observer records problems experienced by different users under selected usage situations. The user actions can be recorded using different methods such as video recordings or other measurements (e.g., outputs of a motion analysis system providing data on locations, velocities, acceleration, posture angles of various body parts of an operator as functions of time), and the records can be subsequently studied by trained analysts. The information gathered during the observations will help in identifying problems of different types (e.g., took long time to find an item, could not find an item, difficulty in reaching to a location, awkward postures, could not comprehend operation of a device) and determining the frequency and severity of occurrences of the problems. In the methods of communication, an interviewer can ask the user to describe the problems encountered during usage of the product (e.g., opening the trunk and replacing a burnt-out tail lamp bulb) and/or ask to provide ratings (e.g., using a 10-point scale, where 10 = very easy and 1 = very difficult) on one or more important issues.

Further, experiments can be set up to evaluate important parameters of the vehicle design along with other comparators (existing or competitor's vehicles), and a number of representative subjects will be asked to perform a number of tasks, and data can be collected from the methods of observation and communication as well as from the measurements on each subject's performance. The analyses of collected data would help determine if improvements can be justified and design superiority can be established.

Additional information on the methods of observation, communications, and experimentation is provided in Chapter 15 on vehicle evaluation methods.

CONCLUDING REMARKS

The vehicle should be designed to make all its user interfaces easy to use and service. Once, a designer came to the author and suggested that the ergonomics engineers should do a study to help design wheel covers that will be easier to clean. The designer who liked to keep his own vehicle spotless complained that he was tired of spending time cleaning each spoke on his beautiful wire wheels with a set of tooth brushes. The point is that no one in the design studio had thought about the problem until other designers realized the need of his colleague. "Cleanability" of wheels is an important consideration to the wheel and wheel cover designers. The ergonomics engineers must therefore interact with the users and find out all their needs, prioritize these needs, and come up with the solutions by applying methods and information provided in this book to eliminate the customer complaints and make the users' experience more delightful.

REFERENCES

Chaffin, D. B., G. B. J. Andersson, and B. J. Martin. 1999. *Occupational Biomechanics*. New York: John Wiley & Sons Inc.
Drury, C. 1983. Task analysis methods in industry. *Applied Ergonomics*, 14, 19–28.
Konz, S., and S. Johnson. 2004. *Work Design: Industrial Ergonomics*. 6th ed. Scottsdale, AZ: Holcomb Hathaway.

10 Automotive Craftsmanship

CRAFTSMANSHIP IN VEHICLE DESIGN

OBJECTIVES

The objective of this chapter is to provide the reader with an understanding of the concept of craftsmanship, its importance, and product characteristics that affect the craftsmanship.

CRAFTSMANSHIP: WHAT IS IT?

Craftsmanship is a relatively new technical area of increasing importance to ergonomics engineers. The whole idea behind craftsmanship is that the vehicle should be designed and built such that the customers will perceive the vehicle to be built by expert craftsmen who apply their skills to enhance the perceptual characteristics of the product such as its looks (appearance—i.e., shape, fit, finish, color, and texture of each part and harmony of appearance between adjacent parts), tactile feel (feel in movements of controls, tactile feel of various interior materials), sound (how the vehicle sounds when you operate its different features—e.g., sound of the engine, sound of a door closing, parking brake engagement clicking sound), smell (the odors—e.g., smell of the leather used in the interiors, new car smell), ease during use (all ergonomic considerations), and other features (that customers associate with craftsmanship). The vehicle should be also perceived by the customers to "belong" to the family of the brand it represents. For example, the customers will expect an expensive vehicle to be extra well made with all the features and overall perception of luxury and quality.

There is no agreement among the automotive experts on what is exactly meant by "craftsmanship." Different customers also expect different product features and characteristics, but when asked based on their internal conceptualization of what constitutes a well-crafted vehicle, they can always tell if a given product looks and feels like it is made by craftsmen.

The attribute of craftsmanship, thus, is based on the following:

1. Customer perceptions of the product after the customer experiences (i.e., sees and uses) the product. (It is assumed that only the customers can tell if a product has superior craftsmanship performance, i.e., it has more upscale characteristics as compared with other similar products in the market.)
2. It is how well the product is executed (i.e., made or perceived by the customers) to possess visual quality (appearance), sound quality, touch and feel quality, smell quality, usability/ergonomics, and features (especially those that delight the customer or create impressions of "wows").
3. It is how the product affects its perception of quality (i.e., image, brand; some examples of semantic differential scales [see Chapter 15, Figure 15.1, scale (f)] that can be used to measure the product perception by using adjective pairs such as solid/flimsy, cheap/expensive, fake/genuineness of materials, quality/shoddy, comfortable/uncomfortable, pleasing/nonpleasing, like/dislike, etc.).

IMPORTANCE OF CRAFTSMANSHIP

The importance of craftsmanship can be better understood by studying the following two models: (1) The Ring Model of Product Desirability and (2) The Kano Model of Quality.

The Ring Model of Product Desirability

The concept presented by Peters (1987) as the Levitt Rings (Levitt, 1980) to describe product service characteristics was applied here to describe product desirability to explain the concept of craftsmanship. In that concept, Levitt described the basic service as "generic," which is enhanced by adding "expected" features and further "augmented" by some unique service features. The model was described by showing the "generic" portion of the service as the core, which was surrounded by two rings—the "expected" ring around the "generic" core and the "augmented" ring around the expected ring. The overall desirability of the service (or product) is indicated by the size of the outer ring.

Figure 10.1 shows a product concept. It is hypothesized that at the core of the product there exists a "functional" design (a mechanism that does some basic product function, e.g., transports four adults). This is shown by the inner core of the product. The functional product design is enhanced by modifying or adding certain features that improve the users' comfort and convenience (i.e., ergonomic characteristics by adding features such as eight-way power seats with four-way power lumbar support, smart headlamps). The "comfort and convenience," in essence, adds a layer (or a ring) on the core of the product. The product's desirability can be further improved by improving its craftsmanship qualities or "pleasing perceptions" by providing (a) materials that have better look and feel characteristics, (b) better fitting parts with smaller gaps between parts, and (c) colors and textures between mating components that look like they were made by the same supplier, etc. The "pleasing perceptions" are also shown in the figure as adding another or second layer (or outer ring) to the product. The overall size of the outer ring is assumed to be proportional to the overall desirability (or value) of the product to the customer. The ergonomics engineers can help in increasing the sizes of both the outer rings.

Figure 10.2 shows product concepts A and B. The size of the inner core of the product concept B is shown larger than the size of the inner core of the product A. This indicates that the product concept B had a better functional design. However, the overall size (with both the rings) of product A is larger than the overall size of product B because the product B design did not offer more comfort and convenience (shown by thinner layer of comfort convenience ring as compared with product A) and had less pleasing-perception-related enhancements (also shown by the thinner outer ring of pleasing perception than product A).

The figure, thus, illustrates that to improve the overall product appeal, it is important that all the three parts (i.e., its core and the two rings) must be carefully designed with attention to all details that are perceived by the customers to make the product more desirable.

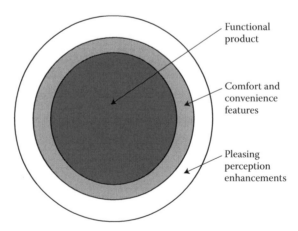

FIGURE 10.1   The ring model of desirability. (Overall desirability (or value) of the product = size of the outer circle = core + comfort convenience ring + pleasing perception ring.)

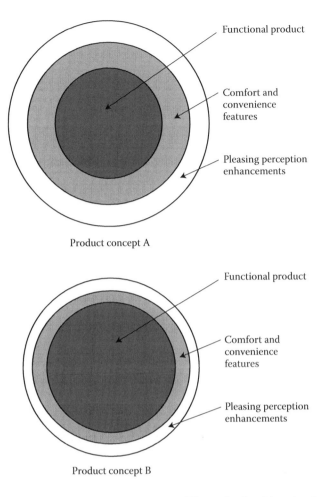

Product concept A

Product concept B

FIGURE 10.2 Illustration of two product concepts with different levels of functional, comfort and convenience, and pleasing perception characteristics.

It is important to realize that as the basic manufacturing technologies (that affect the inner core of the product) are known to all vehicle manufacturers, the task of improving the product appeal by improving its two outer layers, that is, comfort and convenience and pleasing perceptions, is now more important in discriminating between products made by different manufacturers.

It is important to realize that human factors engineers are in a unique situation to help in improving the sizes of both the outer rings. The two outer rings are also largely responsible for increasing product appeal and for brand differentiation.

Kano Model of Quality

The model of quality proposed by Kano conceptualizes that customer satisfaction is affected by three types of product features called (a) the removal of "dissatisfiers" (i.e., providing the "unspoken wants"; otherwise, dissatisfaction arises if the product does not provide what the customer expects), (b) "satisfiers" (giving more of these features increases satisfaction with the product), and (c) "delighters" (that create "wows" when present but do not cause dissatisfaction if not there; Yang and El-Haik, 2003). The Kano model presented in Figure 10.3 shows the effects of these three types of features (represented by the x-axis as the degree of achievement of critical to satisfaction) on the customer satisfaction (shown on the y-axis).

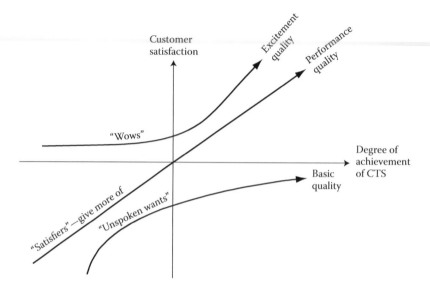

FIGURE 10.3 Kano Model of Quality. (Reproduced from Yang, K., and B. El-Haik, *Design for Six Sigma*, McGraw-Hill, New York, 2003. With permission.)

The model shows that customer satisfaction will increase by the following:

1. Providing product features (or characteristics) that satisfy the unspoken wants (i.e., these are wants the customer expects in the product, and therefore, the customer does not even mention these needs). However, if these unspoken wants are not provided, the customers will be greatly dissatisfied. Furthermore, providing the unspoken product features will not increase the customer satisfaction by much. Thus, the manufacturers consider these unspoken wants as must-have features (e.g., a remote unlocking feature on the key or key fob).
2. Providing more and more of certain types of features called the "satisfiers" that will increase customer satisfaction more and more (e.g., getting more miles per gallon with increased engine horsepower, material with luxury look and feel, illuminated switches with pleasing operational feel and sound).
3. Providing some useful new features that the customer has not seen before and, thus, creating a "wow" response—indicating that the customer is clearly delighted (e.g., providing Internet access with the Bluetooth capability in an economy vehicle). It should be noted that with passage of time, most of the "wow" features become unspoken wants. Thus, the manufacturers must continuously create new "wow" features that were not expected by the customers at the time of their introduction.

A careful study of all the above three types of product features (either available in the market currently or new features that could be developed by studying design and technological trends) and their incorporation in the product is essential to enhance craftsmanship and product quality.

ATTRIBUTES OF CRAFTSMANSHIP

The following characteristics of products have been considered in the automotive industry to be related to perception of quality. The list was prepared by the author through interviews with a number of vehicle owners/users and via discussions with engineers and designers working for different vehicle manufacturers and suppliers.

Visual Quality

1. Excellent fit between any two components in visible regions (characterized by smooth parting lines and small [thin] gaps and nonperceptible mismatches [misalignment] associated with constant gap width along all joints, flush mating surfaces, parallel edges, smooth rounded (nonsharp) edges, low variability/unevenness (warpage or distortions), no see-through areas between joints, etc.)
2. Visual harmony (similarity of look and feel between adjacent components with similar materials based on color, brightness, texture/grain, gloss/reflectivity, finish quality, etc.)
3. No evidence of degradation on visible surfaces (e.g., rust, fading, hazing, fracturing, peeling, wear and scratches)
4. No exposed fasteners (e.g., visible screws, clips, wires, etc.). The underlying concept is that product surfaces with invisible (or unexposed) fasteners should look clean and uninterrupted. Thus, designer should provide the feeling that the product is well made. On the other hand, some screws with a "machined look" can give the impression of precise, well-crafted, and solid joints
5. No annoying visual distractions (e.g., glare from light sources or reflections of other brighter surfaces into glazing or reflective surfaces, waviness, or distortions in reflected or transmitted images)

Touch Feel Quality

1. Vehicle interior surfaces that are often touched by the users (e.g., control grasp areas [buttons, knobs, handles], seats, instrument panels, door trim panels, arm rests, consoles, etc.) should have pleasing feel by considering touch characteristics that can be described and scaled by using adjective pairs such as softness/hardness, smoothness/roughness, textured/nontextured, slippery/sticky, etc.)
2. Pleasing operational feel of switches (e.g., feel or feedback received during switch movements characterized by length of movements [e.g., expected amount of switch travel or deflections, low pressures on the fingertips], nonperceptible vibrations, low slop, sags or looseness at joints, crisp detent feel (measured by force–deflection curves), presence of sound or tactile feedback, etc.)

Sound Quality

1. Pleasing sounds (solid, smooth, nonharsh, or not tinny) emanating from functional equipment (e.g., sound characteristics of engine, door slam, warning signals, auditory feedback on driver executed controls, etc.)
2. Absence of unwanted/annoying sounds (e.g., squeaks, rattles, harshness)

Harmony

1. Harmony across systems, subsystems, and components (e.g., similarities in appearance and operational feel of controls between radio and climate controls) (Note: Some differences are needed to discriminate between functions and reduce driver errors during equipment uses. However, various operational features should provide the impression design consistency, i.e., all systems should look and feel as if they were designed by the same designer, and they should exhibit certain brand characteristics.)
2. Harmony between materials and their finishes within a vehicle of a given brand (to create brand image)
3. Smaller number of dissimilar materials in close proximity (e.g., avoid many components with dissimilar materials with many parting lines placed within a small region as they create perception of clutter, mismatches, unevenness, etc.)

Smell Quality
1. Nonsmelling Materials (avoid materials that produce unpleasant and unsafe odor)
2. Materials that produce pleasing odors (e.g., smell of genuine leather, flowery/fruity, spicy, etc.)

Comfort and Convenience

Ergonomic considerations involved in designing all vehicle systems, subsystems, and components (essentially the entire contents of this book) to assure that the vehicle is comfortable, safe, and easy to use.

MEASUREMENT METHODS

A number of different methods are used in the industry for measuring attributes of craftsmanship listed in the above section. The methods can be categorized as follows:

1. Checklists
2. Objective measurements (e.g., gloss levels, compressibility, roughness, friction)
3. Subjective evaluations using rating scales or comparisons with acceptable reference samples
4. Customer clinics or studies to understand customer preferences and ability to perceive, discriminate, and categorize craftsmanship characteristics

Various applications of the above methods are covered in several chapters of this book (see Chapters 5 and 15).

SOME EXAMPLES OF CRAFTSMANSHIP EVALUATION STUDIES

CRAFTSMANSHIP OF STEERING WHEELS

A study was conducted to evaluate craftsmanship characteristics of steering wheels with a focus on rim and spoke feel. The study involved driving a vehicle in a driving simulator with eight different steering wheels in random orders, and the subjects were asked to provide ratings on the following details on each of the eight production steering wheels:

1. Rim: grasp comfort, thickness/thinness of the rim, rim surface softness/hardness, and rim surface slipperiness/stickiness
2. Top side of upper spoke: surface feel—like/dislike, softness/hardness, and smoothness/roughness
3. Stitches (in the wrapped leather): visual appearance—like/dislike, grasp feel—comfortable/uncomfortable, and touch feel—protrusion/recessing
4. Top spoke size near rim: small/large and easy/difficult to hold
5. Rim grasp area between top and bottom spokes: small/large
6. Seams: grasp feel over the seams—comfortable/uncomfortable and visual appearance—like/dislike
7. Characteristics especially liked or disliked

Figure 10.4 presents sketches of the eight steering wheels used in the study. The steering wheels were obtained from production vehicles made by different automobile manufacturers. The hub areas of the steering wheels were removed and masked with a matte black cardboard to hide the identity of the manufacturers from biasing the subjects.

Figures 10.5 and 10.6 present examples of the outputs of the ratings data obtained from the study involving over 50 drivers of passenger cars.

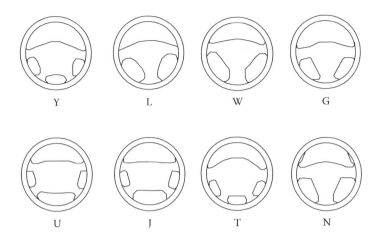

FIGURE 10.4 Eight steering wheels used in the study.

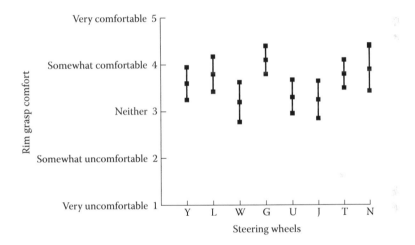

FIGURE 10.5 Mean and 95% confidence intervals of rim grasp comfort ratings of the eight steering wheels.

Figure 10.5 presents the mean and 95% confidence intervals of ratings on how the steering rim grasp comfort was perceived by the subjects while using the eight steering wheels. The data from Figure 10.5 indicated that steering wheels G and N were perceived to be most comfortable.

Figure 10.6 presents the mean and 95% confidence intervals of ratings on how the steering rim hardness was perceived by the subjects on the eight steering wheels. The figure shows that the steering wheels G and N were perceived to be softer than all the other steering wheels.

The relationship plot of mean hardness ratings presented in Figure 10.6 to the objective measure of hardness (called the hardness number) is presented in Figure 10.7. The plot suggested that if the steering wheel rim surface is made with hardness number below 77.5, the drivers will feel that the rim surface is softer. Similar correlation analyses allowed the manufacturer to determine the most preferred characteristics (e.g., rim diameter, rim slipperiness, type of leather, type of stitches) of the steering wheel.

The above example, thus, illustrates how studies can be conducted to obtain preferred characteristics of touch and grasp to improve customer satisfaction.

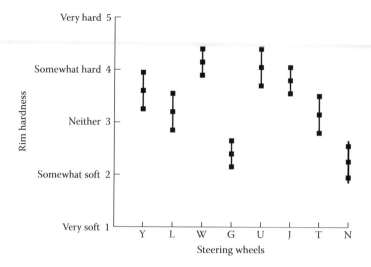

FIGURE 10.6 Mean and 95% confidence intervals of rim surface hardness ratings of the eight steering wheels.

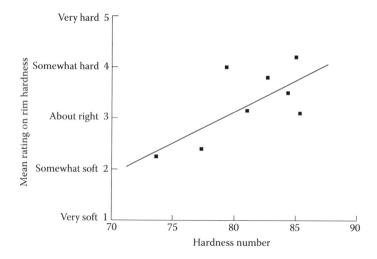

FIGURE 10.7 Relationship between mean ratings on rim surface hardness and hardness.

OTHER STUDIES

Similar studies, to determine customer-pleasing characteristics of interior materials used in armrests, seats, instrument panels, door trim panels, etc., conducted by the author, are available in the literature (Bhise et al., 2006 and 2008; Onkar et al., 2008). In a recent study, Bhise et al. (2009) evaluated effects of seven variables on leather and cloth materials used in automotive interior materials. The variables included (1) compressibility, (2) smoothness (or roughness), (3) shape (flat vs. 30- to 35-mm-diameter round grasp handle), (4) material type (leather vs. cloth), (5) gender of subjects, (6) age of subjects, and (7) sensory modality (visual only vs. visual and tactile). Their results showed that the primary variable that affected the overall pleasing perception of materials was smoothness/roughness of the materials.

CONCLUDING REMARKS

The craftsmanship is relatively a new field, and systematic studies on understanding important variables that affect their pleasing perception are needed. Currently available methods to measure physical characteristics of materials such as compressibility, hardness, coefficient of friction, surface finish, texture, colors, lightness, gloss values, etc., are of limited usefulness in predicting pleasing perception qualities of interior materials. Additional research studies on effects of many objective and subjective characteristics of interior components and their materials and their relationships to craftsmanship are needed. Studies are also needed to determine how these effects are affected by characteristics of customers and users from different market segments.

REFERENCES

Bhise, V., R. Hammoudeh, R. Nagaraj, J. Dowd, and M. Hayes. 2006. Towards development of a methodology to measure perception of quality of interior materials. *Journal of Materials and Manufacturing*, Society of Automotive Engineers Inc., Warrendale, PA. (Also published as SAE Paper 2005-01-0973.)

Bhise, V. D., S. Onkar, M. Hayes, J. Dalpizzol, and J. Dowd. 2008. Touch feel and appearance characteristics of automotive door armrest material. *Journal of Passenger Cars—Mechanical Systems*, SAE 2007 Transactions. (Also published as SAE Paper 2007-01-1217.)

Bhise, V. D., V. Sarma, and P. K. Mallick. 2009. Determining perceptual characteristics of automotive interior materials. SAE Paper 2009-01-0017. Warrendale, PA: Society of Automotive Engineers Inc.

Levitt, T. 1980. Marketing success through differentiation of anything. *Harvard Business Review*. January–February.

Onkar, S., M. Hayes, J. Dalpizzol, J. Dowd, and V. Bhise. 2008. A value analysis tool for automotive interior door trim panel material and process selection. *Journal of Materials and Manufacturing*, SAE 2007 Transactions. (Also published as SAE Paper 2007-01-0453.)

Peters, T. 1987. *Thriving on Chaos*. New York: Harper & Row.

Yang, K., and B. El-Haik. 2003. *Design for Six Sigma*. New York: McGraw-Hill.

11 Role of Ergonomics Engineers in the Automotive Design Process

INTRODUCTION

Designing a new vehicle is generally accomplished by using the systems engineering approach, which involves creating an organization of design teams involving different disciplines. The teams are also colocated at one site or in one building so that they can meet formally and informally to communicate about numerous issues and trade-offs related to interfaces between various systems and subsystems in the vehicle. The ergonomics engineers assigned to the vehicle program follow the vehicle development from its earliest stages of defining the vehicle concept until the vehicle is produced and used by the customers. During each stage of development, the ergonomics engineers conduct a number of tasks to assure that the vehicle being designed will be perceived by its customers to be ergonomically superior. In this chapter, we will review the following: (a) the goal of the ergonomics engineer, (b) methods used to achieve the goal, (c) evaluation measures used, (d) the responsibilities of the ergonomics engineers, and (e) their problems and challenges.

SYSTEMS ENGINEERING MODEL DESCRIBING THE VEHICLE DEVELOPMENT PROCESS

Figure 11.1 presents the Systems Engineering "V" Model (refer to Blanchard and Fabrycky, 2006, for more information). The model shows basic phases of the entire vehicle development project on a horizontal axis, which represents time in months before Job 1. Job 1 in the auto industry refers to the event when the first production vehicle rolls out of the assembly plant. The project (or the vehicle program) generally begins many months prior to Job 1 (typically 12–48 months, depending on the complexity of the program).

In the early stages prior to the official start of a vehicle program, the advanced vehicle planning activity determines the vehicle type (i.e., body style, powertrain type, performance characteristics, etc.), size of vehicle (e.g., size class, number of occupants, and loads [weight and cargo volume]), the intended market (i.e., countries) and a list of reference vehicles (or competitors) that the new vehicle may replace or compete with. A small group of engineers and designers from the advanced design group are selected and asked to generate a few early vehicle concepts to understand design and engineering challenges. A business plan including projected sales volumes, the planned life of the vehicle, program timing plan, facilities and tooling plan, manpower plan and financial plan (including estimates of costs, capital needed, revenue stream, and projected profits) are developed and presented to the senior management along with all other vehicle programs planned by the company (to illustrate how the proposed program fits in the overall corporate plan). The vehicle program, in most automotive companies, begins officially after the approval of the business plan by the company management. This program approval event is considered to occur at x months prior to Job 1 in Figure 11.1.

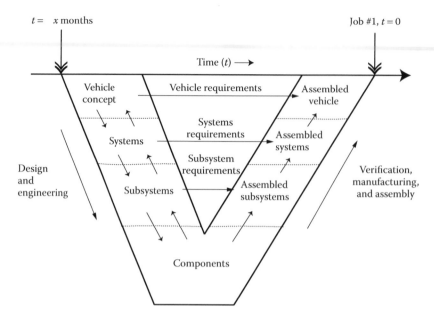

FIGURE 11.1   Systems Engineering "V" Model of the vehicle design process.

At minus x months, the chief vehicle program manager is selected, and each functional group (such as design, body engineering, chassis engineering, powertrain engineering, electrical engineering, climate control engineering, vehicle packaging and ergonomics/human factors engineering, manufacturing engineering, etc.) within the product development and other related activities is asked to provide personnel to support the vehicle development work. The personnel are grouped into teams, and the teams are organized to design and engineer various systems and subsystems of the vehicle. Figure 11.1 shows that the first major task after the team formation is to create an overall product concept. During this phase, the designers (industrial designers) and the vehicle package engineers work with different teams to create the product concept, which involves (a) creating early drawings or computer-aided design (CAD) models of the proposed vehicle, (b) creating computer-generated three-dimensional life-like images or videos of the vehicle (fully rendered with color, shading, reflections, and textural effects), and (c) physical mock-ups (clay models; foam core, wooden, or fiberglass bucks to represent the exterior and interior surfaces). The images and/or models of the proposed vehicle are shown to prospective customers in market research clinics and to the management. Their feedback is used to further refine the product concept.

As the vehicle concept is being developed, each engineering team decides on how each of the vehicle systems can be configured to fit within the vehicle space and how the various systems can be interfaced to work together to meet all the functional and ergonomic requirements of the vehicle. This phase is shown as "Systems" in Figure 11.1. As the systems are being designed, the next phases involve a more detailed design, that is, design of subsystems of each system and components within each of the subsystems. These subsequent phases, straddled in time, are shown as "Subsystems" and "Components." The above phases form the left half of the V and thus represent the time and activities involved in design and engineering.

The right half of the V, moving from the bottom to the top, involves testing, assembly, and verification where the components are individually built and tested to assure that they meet their functional characteristics. The components are assembled to form subsystems which are tested to assure that they meet their functional specifications. Similarly, the subsystems are assembled into systems, and finally, the systems are assembled to create a vehicle. At each of the phases, the corresponding assemblies are tested to assure that they meet the requirements considered during their respective design phases (i.e., the assemblies are verified). These requirements are shown as the horizontal arrows between the left and the right sides of the V in Figure 11.1.

The ergonomics engineers assigned to the vehicle program work throughout all the above phases and continuously evaluate the vehicle design to assure that the vehicle users can be accommodated and they will be able to use the vehicle under all foreseeable usage situations.

VEHICLE EVALUATION

GOAL OF ERGONOMICS ENGINEERS

The primary goal of ergonomics engineers is to work with the vehicle design team to produce ergonomically superior vehicles. Some criteria that can be used to establish ergonomic superiority are the following:

1. Best-in-class or best in the industry (i.e., the product will be perceived by the users to be the best within the selected class of vehicles or in the industry).
2. *x* percent better than products in the class (e.g., all driver interface components of the new vehicle should be at least 10% better than the corresponding items in all other selected reference vehicles).
3. All important and preselected ergonomic requirements and all applicable standards must be met.
4. User accommodation goals set in the early program definition phase must be met. (The goals for accommodation can be set as 90% or 95% of population, 5th percentile female and 99th percentile male, etc.)
5. Minimum acceptable levels of ratings based on attributes such as easy to learn, easy to find and use, nondistracting and safe, comfortable, and convenient must be met.

EVALUATION MEASURES

The following are examples of measures that can be used to evaluate the vehicle:

1. Percentage of ergonomic guidelines and requirements met in each evaluation category (e.g., driver accommodation, field of view, entry/exit, and exterior use items).
2. Weighted sum of ergonomic guidelines met in each category (function or part of the vehicle) (Note: The categories can be weighted by frequency of use and importance of items or vehicle features.)
3. Objective measures: Percentage of users meeting predetermined target values of various task performance measures such as task completion times, error rates, eye involvement times, number of glances, standard deviation of lane position, mean velocity, and standard deviation of velocity.
4. Subjective measures: For example, percentage of users satisfied; percentage of users liked the vehicle (or preferred the vehicle over other benchmarked vehicles); percentage of features rated as expected, pleasing, and delightful to use as compared to features in other benchmarked vehicles; percentage of users liked usability of each feature (control, display, etc.); etc. Other quantitative measures based on ratings such as averages of ratings, percentage of ratings above 8 on a 10-point scale, percentages of ratings below 5 on a 10-point scale, etc. can also be used.

Tools, Methods, and Techniques

A number of tools, methods, and techniques are used for evaluation of vehicles and systems during the vehicle development process (see Chapter 15 for more information). These tools and methods include the following:

1. Benchmarking of selected existing vehicles to understand different designs and ergonomic issues with the designs
2. Quality function deployment: Understanding customer needs and translating them into functional specifications by using quality function deployment (Besterfield et al., 2003).
3. Checklists and scorecards based on ergonomics requirements (in standards) and design guidelines
4. Use of models: For example, physical models (bucks) and mathematical/computer models such as parametric design models for occupant packaging and ergonomic analyses (Bhise et al., 2004a, 2004b; Bhise and Pillai, 2006; Kang et al. 2006), visualization/CAD (packaging studies—three-dimensional models with occupants/manikins and other vehicle systems and components, field-of-view simulations), and other CAE applications to analyze trade-offs between ergonomic and other functional issues (vehicle lighting evaluations, legibility models, occupant crash simulations, aerodynamic drag, air flow and climate control simulations, etc.)
5. Task analysis, failure mode and effects analysis, and cost–benefit analysis
6. Laboratory and/or field studies using methods of observation, communication, and experimentation (e.g., evaluations using programmable vehicle bucks, prototypes, simulators, and instrumented vehicles)

ERGONOMICS ENGINEER'S RESPONSIBILITIES

1. Provide the program teams the needed ergonomics design guidelines, requirements, data/information, and results from analyses and experimental research, scorecards, and recommendations for product decisions at the right point (timing, gateways) in front of the right level of decision makers (teams, program managers, chief engineers, senior management, etc.).
2. Apply methods/models/procedures available in the Society of Automotive Engineers, Inc. (SAE), corporate, and regulatory standards and design guides.
3. Conduct quick-react studies/experiments to resolve issues where sufficient information from prior research or vehicle program is not available.
4. Evaluate product/program assumptions, concepts, sketches/drawings, physical models/mock-ups/bucks, CAD models, mules (other production models equipped with new vehicle components or systems; sometime called mechanical prototypes), prototypes, production vehicles, and competitors.
5. Participate in the development of test drives and market research clinics—with existing leading products (competitors' and your company's) as comparators (or controls).
6. Obtain, review, and act on the customer feedback data (complaints, warranty, customer satisfaction surveys, J. D. Power's ratings/data, inspection surveys with owners, automotive magazines, press, etc.) to improve the product.
7. Prepare ergonomics scorecards and summaries of ergonomic strengths, weaknesses, trade-off considerations, and recommendations.
8. Provide ergonomics consultations within product development on various product issues.
9. In long term, conduct research, translate research results into design guidelines, and develop design tools for application to future vehicle programs.

Steps in Ergonomics Support Process during Vehicle Development

Most automotive companies have a well-developed ergonomics support process that is synchronized with the overall vehicle development timing plan. Therefore, the ergonomics engineers supporting the program must understand the vehicle design process in terms of its phases, the work performed in each phase, functional areas involved in each phase, team structure, people and methods involved in performing different tasks, team communication methods, management review, and the approval process.

The overall flow of a vehicle development program begins with the customers and ends with the customers. In the early stages, even before the program plan is developed, the needs of the customer s are complied and understood by the product planning and the design team. After Job 1, the customer feedback is continuously sought and reviewed to improve the product by removing production defects and planning for future design-related product changes. Thus, basic vehicle development process flow involves the following major steps, many of which are usually conducted concurrently (using what is known as simultaneous engineering or concurrent engineering to reduce time and to avoid redesign/rework):

1. Customers (understanding customers and their needs)
2. Product planning
3. Design (styling and engineering)
4. Detailed engineering
5. Prototyping, testing, and validation
6. Tooling design
7. Plant design and construction
8. Production of vehicles
9. Customers (obtaining feedback after product usages)

STEPS IN THE EARLY DESIGN PROCESS

1. Understand customer needs and translate into vehicle design requirements
   - Review product letter (that describes the proposed product specifications and features). Study new vehicle program assumptions. Predict ergonomic issues. Study customer feedback data.
   - Study/understand the market segment.
   - Benchmark competitive and current products (that will be replaced).
   - Identify the customers and their characteristics, capabilities, and limitations.
   - Determine customer wants.
   - Translate customer wants into product attributes and specifications.
   - Cascade vehicle-level specifications to system, subsystem, and component levels.
   - Evaluate trade-offs between different product attributes.
   - Conduct market research clinics.
2. Understand communications with different teams
   - Learn the team structure (colocated, dedicated, multidisciplinary), team levels, meetings, and procedures for resolutions of issues.
   - Implement simultaneous/concurrent engineering
   - Deliver vehicle-attribute-related information to satisfy customers.
   - Understand vehicle systems, subsystems, components, and interfaces.
   - Implement the Systems Engineering "V" process in program management.
   - Divide the vehicle into parts/modules and functional expertise.
3. Reviewing early product concepts
   - Review early exterior design sketches and renderings (e.g., see Figures 11.2 through 11.4; the vehicle shown in the figures is a low-mass vehicle developed by the students from the University of Michigan-Dearborn [Shulze, 2007]) and mock-ups to identify and resolve issues related to entry/exit, head clearance, field of view (pillar obscurations, windshield, and backlite rake angles), locations of exterior lights, body cut lines (defining body openings and exterior body panels), fuel filler location, trunk opening, loading/unloading, etc.

FIGURE 11.2   Side-view sketches of alternate SUV designs.

FIGURE 11.3   Exterior view of selected SUV concept details. (Reprinted from Shulze, R. ed., *Designing a Low Mass Vehicle*, College of Engineering and Computer Science, University of Michigan-Dearborn, 2007. With permission.)

- Review interior sketches or three-dimensional models (e.g., see Figure 11.5) for driver accommodation issues (e.g., seat positioning, eye location, instrument panel layout, etc; Bhise et al., 2004a and 2004b).
- Conduct package analyses (e.g., use parametric model [Bhise et. al., 2004b] to evaluate occupant accommodation; see Figure 11.6) and conduct customer clinics.
4. Possible steps after a vehicle concept is selected for production are the following:
   - Verify application of the SAE practices related to placement of seat track, location of eyellipse, reach and head clearance contours, visibility through the steering wheel, controls locations, direction of motion stereotypes, labeling, exterior lighting photometric requirements, etc.
   - Conduct task analyses on items and issues related to different vehicle usages.
   - Conduct studies to evaluate special issues (e.g., legibility of labels and graphics, evaluation of unwanted interior reflections, new center stack units, entry/exit, obscurations

FIGURE 11.4 Fully surfaced vehicle with shading and reflections details.

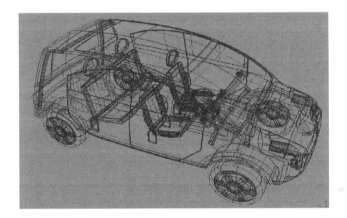

FIGURE 11.5 Wire-frame diagram showing interior details. (Reprinted from Shulze, R. ed., *Designing a Low Mass Vehicle*, College of Engineering and Computer Science, University of Michigan-Dearborn, 2007. With permission.)

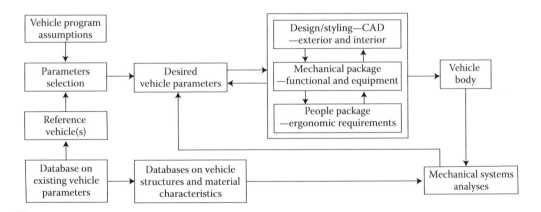

FIGURE 11.6 Flow diagram of a parametric model developed for vehicle packaging. (Redrawn from Shulze, R. ed., *Designing a Low Mass Vehicle*, College of Engineering and Computer Science, University of Michigan-Dearborn, 2007. With permission.)

due to pillars, operation of navigation units, entertainment systems, exterior lighting photometric issues, etc.).
- Follow design refinements and conduct additional evaluations or studies as needed.
- Evaluate hardware as prototype parts and vehicle models.
- Provide ergonomics assessments/maintain score cards (e.g., smiley face charts) on controls, displays, field of view, entry/exit, craftsmanship, etc.

TRADE-OFFS IN THE DESIGN PROCESS

During the entire design process of an automotive product, designers and engineers are continually faced with the problems of trade-offs between different vehicle features to meet many design requirements. For example, the designers could like highly sloped windshields for styling and aerodynamic reasons. But the windshield glass, wiper design, and climate control engineers would have a more difficult task because the higher windshield rake angle will require larger glass (due to increased height), longer wipers, more powerful windshield wiper motor, large capacity air-conditioner (due to larger sun load through the windshield). The glass engineer would need to make the windshield glass thicker to maintain the rigidity of the windshield. The thicker windshield will decrease the light transmissivity which in turn will reduce visibility distances during night driving. The higher raked windshield may also reduce the driver's head clearance to the header and may also require the tall driver and the front seat passenger to bend their heads and torsos during entry and exit. Thus, one change in a design need may cascade a series of design changes and challenges in a number of affected vehicle systems.

PROBLEMS AND CHALLENGES

The ergonomics engineers also face many problems and challenges during the vehicle design process. Some major issues related to the problems and challenges are the following:

- Insufficient data available (constantly changing/evolving designs and unavailability of sufficient design details) during early design phases
- Insufficient time and resources available to thoroughly research problems
- Need to consider many users (and markets), usages, needs, and expectations (difficult to study many issues, prioritize or weigh, and trade-off different considerations)
- Fierce competition (competing in fast changing and chaotic environment)
- High cost of research studies
- Difficulty to predict future design and technology trends
- Difficulty to predict future plans of your competitors (i.e., "futuring"—what would your competitors do?)
- Need to know limits of applicability of your tools, procedures, data, etc.
- Insufficient understanding of trade-offs between different vehicle attributes

Many of the above challenges can be overcome through in-house and continual training of ergonomists, continual communication within members of the ergonomics departments on various projects under study and completed by various design teams, information exchanges with other engineering departments on methods and completed projects, planning activities of future research projects (see Chapter 17), and attending meeting and conferences where ergonomic research methods and studies are presented (e.g., SAE Technical Committee Meetings, SAE International Congress, Transportation Research Board Annual Meetings, Human Factors and Ergonomics Society Meetings).

CONCLUDING REMARKS

The product development process is rarely smooth because of many unknown issues involving hundreds of interfaces between many systems within the vehicle need resolution within the tight program timing plan. In many major automotive companies, more experienced ergonomics experts are generally available to guide the ergonomics engineers working in the teams. In addition, the ergonomics process and tools to be used in product development are also well documented. An extensive library of previous research studies, literature, and reference books are maintained within most ergonomics departments. The available information is also placed on internal websites of most automotive companies and incorporated in many computer-integrated design and engineering analysis tools.

REFERENCES

Besterfield, D. H., C. Besterfield-Michna, G. H. Besterfield, and M. Besterfield-Scare. 2003. *Total Quality Management*. Upper Saddle River, NJ: Prentice Hall.

Bhise, V., R. Boufelliga, T. Roney, J. Dowd, and M. Hayes. 2006. Development of innovative design concepts for automotive center consoles. SAE Paper 2006-01-1474. Presented at the SAE 2006 World Congress, Detroit, MI.

Bhise, V., G. Kridli, H. Mamoola, S. Devraj, A. Pillai, and R. Shulze. 2004b. Development of a parametric model for advanced vehicle design. SAE Paper 2004-01-0381. Also in SAE Special Publication SP-1858, SAE International, Warrendale, PA.

Bhise, V., and A. Pillai. 2006. A parametric model for automotive packaging and ergonomics design. Proceedings of the International Conference on Computer Graphics and Virtual Reality, Las Vegas, NV.

Bhise, V., R. Shulze, H. Mamoola, and J. Bonner. 2004a. Interior design process for UM-D's low mass vehicle. SAE Paper 2004-01-1709. Also in SAE Special Publication SP-1848, SAE International, Warrendale, PA.

Blanchard, B. S., and W. J. Fabrycky. 2006. *Systems Engineering and Analysis*. 4th ed., Upper Saddle River, NJ: Pearson Prentice Hall.

Diehl, C., and V. Bhise. 2006. Design concepts for vehicle center stack and console areas for incorporating new technology devices. Presented at the Society of Information Display Vehicle and Photons Symposium, Dearborn, MI.

Kang, H., N. Bhat, and V. Bhise. 2006. Parametric approach to the development of a front bucket seat frame. SAE Paper. 2006-01-0336. Presented at the SAE 2006 World Congress, Detroit, MI.

Shulze, R. (ed.). 2007. *Designing a Low Mass Vehicle*. Dearborn: College of engineering and computer science, University of Michigan-Dearborn.

Part II

Advanced Topics, Measurements, Modeling, and Research

12 Modeling Driver Vision

USE OF DRIVER VISION MODELS IN VEHICLE DESIGN

The objectives of this chapter are to present driver vision models and discuss their applications for assessing driver's visibility-related problems encountered during the vehicle design process. The problems covered in this chapter are determination of (a) visibility of targets under headlamp illumination, (b) legibility of displays, and (c) veiling glare caused by sun reflections.

The driver vision models considered in this chapter are based on determinations of what the driver can see and read. The models are primarily used to evaluate detection of targets on the roadway due to illumination from vehicle lighting systems, legibility of displays, and evaluation of daytime visibility under situations of veiling glare caused by reflections of the top of the instrument panels into the windshields.

This chapter will begin with a discussion of the visual contrast thresholds of the human eye. During his research work on driver vision, the author found that the visual contrast threshold curves developed by Blackwell (1952) could be used to predict visibility and legibility under a variety of daytime and nighttime conditions. The prediction capabilities of these models were compared with the visibility distances obtained in seeing distance tests while driving in both the absence and presence of glare caused by the oncoming vehicle headlamps (Bhise et al., 1977a, 1977b). The discomfort glare evaluation model developed by DeBoer (1973) was also found to be useful by comparing the predicted values of the DeBoer discomfort index with the dimming request behavior of the drivers on public roads (Bhise et al., 1977b). The visual contrast thresholds were also used to predict the legibility of labels and numerals under day, dawn/dusk, and night-driving conditions of automotive displays (Bhise and Hammoudeh, 2004). Rockwell et al. (1988) also used a similar model to predict legibility of electronic displays. The veiling glare caused by the reflections of the top of the instrument panels in the windshields were modeled recently to predict veiling glare experienced by the drivers and their effects on target visibility (Bhise and Sethumadhavan, 2008a, 2008b).

SYSTEMS CONSIDERATIONS RELATED TO VISIBILITY

Determining the level of visibility of an object is a systems problem because it depends on characteristics of a number of components of the highway transportation system. The visibility of an object (a target or a visual detail) depends primarily on the photometric and geometric characteristics of the object, the visual environment, the vehicle, and the observer's (driver's) eyes and his or her visual information-processing capabilities. The important characteristics related to visibility are given below.

1. Target characteristics

   a. Size (i.e., the angular size or the visual angle subtended by the target at the driver's eyes)
   b. Reflectance
   c. Shape (e.g., length-to-width ratio)
   d. Location of the target on the roadway
   e. Target orientation (stand-up target or horizontal target on the roadway, e.g., lane marking)
   f. Motion/movement of the target
   g. Temporal characteristics of the target luminance (e.g., flash rate)
   h. Color of the target

2. Visual environment characteristics

   a. Ambient lighting conditions (i.e., illumination from external sources)
   b. Road geometry (e.g., lane configurations, curvatures, grades)
   c. Background against which a target is seen (i.e., luminance or reflectance of background material)
   d. Weather conditions (e.g., transmission and scattering of light through fog, rain, and snow)
   e. Glare sources (i.e., their locations with respect to observer's eyes and line of sight, luminous intensity)

3. Vehicle characteristics

   a. Driver's (observer's) eye locations in the vehicle (e.g., coordinates of the eyellipse centroids)
   b. Vehicle components causing obstructions in the driver's field of view (e.g., pillars, mirrors, headrests)
   c. Headlamp locations (i.e., coordinates of headlamp centers)
   d. Headlamp beam patterns (i.e., distribution of luminous intensities of left and right headlamps)
   e. Headlamp aim (i.e., angular locations of the headlamp axes with respect to the longitudinal axis of the vehicle)
   f. Glazing (glass) materials and installation angles (i.e., light transmissivity at installed angle of the vehicle glass through which the target is viewed)
   g. Vehicle location on the roadway and speed
   h. Graphics in visual displays (i.e., sizes of visual details, their background luminance, visual contrast, colors)

4. Observer's (driver's) characteristics

   a. Observer's age (OA) (Note: The visual contrast thresholds, disability glare, and discomfort glare increase with an increase in the OA.)
   b. Individual differences in visual capabilities of drivers (e.g., percentile level of contrast threshold, eye defects, corrections in eye glasses)
   c. Seating position and eye locations (with respect to the locations of the seating reference point and the eyellipses)
   d. Eye and head movement behavior (i.e., eye and head-turn angles)
   e. Psychological and physiological state (e.g., attentiveness of the driver)

LIGHT MEASUREMENTS

This section provides definitions of light measurement units that are used to compute photometric characteristics of targets and their backgrounds which, in turn, are used to predict visibility and legibility.

LIGHT MEASUREMENT UNITS

Light is defined as visually sensed radiant energy. Radiant energy is the total energy emitted by a source. Only some part of the radiant energy can be sensed by the human visual system and is perceived as light. Thus, visible light is a part of radiant energy. The light measurement units used to measure the outputs of light sources and resulting luminance of targets and their backgrounds are defined as follows:

1. Luminous energy (Q) or quantity of light is defined as visually sensed radiant energy. It is measured in Talbots (T) with the units lumen-second (lm s).
2. Luminous flux (Φ) is light power. It is the time rate of flow of light energy, and it is defined as $\Phi = Q / t$ and is measured in lumen (lm), where t = time in seconds.
3. Luminous intensity (I) is defined as the amount of luminous flux emitted (Φ) in a given solid angle (ω). The unit of luminous intensity is candela (cd).

   Thus, $I = \Phi / \omega$, where ω = solid angle measured in steradians (sr). The solid angle is defined as $\omega = S / r^2$, where S = elemental area (through which the light flux is emitted) on the surface of a sphere of radius r (through which the light flux is emitted).
4. Illuminance or illumination (E) is the density of luminous flux incident upon a surface. It is the quotient of flux (Φ) divided by the projected area A of the illuminated surface when the flux is averaged over the area.

   Thus, $E = \Phi / A$. The illumination is measured in the SI metric system in lux (lx), which has the units of lumens per square meter.. The measure of illumination in the English units is foot-candle (fc), which has the units of lumens per feet squared. The conversion equations are 1 fc = 10.76 lx, and 1 lx = 0.0929 fc.

   $E = (I / d^2) * \cos \alpha$, where I = luminous intensity, d = distance from the source to the receiving plane, and α = the angle of the normal to the plane with respect to the incident light. The inverse square law thus applies here, stating that illumination falling on a surface is directly proportional to the inverse of the square of the distance from its source to the surface.
5. Luminance (L) or photometric brightness is measured in nit or candela per meter squared in the SI metric system or in foot-Lambert (fL) in the English system. The conversion equations are 1 nit = 1 cd/m^2 = 0.2919 fL; 1 fL = 3.426 cd/m^2; $1/\pi$ cd/ft^2 = 1 fL, where $\pi = 3.142$.

   For a surface viewed under direct illumination: The luminance of a surface (L) due to illumination (E) incident on the surface (of a target or a background of a target) can be calculated as

$$L = [(\text{reflectance of the surface}) \times (\text{illumination})] / \pi = [R \times E] / \pi$$
$$= R \times E \text{ (fc)} = L \text{ (fL)}$$
$$= [R \times E \text{ (lx)}] / \pi = L \text{ (cd/m}^2)$$

   Note: Reflectance (R) is a dimensionless number ranging between 0 and 1 (or specified as percentage). The reflectance of white matte paint is about 0.85, and the reflectance of matte black paint is about 0.04.

   For a surface viewed under transmitted illumination: Luminance of a surface (L) due to an illumination (E) incident on the back surface (of a translucent surface of a target or a background of a target) can be calculated as

$$L = [(\text{transmittance of the surface}) \times (\text{illumination})] / \pi = [T \times E] / \pi$$
$$= T \times E \text{ (fc)} = L \text{ (fL)}$$
$$= [T \times E \text{ (lx)}] / \pi = L \text{ (cd/m}^2)$$

   Note: Transmittance (T) is a dimensionless number ranging between 0 and 1 (or specified as percentage).
6. The total light energy falling on a given material will be transferred as the sum of reflected energy, transmitted energy, and absorbed energy. In photometry, we are concerned with the measurement of the reflected or transmitted visible light energy, which is sensed by the human eye as the luminance.

PHOTOMETRY AND MEASUREMENT INSTRUMENTS

Photometry is defined as the science of measuring visible light based on the response of an average human observer. A photometer is an instrument used to measure photometric quantities such as luminance, illuminance, luminous flux, and luminous intensity. For luminance measurements, a photometer projects an image of a visual detail (e.g., target or background) on a photo cell. For illuminance or illumination measurement, it measures light flux falling on a white-light-gathering lens placed in front of the photo cell. A spectroradiometer is an instrument for measuring the spectral energy radiated by a source. The spectral data can be used to calculate photometric and colorimetric parameters.

VISUAL CONTRAST THRESHOLDS

Blackwell (1952) developed a basic model of the visual detection capabilities of the human eye based on the concept of visual contrast thresholds. A visual target is seen by an observer because of the visual contrast between the target and its background. Visual contrast (C) is defined as follows:

$$C = \frac{|L_t - L_b|}{L_b}$$

where L_t = luminance of the target and L_b = luminance of the background.

Figure 12.1 presents three circular targets with different visual contrast values. The two targets on the left have the same luminance of the target (L_t). The middle target is more visible because its background luminance (L_b) is lower than the background of the target on the left. The right target has the same luminance as the background luminance of the left target, and its background luminance is the same as the middle target. However, the right target is not as visible as the middle target because the luminance and the contrast of the right target are lower.

The contrast (C) must be greater than the threshold contrast (C_{th}) for the target to be visible. The threshold contrast levels were determined by Blackwell (1952) by conducting laboratory experiments. The experiments involved presenting to the subjects a number of circular targets with different diameters and luminance values against backgrounds with different luminance values to create different contrasts levels. Each subject sat in front of a screen, and the subject was presented with a different target in each trial. The targets were presented in tachistoscopic exposures ranging from 0.01 to 1 s. (The tachistoscopic exposures were accomplished by using an electromechanical shutter in front of the lens of a slide projector.) The task of the subjects was to determine if a target presented in each trial was visible or not visible. The trials included combinations of target size, background luminance, contrast, and exposure duration. From these data, Blackwell developed contrast threshold curves for target sizes ranging from 1 to 64 min in the space defined by the logarithm of adaptation luminance on the x-axis and logarithm of threshold contrast on the y-axis. The contrast threshold curves obtained under 1/30th of a second exposure are presented in Figure 12.2.

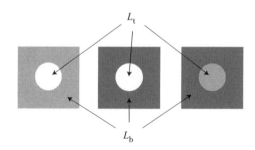

FIGURE 12.1 Illustration of three circular targets with different combinations of target luminance (L_t) and background luminance (L_b).

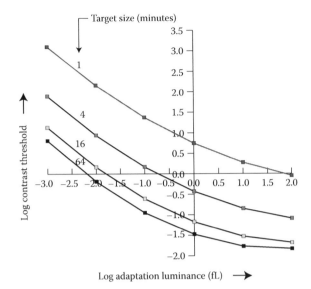

FIGURE 12.2   Visual contrast threshold curves for the 1/30th second target exposure used in the model.

The adaptation luminance (L_a) is the luminance level at which the observer's eyes are adapted. The adaptation luminance depends on the background luminance of the target being viewed, the adaptation level in the previous fixations (due to an effect called the transient adaptation), and the veiling glare experienced by the observer from brighter areas and other light sources in the observer's visual field. For an observer looking for a target in a relatively uniform luminance background, the adaptation luminance can be assumed to be equal to the luminance of the background.

BLACKWELL CONTRAST THRESHOLD CURVES

The visual contrast threshold data developed originally by Blackwell (1952) and validated for night-driving situations by Bhise et al. (1977a) were used to model basic target detection situations in night-driving (Bhise and Matle, 1989; Bhise, Matle, and Farber, 1988). It should be noted that the contrast threshold curves presented in Figure 12.2 were obtained by fitting Blackwell's threshold data obtained under 1/30th of a second exposure. The log values used in the abscissa and ordinate of the graphs are to the base value of 10. The 1/30th-second-exposure duration is much smaller than the typical eye fixation durations (of about 1/3 s) made by the drivers during driving. However, since Bhise et al. (1977a) found that the higher contrast thresholds for 1/30th second (as compared with that at 1/3-s exposure) predicted visibility distances to stand-up targets more accurately under the more difficult actual dynamic driving conditions, the Blackwell's contrast thresholds obtained under 1/30th of a second were used here. Overall, the field-observed seeing distances of stand-up and delineation line targets (i.e., painted lane markings) could be predicted within about 13% accuracy by the use of the Blackwell thresholds.

COMPUTATION OF CONTRAST VALUES

The target contrast (C) was computed as follows: $C = [|L_t - L_b|] / L_b$

where L_t = target luminance (fL)
$\quad\quad L_b$ = background luminance (fL)
$\quad\quad L_t = R_t \times (E_{lt} + E_{rt}) + L_{at}$
$\quad\quad L_b = R_b \times (E_{lb} + E_{rb}) + L_{ab}$
$\quad\quad R_t$ = reflectance of the target (fL/fc)

R_b = reflectance of the background (fL/fc)

(Note: Typical retroreflectance values of dry pavements range from about 0.03 to 0.10 [Bhise et al., 1977b]. Wet pavement retroreflection values range from about 0.005 to 0.04 [Bhise et al., 1977b]. Pedestrian summer and winter clothing reflectance range typically from 0.02 to 0.50 and 0.02 to 0.30, respectively [Bhise et al., 1977b].)

E_{lt} = illumination from the left headlamp falling on the target (fc)
E_{rt} = illumination from the right headlamp falling on the target (fc)
E_{lb} = illumination from the left headlamp falling on the target background (fc)
E_{rb} = illumination from the right headlamp falling on the target background (fc)

(Note: The above illumination levels are computed from the luminous intensity [cd] values directed at each target/background point from each headlamp and dividing them by squared values of the corresponding distances from the headlamp to the target/background point.)

L_{at} = ambient luminance of the target (fL)
L_{ab} = ambient luminance of the background (fL)

(Note: Typical ambient luminance values in rural nonilluminated areas range from about 0.0001 to 0.01fL, and those in urban illuminated areas range from about 0.001 to 0.10 fL [Bhise et al., 1977b].)

COMPUTATION OF THRESHOLD CONTRAST AND VISIBILITY DISTANCE

The contrast threshold (C_{th}) was modeled as a function of adaptation luminance (L_a) and target size (θ) in minutes of arc at the eye, and contrast multipliers were used to account for the effects of driver's age, confidence in detection judgment, driver alertness level, and attention-getting value of the target (conspicuity level). The C_{th} was computed by using the following expression:

$$\log_{10} C_{th} = \frac{B_0 + 10B_1 L + 100B_2 L^2}{10} + \log_{10}(T_M)$$

where $L = \log_{10} L_a$
   $L_a = W_T \times L_b$
   $B_0 = 7.4935 - 6.97678\ S + 0.544938\ S^2$
   $B_1 = -0.55315 + 0.021675\ S + 0.0003125\ S^2$
   $B_2 = 0.007721 + 0.000558\ S + 0.0000175\ S^2$
   $S = \log_2 \theta$ = logarithm to the base 2 of target size in minutes
   TS = target size (feet) (the diameter of a circle that has the same area as the target)
   θ = target size (minutes) = $[\tan^{-1}(TS\ /\ VD)] \times (180.0\ /\ \pi) \times 60$
   VD = viewing distance (in feet)
   W_T = windshield transmittance (typical value is about 0.65 or 65%)
   $T_M = M_a \times M_c \times M_{at}$ = contrast multiplier to account for the effects of observer age, confidence in detection judgment, and level of attention getting characteristic (or conspicuity) of the target. The three multipliers are described below:
   M_a = multiplier for contrast to account for effects of OA (Bhise et al., 1977b; Blackwell and Blackwell, 1971)
      = $-0.3796391 + 0.1343982\ OA - 0.0044422\ OA^2 + 0.0000550484\ OA^3$

Figure 12.3 shows a plot of the contrast multiplier (M_a) as a function of the OA.

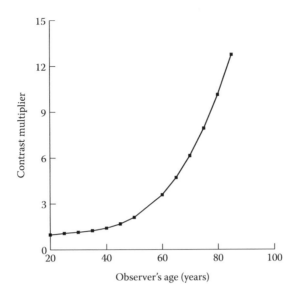

FIGURE 12.3   Contrast multiplier to account for the observer's age.

M_c = multiplier to account for the observers confidence (or the probability) in detecting the target. The multipliers were derived from the variability in the contrast threshold reported by Blackwell (1952).

= 1.00 for 50% confidence

= 1.58 for 90% confidence

= 1.78 for 95% confidence

= 2.24 for 99% confidence

M_{at} = multiplier to account for attention-getting level (or conspicuity) of the target

= 1.0 for "just-detectable" target

= 2.5 for "easy-to-see" target

= 5.0 for an "unalerted" driver (i.e., to model a driver who is unaware that a target will appear as compared with an alerted driver who is expecting a target to appear in a experimental situation)

= 10.0 for attention getting

The target is considered visible if the computed target contrast C is greater than the threshold contrast (i.e., $C > C_{th}$).

To determine the maximum visibility distance, the above procedure should be iterated by increasing VD in steps of 10–100 ft until a previously visible target becomes invisible or by decreasing VD in steps of 10–100 ft until a previously invisible target becomes visible.

The above contrast multipliers were developed by the author during the experiments and analyses conducted in the headlighting research program at the Ford Motor Company (Bhise et al., 1977b). The approach of using the multipliers is somewhat controversial as the selection of values for the multipliers requires the researcher's judgment and some calibration with actual detection data collected under known field conditions. However, for purposes of sensitivity analysis and to determine relative changes in visibility with respect to certain reference or baseline conditions, the application of multipliers is a useful approach.

EFFECT OF GLARE ON VISUAL CONTRAST

When one or more light sources are present in a driver's (or observer's) field of view, the illumination from the light sources can enter the driver's eyes and become scattered inside the eyes. The scattered

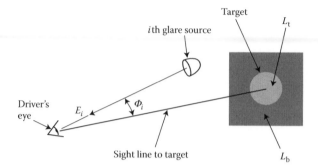

FIGURE 12.4 Target detection situation in the presence of a glare source directing glare illumination into the driver's eye.

light is superimposed on the image seen by the driver. Figure 12.4 shows a target viewed by the driver in the presence of a glare source (called the ith glare source) located at an angle of Φ_i from the line of sight to the target. The glare illumination entering the observer's eye is shown as E_i.

The additional luminance superimposed on the image of the target and its background due to the internal scattering of the illumination E_i is called the veiling luminance, and it can be computed using a formula developed by Fry (1954). The veiling luminance caused by the ith glare source is defined as L_{vi} and can be computed from Fry's veiling glare formula as follows:

$$L_{vi} = 10 \, p \frac{(E_i \cos q_i)}{(\Phi_i + 1.5) \times \Phi_i} \quad \text{(measured in footlambert)}$$

where E_i = illumination from the ith glare source (fc) and Φ_i = glare angle in degrees

If there is more than one glare source, the formula for the total veiling luminance (L_v) obtained by summing the veiling luminance from each of the glare sources can be stated as follows:

$$L_v = 10 \, p \sum \frac{(E_i \cos \Phi_i)}{(\Phi_i + 1.5) \times \Phi_i}$$

The observer's adaptation luminance (L_a), due to the superimposed veiling luminance, will increase as shown below.

$$L_a = L_b + L_v$$

The luminance contrast of the target will be modified due to the addition of the veiling glare to the luminance of the target and the background as follows:

$$C = \frac{|(L_t + L_v - L_b - L_v)|}{L_b + L_v} = \frac{|L_t - L_b|}{L_b + L_v}$$

The above equation, thus, shows that in the presence of veiling glare, the value of the contrast will always decrease from the value obtained under no-glare situation (i.e., without the presence of illumination from any glare source).

STEPS IN COMPUTING VISIBILITY OF A TARGET

Step 1: Measure the distance (D_t) of the target from the driver's eye location

The distance should be measured using the 95th percentile eyellipse (defined in Society of Automotive Engineers Inc. [SAE] standard J941, SAE 2009). Using the midpoint of the rearmost eyes on the 95th percentile eyellipse will provide the farthest eye location to measure the distance to the target. This distance will cover eye locations of most drivers and provide a conservative estimate of visibility.

Step 2: Determine the projected target size

The projected area of the target (A_p) at the driver's eye is equal to $A \cos \alpha$.
where α = angle between normal to the target surface and the sight line from the selected eye point to the target and A = target area.

Determine target size (TS) = diameter of a circular target of area A_p

$$= \left(\frac{4A_p}{p} \right)^{0.5}$$

Target size (θ) in minutes = $[\tan^{-1}(\text{TS} / D_t)] (180 \times 60)/\pi$.

Step 3: Determine target luminance

The target luminance (L_t) can be measured by using a photometer or calculated by knowing target reflectance and the illumination (E_t) incident on the target as $L_t = [(R_t \times E_t) + L_{at}]$. (Note: It is assumed here that the illumination [E_t] is from a single source).

Step 4: Determine luminance of the background of the target

The target background luminance (L_b) can be measured by using a photometer or calculated by knowing the background reflectance and illumination incident on the background as $[(R_b \times E_b) + L_{ab}]$. (Note: It is assumed here that the illumination [E_b] is from a single source.)

Step 5: Compute veiling luminance

The veiling luminance (L_v) can be measured directly by using a photometer equipped with a Fry veiling glare lens adaptor or it can be computed by using the formula given above, by knowing the glare angle and glare illumination from each glare source.

Step 6: Compute luminance contrast and its logarithm

Luminance contrast $C = (|L_t - L_b|) / (L_b + L_v)$
Compute logarithm of $C = \log_{10} C$.

Step 7: Compute logarithm of adaptation luminance

Logarithm of adaptation luminance = $\log_{10} (L_b + L_v)$.

Step 8: Plot point in the Blackwell space

Plot the point with the coordinates ($\log_{10} [L_b + L_v]$, $\log_{10} C$) in the Blackwell space (on the plot of Blackwell curves, Figure 12.2).

Step 9: Determine required contrast threshold curve for the target of size (θ)

For the value of target size θ (calculated in Step 2), using Figure 12.2 estimate (or interpolate the location of) the threshold contrast curve corresponding to θ. The threshold contrast curve should be adjusted (shifted up/down) by the magnitude of $\log_{10} (T_M)$, or compute $\log_{10} C_{th}$ by using the formula given above.

Step 10: Determination of target visibility

If the point computed in Step 8 is located above the adjusted threshold curve obtained in Step 9, then the target is visible.

Example 1: Target Visibility without Glare

Determine if a 0.3-m (1-ft) diameter target of 0.07 reflectance placed at 63.4 m (208 ft) from the driver's eyes would be visible against a background with 0.03 reflectance placed at 152 m (500 ft) from the driver's eyes illuminated by a 10,000-cd intensity headlamp placed at 61 m (200 ft) from the target. Assume that the ambient luminance of the target and the background is 0.001 fL and the driver is 20 years old.

The visibility model available at the publisher's website (called "Visibility Prediction Model Jan 3 2011") was used to input the data described in the above problem. The output of the model presented in Table 12.1 below shows that the driver could see the target because the bottom line of the table shows that the value of C/C_{tha} is greater than 1.0.

TABLE 12.1
Application of the Visibility Model for Example 1

Inputs		Value	Units
1	Luminous intensity of the source directed at the target	10,000	cd
2	Distance of the target from the light source	200	Ft
3	Distance of the background from the light source	500	Ft
4	Reflectance of the target $(0 < R_t < 1)$	0.07	fL/fc
5	Reflectance of the background $(0 < R_b < 1)$	0.03	fL/fc
6	Viewing distance (observer's eyes to target)	208	ft
7	Target size (diameter)	1	ft
8	Ambient luminance of the target	0.001	fL
9	Ambient luminance of the background of the target	0.001	fL
10	Glare source luminous intensity	0	cd
11	Distance of the glare source from the observer's eyes	400	ft
12	Angle of the glare source from the sight line to the target	2	Degrees
13	Observer's age	20	Years
14	Contrast multiplier to account for confidence, conspicuity, etc.	1	

Outputs		Value	Units
1	Illumination at the target (E_t)	0.2500	fc
2	Luminance of the target $(L_t + L_v)$	0.0185	fL
3	Illumination at the background (E_b)	0.04	fc
4a	Luminance of the background $(L_b + L_v)$	0.0022	fL
4b	Veiling glare luminance due to the glare source (L_v)	0	fL
5	Contrast of the target against background	7.40909091	
6	Log contrast	0.86976492	Log contrast
7	Log background luminance	−2.6575773	Log luminance
8	θ = target size (angle subtended at observer's eye)	16.5275014	Minutes
9	$S = \log_2 \theta = \log$ of target size to the base 2	4.04679673	$\log_2 (\theta$ in minutes)
10	$B_0 = 7.4935 − 6.97678\ S + 0.544938\ S^2$	−11.815899	Coefficient
11	$B_1 = −0.55315 + 0.021675\ S + 0.0003125\ S^2$	−0.460318	Coefficient
12	$B_2 = 0.007721 + 0.000558\ S + 0.0000175\ S^2$	0.0102657	Coefficient
13	$\log C_{th} = \log$ of Blackwell threshold contrast	0.76677836	
14	Age multiplier (M_a)	0.9718321	
15	$\log C_{tha} = \log$ threshold contrast with age and other multipliers	0.7543696	
16	Difference between actual and required log contrast = $\log C − \log C_{tha}$	0.11539532	
17	Ratio of C/C_{tha} (target is visible if the ratio is greater than or equal to 1)	1.30435354	

Example 2: Target Visibility in the Presence of a Glare Source

Determine the target visibility in Example 1 above if a glare source of 2000 cd is located at 122 m (400 ft) from the driver and at 2 degrees from the driver's line of sight to the target.

The output of the model presented in Table 12.2 shows that since the value of C/C_{tha} is less than 1.0, the target will not be visible to the driver.

Comparing values provided in Table 12.2 with the corresponding values in Table 12.1, the following observations can be made:

1. The contrast value (given in line 5 of the above output tables) changed from 7.409 when the glare source was absent to 0.041 when the glare source was present.
2. The driver's adaptation luminance (given in line 4a of the above output tables) changed from 0.0022 fL when the glare source was absent to 0.394 fL when the glare source was present.
3. When target distance was increased in increments of 3 m (10 ft) in Example 1 (after iterating the model), the maximum distance at which the target was visible was 66.5 m (218 ft) from the driver.
4. When target distance was decreased in increments of 3 m (10 ft) in Example 2 (after iterating the model), the maximum distance at which the target was visible was 45 m (148 ft) from the driver.

DISCOMFORT GLARE PREDICTION

The perception of level of discomfort that a driver will experience due to the presence of glare sources can be quantified by asking the driver to rate discomforting sensation using the 9-point discomfort glare scale developed by DeBoer (1973) given below. The scale rating (W) is called the DeBoer index.

$W = 1$—unbearable glare
  $= 3$—disturbing glare
  $= 5$—just-acceptable glare
  $= 7$—satisfactory glare
  $= 9$—just-noticeable glare

It should be noted that the value of the DeBoer index (W) decreases with increasing discomfort.

The value of the discomfort glare rating (W) can be also predicted by using an equation developed by DeBoer (1973). The DeBoer equation to predict discomfort glare index (W) caused by multiple glare sources and measured on the 9-point scale is given below.

$$W = 2 \log_{10}(1 + 269.0966\, L_a) - 2 \log_{10}(\Sigma E_i / F_i^{0.46}) - 2.1097$$

where L_a = adaptation luminance (fL)
  E_i = illumination from the ith glare source into the observer's eyes (fc)
  Φ_i = glare angle between the observer's line of sight and the line from the observer's mid eye to the ith glare source (minutes).

The above equation was found to be useful by Bhise et al. (1977b) and Bhise and Matle (1985) in evaluating glare from oncoming headlamps and signal lamps (Hoffmeister and Bhise, 1978). The above index (W) was developed under static conditions, and therefore, the magnitude of sensation of discomfort experienced by the driver under dynamic passing situation will be different. Based on the DeBoer index, Bhise et al. (1977b) developed a model to predict the probability of an oncoming driver making a dimming request (i.e., flashing his or her high beams to signal the opposing driver to dim down to the low beam). The dimming requests made by the oncoming drivers were measured by using a glare vehicle equipped with a variable intensity headlamp system on a two-lane highway. Bhise et al. (1977b) used the DeBoer index to account for the effect of adaptation luminance and glare illumination and used the oncoming driver's dimming requests as a measure of unacceptable

TABLE 12.2
Application of the Visibility Model for Example 2

Inputs		Value	Units
1	Luminous intensity of the source directed at the target	10,000	cd
2	Distance of the target from the light source	200	ft
3	Distance of the background from the light source	500	ft
4	Reflectance of the target $(0 < R_t < 1)$	0.07	fL/fc
5	Reflectance of the background $(0 < R_b < 1)$	0.03	fL/fc
6	Viewing distance (observer's eyes to target)	208	ft
7	Target size (diameter)	1	ft
8	Ambient luminance of the target	0.001	fL
9	Ambient luminance of the background of the target	0.001	fL
10	Glare source luminous intensity	2000	cd
11	Distance of the glare source from the observer's eyes	400	ft
12	Angle of the glare source from the sight line to the target	2	Degrees
13	Observer's age	20	Years
14	Contrast multiplier to account for confidence, conspicuity, etc.	1	

Outputs		Value	Units
1	Illumination at the target (E_t)	0.2500	fc
2	Luminance of the target $(L_t + L_v)$	0.41101069	fL
3	Illumination at the background (E_b)	0.04	fc
4a	Luminance of the background $(L_b + L_v)$	0.39471069	fL
4b	Veiling glare luminance due to the glare source (L_v)	0.39251069	fL
5	Contrast of the target against background	0.04129607	
6	Log contrast	−1.3840913	Log contrast
7	Log background luminance	−0.4037211	Log luminance
8	θ = target size (angle subtended at observer's eye)	16.5275014	min
9	$S = \log_2 \theta$ = Log of target size to the base 2	4.04679673	$\log_2$ (θ in minutes)
10	$B_0 = 7.4935 − 6.97678\ S + 0.544938\ S^2$	−11.815899	Coefficient
11	$B_1 = −0.55315 + 0.021675\ S + 0.0003125\ S^2$	−0.460318	Coefficient
12	$B_2 = 0.007721 + 0.000558\ S + 0.0000175\ S^2$	0.0102657	Coefficient
13	$\log C_{th}$ = log of Blackwell threshold contrast	−0.9790176	
14	Age multiplier (M_a)	0.9718321	
15	$\log C_{tha}$ = log threshold contrast with age and other multipliers	−0.9914264	
16	Difference between actual and required log contrast = $\log C − \log C_{tha}$	−0.3926649	
17	Ratio of C/C_{tha} (target is visible if the ratio is greater than or equal to 1)	0.40488818	

discomfort. The probability that a driver will make a dimming request due to the glare caused by the opposing headlamps can be computed as follows:

$$P_D(t) = \frac{-7.622 - 0.099t_g^2 + 6.34t_g - 1.056\ [W(t)]^2}{100}$$

where $P_D(t)$ = probability that a driver will make a request to an oncoming driver to switch to low beam at time t (in seconds) before passing the oncoming vehicle

  = 0 if $t \le 2.5$ s

  = 0 if $W \ge 7$

   t_g = potential glare exposure (s) (total time during which the oncoming driver will be exposed to the glare headlamps during a meeting situation)

  $W(t)$ = DeBoer discomfort index computed at time t before passing of the oncoming vehicle

TABLE 12.3
DeBoer Discomfort Glare Index and Dimming Request Probability Model Inputs and Outputs

Inputs		Value	Units
1	Luminous intensity of source 1 directed at the observer's eyes	10,000	cd
2	Luminous intensity of source 2 directed at the observer's eyes	10,000	cd
3	Distance of source 1 from the observer's eyes	500	ft
4	Distance of source 2 from the observer's eyes	500	ft
5	Angle of source 1 from the observer's line of sight	2	Degrees
6	Angle of source 2 from the observer's line of sight	2	Degrees
7	Adaptation luminance of the observer's eyes	0.01	fL
8	Potential glare exposure (t_g)	25	s

Outputs		Value	Units
1	DeBoer Discomfort Glare Index (W)	3.1	Index
2	Probability of observer making a dimming request	0.8	

An Excel model is available at the publisher's website to solve the above two equations (see file named "Discomfort Glare and Dimming Request Prediction Model Jan 3 2011"). The inputs and outputs of the model are illustrated in Table 12.3.

LEGIBILITY

Legibility can be defined as the ability of users to read or decipher the text, graphics, or symbols of a display. Legibility is measured by determining the maximum distance at which the display can be read by the user or by determining the characteristics of an observer (e.g., age, visual acuity) who can read a given display from a given distance. Legibility depends upon the user's visual ability to resolve and discriminate key (or critical) visual details required for acquisition of the displayed information. Thus, legibility assumes some level of processing of information from the display after the image of the display is sensed by the observer's eyes.

For the purpose of predicting the legibility of a display, the problem can be simplified into determining if the user can see a key visual detail in a letter, numeral, or a graphic character in the visual display. The key element in reading text is generally considered to be the smallest element such as a stroke or a gap between the strokes of a complex letter such as an E. To recognize the letter E, the reader needs to visually discriminate between the following five horizontal details (i.e., see each detail separately; see Figure 4.10): (1) the upper horizontal stroke of the E, (2) the gap between the top and the middle horizontal strokes, (3) the middle horizontal stroke, (4) the gap between the middle and bottom horizontal stokes, and (5) the bottom horizontal stroke. Thus, assuming that the above five details are equal in height (i.e., 1/5th the height of the letter) and the smallest visual detail that a person with normal vision can read in a black letter on a white background in the photopic vision is 1 min of visual arc subtended at the eye in size, the letter height should subtend at least 5 min of visual angle. It should be noted that the width of the vertical stroke of letter E is considered to be the same in any of the three horizontal strokes, and thus, its 1-min width will be discriminable.

FACTORS AFFECTING LEGIBILITY

The legibility of a display will depend on the characteristics of the display, the driver, the vehicle interior, and the visual environment inside the vehicle. Thus, legibility evaluation is a systems problem and it should be evaluated by understanding and selecting a proper combination of characteristics of all the components of the system.

1. Geometric displays characteristics

    a. Character (letter, symbol) height (e.g., height of the heated backlight defroster symbol on a push button in a climate control)

    b. Width of letter or symbol

    c. Stroke width (or size of smallest visual detail required to read correctly)

    d. Height-to-width ratio

    e. Font

    f. Horizontal spacing between characters or numerals

    g. Vertical spacing between lines

2. Photometric display characteristics

    a. Luminance of the visual display element(s) (i.e., target)

    b. Luminance of the background (e.g., luminance of the background of a gauge or screen)

    c. Luminance variations (i.e., nonuniformity in the luminance of a letter and/or its background due to lighting variations or sunlight and shadows can also create uneven luminance)

    d. Color characteristics of display elements and their background

3. The vehicle interior or package characteristics

    a. Locations of driver's eyes (e.g., eyellipses)

    b. Location and orientation of the display plane (i.e., viewing distances and viewing angles)

    c. Obstructions of the display caused by the steering wheel, stalks, etc.

    d. Luminance of interior surfaces

    e. Illumination from glare sources falling into the driver's eyes and angular locations of glare sources with respect to the sight lines to the displays (e.g., sunlight reflected from chrome bezel surrounding the display)

4. The environmental characteristics

    a. Ambient lighting conditions

    b. External illumination (e.g., sunlight, street lights)

    c. Adaptation luminance

    d. Dynamic/transient aspects of lighting conditions (changes in external lighting, approaching oncoming vehicle headlamps, reflections through mirrors, chrome/shiny surfaces, etc.)

Many research studies on legibility have been reported in the driver vision literature. Cai and Green (2005) reviewed some 112 legibility (or minimum letter height required to read) prediction equations proposed by different authors and found that 22 of them examined automotive displays and screens. They found that the range of letter heights predicted by different equations differed considerably. Only two models that they reviewed were based on visual contrast and adapting luminance. However, none of the models provided the ability to incorporate many variables (required to define the viewing situation and display characteristics) and allowed to input adaptation luminance as a continuous variable.

    The legibility prediction model presented in the next section takes into account many photometric and geometric variables described above.

MODELING LEGIBILITY

Many of the variables that affect legibility are the same variables that affect visibility of targets. Therefore, the legibility of a display can be predicted by using the visibility model described earlier. The basic assumption in predicting legibility is that the researcher (or the model user) is able to determine the key visual detail that must be visible to the observer. If that key detail is visible, then it is further assumed that the observer will be able to read the content of the display. Thus, to read letters, numerals, or symbols, we can determine if the smallest visual detail such as the width of a stroke (in a letter or a symbol) can be visible.

A version of the visibility model described earlier in this chapter was created by Bhise and Hammoudeh (2004) to predict legibility of displays. The visual contrast threshold data originally published by Blackwell (1952) (described earlier) and later validated for legibility predictions by Rockwell et al. (1988) were used to model the basic contrast thresholds as functions of adapting luminance and target size. For legibility computations, the model provides options to evaluate externally illuminated or backlighted displays. The user is provided three options to input illumination data by inputting (1) light source intensity and distance from the light source to target (letters, numerals, or symbols on the display plane) in English or metric units; (2) illumination directed at the target and background and their reflectances, or (3) luminances of target (i.e., numerals or letters) and background. Depending on the above-selected method of data entries, the program shows open data entry boxes (or grayed out the boxes when not needed) for required reflectance or transmittance values. The program also requires other inputs such as letter height, letter-height-to-stroke-width ratio, viewing distance, OA, and level of confidence and determines if the letter or numeral can be read by the observer.

The legibility computation program developed by Bhise and Hammoudeh (2004) is provided at the publisher's website for the reader's use. The legibility program was exercised using the following inputs for a backlit display: (a) background luminance = 1 cd/m^2; (b) letter luminance = 3, 5, and 9 cd/m^2 to obtain contrast of the letters with the background = 2, 4, and 8, respectively; (c) letter height = 2 to 25 mm; (d) letter-height-to-stroke-width ratio = 5.0; (e) viewing distance = 900 mm; (f) ease of reading = 95% confidence; and (g) driver's age from 20 to 80 years in increments of 10 years. Figure 12.5 presents the relationship between minimum letter height (millimeters) and driver's age for the three contrast levels used in the above exercise by repeated applications of the model. The exercise involved changing the letter height value and determining the smallest letter height that was legible under each combination of input variables provided above. (Note: Most automotive speedometers have major scale numerals [i.e., MPH numerals in the United States] printed in about 6-mm-high white letters on a black background).

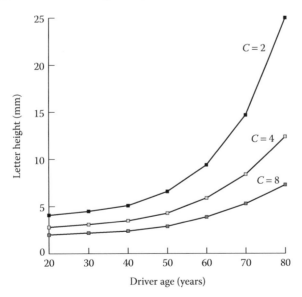

FIGURE 12.5 Illustration of outputs of the legibility model showing minimum letter height required as functions of driver's age and contrast (C) of the letters with their background.

It is important to note that the contrast ratios of displays are affected considerably by the ambient light reflected and scattered inside the display optics or lenses. Thus, the ergonomics engineer must be careful when determining the contrast ratio for inputting in the program. The author found out that many speedometer graphic elements, which had measured values of contrast ratio above 8 in the studio or laboratory environment, had much lower contrast ratios of about 1.5–3 when measured outdoors in the daytime with a photometer. The reduction in contrast due to sunlight falling on the display surface and scattering inside the display elements is also a notable problem with many new technology displays (e.g., LCD) Bhise (2007).

VEILING GLARE CAUSED BY REFLECTION OF THE INSTRUMENT PANEL INTO THE WINDSHIELD

In designing vehicles, especially with more sloping windshields, it is critical that the right combination of windshield slope or rake angle, instrument panel angle, and low reflectance materials on the instrument panel is selected to avoid the problem of degrading the driver's forward visibility. The windshield rake angle is defined as the angle of the windshield surface from the vertical Z-axis measured in the longitudinally located vertical plane at the vehicle centerline ($Y = 0$). The visibility degradation occurs due to the veiling glare caused by the reflection of the sun-illuminated instrument panel at the windshield. The term "veiling glare" is defined here as the light that is reflected or scattered from the vehicle windshield into the driver's eyes. It is called the "veiling glare" because it creates a veil that is superimposed as unwanted luminance in the driver's view and thus reduces the driver's visibility. When sunlight falls on the windshield, it illuminates the top part of the instrument panel, and its reflection in the windshield is seen by the driver as a veil. The factors that affect this visual effect are as follows: (a) the windshield angle (defined in SAE standard J1100 as dimension A121-1 "windshield slope angle"; refer to SAE, 2009); (b) windshield type or material; (c) the angle between the instrument panel and the windshield; (d) the source of incident light, that is, the sun during the daytime; (e) source (or sun) angle; (f) source illuminance (i.e., illumination falling on the windshield); (g) instrument panel (top surface) material characteristics such as gloss (or reflectance), texture, color; and (h) parting line geometry (joints or discontinuities in the top parts of the instrument panel).

Figure 12.6 presents a driving situation in which the effect of veiling glare will be very critical. The situation involves a driver approaching a dimly lit area such as a tunnel (or a parking structure) and is looking for an object (or a target) in his or her path inside the tunnel while the sunlight is falling on his or her windshield. The visibility of the target inside the tunnel will be reduced by the veil created by the reflection of the of the sun-illuminated instrument panel into his or her windshield. The visibility of the target will also depend on other factors such as the size and reflectivity of the target, its location in the tunnel, tunnel lighting and ambient illumination falling on the target, and the driver's age.

Figure 12.7 presents the ray geometry of the incident sunlight as it is reflected into the windshield and viewed by the driver while detecting (observing or looking for) a target. During the early design phases of a vehicle, a vehicle package engineer will generally draw such ray geometry diagrams from various different driver eye points on the eyellipses to assess the windshield and instrument panel architecture for potential veiling glare effects.

A DESIGN TOOL TO EVALUATE VEILING GLARE EFFECTS

Bhise and Sethumadhavan (2008a, 2008b) measured the veiling glare characteristics of windshield reflections and modified the visibility prediction model (described earlier in this chapter) to evaluate the effects of the veiling reflections on the driver's forward visibility. The visibility model can, thus, be used as a design tool to eliminate the distracting and visibility degrading effects of the veiling glare during the early stages of designing a new vehicle.

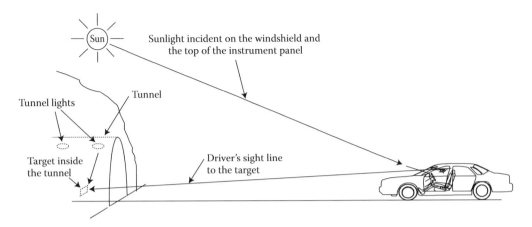

FIGURE 12.6 Veiling glare situation experienced by a driver approaching a target inside a darker tunnel when the sunlight falls on the windshield. (Reprinted from Bhise, V. D., and S. Sethumadhavan, Predicting Effects of Veiling Glare Caused by Instrument Panel Reflections in the Windshields, SAE Paper 2008-01-0666, *International Journal of Passenger Cars—Electronics Electrical Systems*, 1(1), 275–281, Society of Automotive Engineers Inc. Warrendale, PA, 2008. Copyright 2008 SAE International. With permission.)

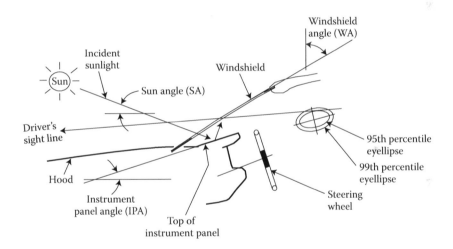

FIGURE 12.7 Ray geometry in the veiling glare situation. (Reprinted from Bhise, V. D., and S. Sethumadhavan, Predicting Effects of Veiling Glare Caused by Instrument Panel Reflections in the Windshields, SAE Paper 2008-01-0666, *International Journal of Passenger Cars—Electronics Electrical Systems*, 1(1), 275–281, Society of Automotive Engineers Inc. Warrendale, PA, 2008. Copyright 2008 SAE International. With permission.)

To develop the model to predict driver visibility under the veil created by reflections in the windshield, Bhise and Sethumadhavan (2008a, 2008b) conducted a two-phase reaserch study. In the first phase, a miniature veiling glare simulator was developed to simulate and measure the veiling glare luminance (using a photometer) on the windshield caused by reflection of the instrument panel illuminated by simulated sunlight. The measured veiling glare luminance data were used to develop a linear regression model to predict the veiling glare coefficient (VGC) in cd / (m² × lx) as a function of the geometric and photometric variables asociated with the veiling glare situation and the instrument panel material characteristics. The second phase involved modification of the visibility model to incorporate the veiling glare prediction equation to predict visibility distances to targets in veiling glare situations similar to the tunnel-approaching situation shown in Figure 12.6.

The VGC (defined as the veiling luminance divided by the illumination incident on the windshield) was used as the response measure for regression analysis. A number of stepwise linear regression models to predict the VGC as linear and quadratic functions of the following variables were developed: (a) the windshield angle (WA), (b) instrument panel angle (IPA), (c) sun angle (SA), and (d) gloss value of the instrument panel material (G). The angles WA, IPA, and SA are defined in Figure 12.7 and specified in degrees. The gloss (reflectance) value (expressed in percent) was measured by a commercially available gloss meter which measured the value at incidence and reflection angles of 60 degrees.

The best-fitting linear model to predict VGC is given below:

$$VGC = -0.09276 + 0.00216 \text{ WA} + 0.00062 \text{ IPA} - 0.0099 \text{ SA} + 0.00561 \text{ } G$$

The above equation thus shows that the veiling glare effect (or the VGC) increases as the windshield angle, instrument panel angle, and the gloss of the instrument panel top are increased. The veiling effect decreases as the sun angle is increased.

Veiling Glare Prediction Model

The situation shown in Figure 12.7 was modeled by using the following variables:

WA = windshield angle (measured in degrees with respect to the vertical)
IPA = instrument panel angle (measured in degrees with respect to the horizontal plane)
SA = sun angle (measured in degrees with respect to the horizontal plane)
TS = target size (diameter in feet of a circle having the same area as the target)
TR = target reflectance (value between 0 and 1)
BR = reflectance of target background (value between 0 and 1)
W_T = windshield transmission (value between 0 and 1)
SI = sun (i.e., source) illumination incident on the windshield (measured in lux in the plane normal to the sun rays)
TI = tunnel illumination (measured in lux in the horizontal direction)
C = target contrast

$$= \frac{|(TR \times TI) - (BR \times TI)|}{(BR \times TI) + (VGC \times SI)}$$

(Note: All basic calculations in the model were conducted using English units. The illumination and VGC input values were in metric units.)

θ = target size in minutes = $[\tan^{-1} (TS / VD)] \times (180.0 / \pi) \times 60$
TS = target size in feet (the diameter of a circle that has the same area as the target)
VD = viewing distance to the target in feet

The visual contrast threshold data developed originally by Blackwell (1952) were used to model basic target detection using the same equations provided earlier.

The target was considered visible if the computed target contrast C was greater than the threshold contrast (i.e., $C > C_{th}$), and the value of the answer parameter (ANS) was set equal to 1. Thus, ANS = 1 if the target is visible (detectable) and 0 if not visible.

The visibility distance was determined by iterating the model by using different values of VD. To compute the visibility distance, the VD value was first set low, and then it was incremented by 3 or 15.2 m (10 or 50 ft) in each iteration until the farthest distance beyond which the value of ANS became 0.

The visibility prediction model was programmed using the Microsoft Excel application. The interface screen showing the inputs and outputs of the model is presented in Table 12.4.

TABLE 12.4
Model Input Variables and Outputs Showing the Values of the Baseline Situation

Sr. no.	Variable	Inputs for Daytime Driving Condition	Baseline
1	SI	Sun illumination on windshield (lx)	10,000
2	SA	Sun angle from the horizontal plane (degrees)	20
3	TI	Illumination on the target (lx)	5,000
4	TR	Reflectance of the target (0 < TR < 1)	0.1
5	BR	Reflectance of the background (0 < BR < 1)	0.05
6	VD	Viewing distance (ft)	1,600
7	TS	Target size in equivalent diameter (feet)	2
8	OA	Observer's age (years)	45
9	CME	Contrast multiplier for confidence or ease	1
10	AC	Adjustment to contrast thresholds (in log C)	0
11	IPA	Instrument panel top angle from horizon (degrees)	0
12	G	Material gloss (0.8 to 2.65)	1.8
13	WA	Windshield angle from vertical (degrees)	64
14	W_T	Windshield transmission along normal to the glass	0.7
15	WR	Windshield reflectance at 20 degrees (considered in VGC)	1
16		**Outputs**	
17	VGC	Veiling glare coefficient of instrument panel material (L_v in cd/[m² × lx])	0.035778
18	LT	Target luminance (fL)	74.22373
19	LB	Background luminance (fL)	37.11187
20	L_v	Veiling luminance (fL)	104.436
21	LA	Adapting luminance (fL) = LB + L_v	141.5478
22	CA	Contrast of the target against adapting luminance	0.262186
23	LCA	Log contrast	−0.58139
24	LLA	Log adapting luminance	2.150903
25	TA	Target subtended angle (minutes) at the observer's eye	4.297181
26	LLAU	Log adapting luminance upper bound	2
27		Contrast multipliers	
28	AM	Age multiplier	1.68911
29	CM	Confidence multiplier	2
30	EM	Exposure multiplier	1
31	TCM	Total contrast multiplier	3.378221
32		Blackwell model calculations	
33		Theta	2.103391
34		B_0	−4.77045
35		B_1	−0.50618
36		B_2	0.008972
37	LTHC	Log threshold contrast with above total multiplier	−0.60183
38	THC	Threshold contrast with above total multiplier	0.250135
39		**Result**	
40		Detectable at the 1/30th second Blackwell threshold contrast	
41	ANS	With the above total multiplier (1 = yes; 0 = no)	1

Source: Reprinted from Bhise, V. D., and S. Sethumadhavan, Predicting Effects of Veiling Glare Caused by Instrument Panel Reflections in the Windshields, SAE Paper 2008-01-0666, *International Journal of Passenger Cars—Electronics Electrical Systems*, 1(1), 275–281, Society of Automotive Engineers Inc., Warrendale, PA, 2008. Copyright 2008 SAE International. With permission.

MODEL APPLICATIONS ILLUSTRATING EFFECTS OF DRIVER'S AGE, SUN ILLUMINATION, AND VEHICLE DESIGN PARAMETERS

Baseline Situation: To illustrate the capabilities of the model in predicting visibility distances to a target located in the tunnel as the driver approaches the tunnel with sunlight falling on his or her vehicle windshield, a baseline situation was created. The primary parameters of the baseline situation are shown in rows numbered 1–15 of Table 12.4. Thus, the baseline situation can be described as a 45-year-old driver in a vehicle with a 64-degree windshield angle, 0-degree instrument panel angle and 1.8% gloss, 10,000-lx illumination falling on the vehicle windshield at 20-degree sun angle, approaching a 0.61-m (2-ft) diameter target with 10% reflectance placed in the tunnel. The target is illuminated with 5000 lx (from tunnel and ambient lighting), and the background of the target is assumed to have 5% reflectance. In this baseline situation, the predicted visibility distance of the target was 488 m (1600 ft).

Effect of Incident Sunlight and Driver's Age: The effect of sunlight illumination incident on the windshield was evaluated by conducting four additional prediction runs under the above baseline situation by changing the sun illumination from 10,000 to 5,000, 15,000, 20,000, and 25,000 lux and using driver's ages of 25, 45, and 65 years. The maximum viewing distances at which the target was visible (i.e., ANS = 1, as the viewing distances were incremented in steps of 15.2 m [50 ft]) are shown in Figure 12.8 as the visibility distances to the target. The visibility distances decreased as the sun illumination level increased (due to increased veiling luminance). The visibility distances also decreased as the driver's age increased due to increase in the contrast thresholds with increase in the driver's age. The visibility distances of the 65-year-old driver were about 457 m (1500 ft) shorter than those of the 25-year-old driver under the same situation (i.e., the separation between the visibility distance curves of 25- and 65-year-old drivers in Figure 12.8). The veil at 25,000 lx reduced the visibility for a 65-year-old driver to "zero" distance.

Effect of Windshield Angle and Sun Angle: Figure 12.9 shows the effect of the windshield angle and the sun angle on the baseline situation (10,000-lx sun illumination) as the windshield angle was increased from 55 to 70 degrees from the vertical and the sun angle was increased from 10 to 30 degrees above the horizontal plane. The luminance of the veil increased with an increase in the

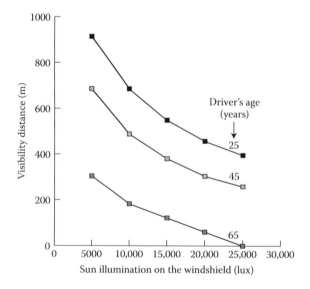

FIGURE 12.8 Effect of sun illumination and driver's age on visibility distance to a 0.61-m (2-ft) diameter target under the baseline situation. (Reprinted from Bhise, V. D., and S. Sethumadhavan, Predicting Effects of Veiling Glare Caused by Instrument Panel Reflections in the Windshields, SAE Paper 2008-01-0666, *International Journal of Passenger Cars—Electronics Electrical Systems*, 1(1), 275–281, Society of Automotive Engineers Inc. Warrendale, PA, 2008. Copyright 2008 SAE International. With permission.)

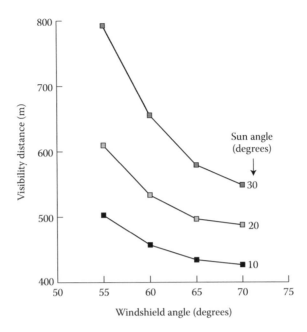

FIGURE 12.9 Effect of windshield angle and sun angle on visibility distance to a 0.61-m (2-ft) diameter target under the baseline situation. (Reprinted from Bhise, V. D., and S. Sethumadhavan, Predicting Effects of Veiling Glare Caused by Instrument Panel Reflections in the Windshields, SAE Paper 2008-01-0666, *International Journal of Passenger Cars—Electronics Electrical Systems*, 1(1), 275–281, Society of Automotive Engineers Inc. Warrendale, PA, 2008. Copyright 2008 SAE International. With permission.)

windshield angle and with a decrease in the sun angle, which in turn reduced the visibility distance. Thus, the worst visibility distance was obtained when the windshield rake angle was at 70 degrees and the sun angle was at 10 degrees.

Effect of Vehicle Design Variables: The design and appearance of the vehicle can be directly affected by the windshield angle, the instrument panel angle, and the material on the top of the instrument panel. Therefore, a sensitivity analysis on visibility distance was conducted by using the above three variables. The results of the analysis are presented in Figure 12.10. The dashed and solid-line curves in this figure are for gloss values of 2.7 and 0.8, respectively. The results are especially useful in the considerations of trade-offs between the interior and exterior design variables of the vehicle. For introducing higher raked windshields, the instrument panel angle can be decreased and/or the top of the instrument panels can be made (or covered) with materials with lower gloss values. For example, Figure 12.10 shows that visibility with "0 deg. instrument panel angle at 0.8 gloss value" will be similar to "20 deg. instrument panel angle at 2.7 gloss value." Thus, if materials of higher gloss values are selected for the top of the instrument panel, the top surface of the instrument panel should be sloped more down (i.e., away from the windshield).

CONCLUDING REMARKS

The models presented in this chapter were found by the author to be very useful in teaching concepts and understanding variables related to visibility and legibility to students enrolled in the Automotive Systems Engineering program. The models presented in this chapter can be downloaded from the publisher's website. The reader is encouraged to exercise the models to gain a better understanding into the sensitivity of different photometric-, geometric-, and observer-related factors. The models are useful not only as educational tools, but they can also serve as guides in evaluating trade-offs between many variables related to the vehicle exterior and interior designs.

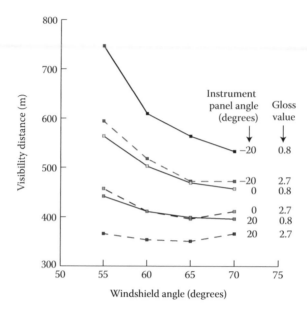

FIGURE 12.10   Effect of windshield angle, instrument panel angle, and the gloss value of the instrument panel material. (Reprinted from Bhise, V. D., and S. Sethumadhavan, Predicting Effects of Veiling Glare Caused by Instrument Panel Reflections in the Windshields, SAE Paper 2008-01-0666, *International Journal of Passenger Cars—Electronics Electrical Systems*, 1(1), 275–281, Society of Automotive Engineers Inc. Warrendale, PA, 2008. Copyright 2008 SAE International. With permission.)

REFERENCES

Blackwell, H. R., 1952. Brightness discrimination data for the specification of quantity of illumination. *Illuminating Engineering*, 47(11).

Blackwell, O. M., and R. H. Blackwell. 1971. Visual performance data for 156 normal observers of various ages. *Journal of Illuminating Engineering Society*, 1(1).

Bhise, V. D., E. I. Farber, and P. B. McMahan. 1977a. Predicting target detection distance with headlights. *Transportation Research Record*, 611, Transportation Research Board, Washington, DC.

Bhise, V. D., E. I. Farber, C. S. Saunby, J. B. Walnus, and G. M. Troell. 1977b. Modeling vision with headlights in a systems context. SAE Paper 770238, 54 pp. Presented at the 1977 SAE International Automotive Engineering Congress, Detroit, MI.

Bhise. V. D., and C. C. Matle. 1989. Effects of headlamp aim and aiming variability on visual performance in night driving. *Transportation Research Record*, 1247, Transportation Research Board, Washington, DC.

Bhise, V. D., C. C. Matle, and E. I. Farber. 1988. Predicting effects of driver age on visual performance in night driving. SAE paper no. 881755 (also no. 890873). Presented at the 1988 SAE Passenger Car Meeting, Dearborn, MI.

Bhise, V. D., and C. C. Matle. 1985. Review of driver discomfort glare models in evaluating automotive lighting. Presented at the 1985 SAE International Congress, Detroit, MI.

Bhise, V. D., and R. Hammoudeh. 2004. A PC Based model for prediction of visibility and legibility for a human factors engineer's tool box. *Proceedings of the Human Factors and Ergonomics Society 48th Annual Meeting*, New Orleans, LA.

Bhise, V. D., 2007. Effects of veiling glare on automotive displays. *Proceedings of the Society of Information Display Vehicle and Photons Symposium*, Dearborn, MI.

Bhise, V. D., and S. Sethumadhavan. 2008a. Effect of windshield glare on driver visibility. *Transportation Research Record (TRR), Journal of the Transportation Research Board*, 2056, Washington, DC.

Bhise, V. D., and S. Sethumadhavan. 2008b. Predicting effects of veiling glare caused by instrument panel reflections in the windshields. SAE Paper 2008-01-0666. *International Journal of Passenger Cars—Electronics Electrical Systems*, 1(1), 275–281. Warrendale, PA: Society of Automotive Engineers Inc.

Cai, H., and P. Green. 2005. Range of character heights for vehicle displays as predicted by 22 equations. *Proceedings of the 2005 SID Vehicle Display Symposium*, Dearborn, MI.

DeBoer, J. B. 1973. Quality criteria for the passing beam of motorcar headlights. Paper presented at the CTB Meeting, Walldorf, Germany.

Fry, G. A. 1954. Evaluating disabling effects of approaching automobile headlights. *Highway Research Board Bulletin*, 89.

Hoffmeister, D. H. and V. D. Bhise. 1978. A driver glare-discomfort model to evaluate automotive stop lamp brightness. *Proceedings of the 1978 Annual Meeting of the Human Factors Society*, Detroit, MI.

Rockwell, T. H., A. Augsburger, S. W. Smith, and S. Freeman. 1988. The older driver: A challenge to the design of automotive electronic displays. *Proceedings of the Human Factors Society, 32nd Annual Meeting*, Anaheim, CA.

Society of Automotive Engineers Inc. 2009. *SAE Handbook*. Warrendale, PA: Society of Automotive Engineers Inc.

13 Driver Performance Measurement

INTRODUCTION

To evaluate different vehicle designs, vehicle features, and the effects of changes or improvements made in vehicle designs, an ergonomics engineer should be able to measure and demonstrate how well the driver performs in different tasks while using the vehicle. Currently, there are very few standardized measures (or variables) and methods for measurement of driver behavior and driver performance. A number of researchers have used many different measures, and there is some agreement on the general approaches for performance measurements. However, many differences in defining even commonly used measures such as task completion times, variability in lane position, errors. exist due to differences in instrumentation, experimental procedures, and data collection techniques. Thus, the measurement problem occurs due to inconsistencies between researchers in determining what to measure and how to measure in any given driving or vehicle usage situation.

Therefore, objectives of this chapter are (a) to review various variables used in measuring driver performance in various tasks involved in vehicle uses, (b) to provide the reader a better understanding into issues and problems associated with the measures, and (c) to develop a background in evaluation of vehicle designs by measuring driver and vehicle outputs.

CHARACTERISTICS OF EFFECTIVE PERFORMANCE MEASURES

For any measure to be acceptable and useful, it should meet certain key characteristics. The following characteristics were based on the characteristics of effective safety performance measures presented by Tarrants (1980).

1. Administrative feasibility: The measuring system or measuring instrumentation used to obtain the value of the measure must be practical, that is, we must be able to construct it and use it quickly and easily without excessive costs. Thus, an ergonomics researcher should be able to use the measurement system to make the necessary measurements, and the vehicle development team should be able to set targets by using the measure and determine if the ergonomic goals have been reached.
2. Interval scale: The measurement system should be able to provide the measure by using at least an interval scale. The interval scale should be graduated with equal and linear units, that is, the difference between any two successive point values should be the same throughout the scale.

    It should be noted that there are four types or orders of measurement scales: nominal, ordinal, interval, and ratio scales—with ascending order of power to perform mathematical operations. The nominal scale (the most primitive) is used for categorizing or naming (e.g., football jersey numbers, car model numbers) of items. We can only analyze the data obtained by using the nominal scale by counting frequencies or percentages of values in each category. The ordinal scale is used to order or rank items. However, the distance measured on a scale between ranked items may not be equal. Thus, in ranked items, we can only conclude that an item with a higher rank is better than another item with a lower rank (e.g., the Mohs hardness scale in geology). The interval scale has equal intervals, but the zero point on the interval scale is arbitrary (e.g., like the Fahrenheit or Celsius temperature scales). On an interval scale, the difference between any two items measured on a scale can be determined by the difference between their two respective scale values. The ratio scale is the most informative (or quantitative) as ratios of quantities defined by the scale values can be constructed. For example, a 10-lb weight is two times heavier than a 5-lb weight. The ratio scale also contains an absolute zero (i.e., the point of "no amount"). Thus, we should make sure that the measure that we select should use the highest possible order of scale—with the interval scale as the minimum order of acceptable scale.

3. Quantifiable: A quantitative measure will allow comparison between any two values in terms of at least a difference on an interval scale. The quantitative measure should permit application of more sensitive statistical inference techniques. (Note: A nonquantitative, i.e., a qualitative, measure limits statistical inference [due to use of data on nominal and/or ordinal scales] and opens the way for individual interpretation. For example, if the result of a speedometer comparison study states that "the analog speedometers are better than digital speedometers," the reader does not have sufficient information on the magnitude of improvement gained by the use of an analog speedometer.)

4. Sensitivity: The measurement technique should be sensitive enough to detect changes in a product characteristic on the product or user performance to serve as a criterion for evaluation. (Note: A tiny diamond cannot be measured on a cattle scale.)

5. Reliability: The measurement technique should be capable of providing the same results for successive applications in the same situations.

6. Stability: If a process does not change, the performance level obtained from the measure at any another time should remain unchanged.

7. Validity: The measure should produce information that is representative of what is to be inferred in the real world. This is particularly important because in the ergonomic research many different types of measures in a wide range, from indirect and surrogate measures to direct measures, can be used. This issue is discussed later in this chapter.

8. Error-free results: An ideal measuring instrument should yield results that are free from errors. However, in general, any measurement will have some constant and random errors. The ergonomics engineer needs to understand the sources of such errors and minimize the errors. The errors can also be statistically isolated and estimated by their sources (or effects of the sources).

DRIVING AND NONDRIVING TASKS

It is important to understand that the driver performs a number of tasks while driving. These tasks can be classified as follows:

1. Lateral control of the vehicle: maintaining lateral control (left–right movements) of the vehicle within a given driving lane

2. Longitudinal control: maintaining longitudinal (fore–aft) control of the vehicle on the roadway and maintaining a safe distance from the lead vehicle in the same lane

3. Roadway monitoring: acquiring information about the state of the roadway and traffic (e.g., pavement surface characteristics; viewing signs, signals, and traffic control devices; detecting/monitoring other vehicles and objects on or on the sides of roadway; using inside and outside mirrors to view traffic in adjacent lanes, etc.)

4. Crash avoidance: avoiding a collision with objects in the vehicle path

5. Route guidance: obtaining information from route guidance signs or memorized roadway landmarks to follow a route to the intended destination

6. Using in-vehicle controls and displays: acquiring information from displays and control settings to understand vehicle state and operating controls

7. Other nondriving tasks: performing tasks that are not required to drive the vehicle, for example, conversing with passengers, using car phones or other devices, eating, reading materials, searching/grasping objects (e.g., cups, coins, maps, papers, etc.)

The driver's actions and performance in each of the above mentioned tasks can be measured as functions of time and/or distance and occurrence frequencies (or rates) of many predefined events can be also measured. The measurements can provide information on (a) state of the driver, (b) driver outputs, (c) driver behavior, (d) driver performance, (e) driver preferences, (f) driver-encountered problems and difficulties, (g) state of the vehicle, and (h) vehicle motion with respect to the highway.

DETERMINING WHAT TO MEASURE

The problem of what to measure depends on the researcher's understanding of the driver's tasks, the purpose of the research, and resources available. When a driver takes a vehicle out for a trip, the researcher assumes that the driver has certain objectives in making the trip under the driving situation. Figure 13.1 presents a flow diagram to help understand links between the driver's objectives and different types of measures that can be used to evaluate the driver's performance.

The driver's objectives that are related to how he or she wants to make the trip depend on the driver's understanding of his or her capabilities, desires, and the characteristics of his or her vehicle and the driving situation (see Figure 13.1). For example, the driver may be late for a job interview and would like to make the trip of 50 km within 0.5 h while also making sure that he or she drives safely near the speed limit. The driver–vehicle combination is then driven on the trip route, which has its characteristics such as road geometry, traffic conditions, pavement surface. Thus, the driver behavior (i.e., how he or she will act, do, or behave during the trip) will depend on the characteristics of the driver, the vehicle, the roadway, and the driving situation. While driving on the route, the driver gets

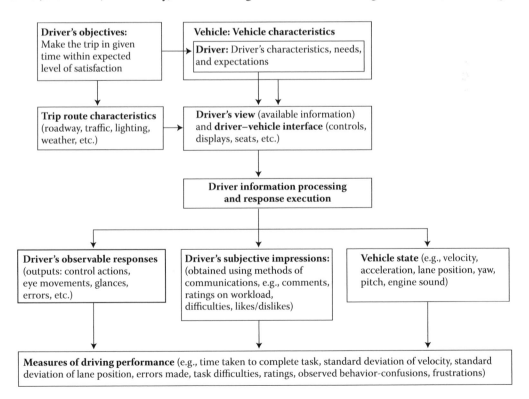

FIGURE 13.1 Flow diagram illustrating links between driver's objectives and driving performance measures.

his or her visual information from the fields of view available from the vehicle and interior displays and operates controls to follow the roadway (see Driver's view box in Figure 13.1).

To measure how well the driver is performing his or her driving tasks during the trip, we can measure the following (see lower half of Figure 13.1):

1. Driver's observable responses: We can observe driver's responses such as his or her visual information acquisition behavior through measurements of eye movements, head movements, eye glances, time spent in viewing different objects, and his or her control movements from measurements of hand and foot movements while operating the steering wheel and the pedals. We can also measure the physiological state of the driver by measuring the driver's heart rate.
2. Driver's subjective responses: We can also develop a structured questionnaire and ask the driver a number of questions at different points in the route (if an experimenter is present) or at the end of the trip to understand the driver's problems, difficulties, confusions, frustrations, and situational awareness issues and also ask the driver to provide ratings on his or her workload, comfort, ease of usages of different controls and displays, etc.
3. Vehicle state: We can also record the state of the vehicle by installing measuring instruments in the vehicle to measure vehicle outputs (as functions of time) such as steering wheel position, accelerator and brake pedal positions, distance traveled, lateral position in the lane, vehicle speed, vehicle accelerations, heading angle of the vehicle with respect to the roadway.

DRIVER PERFORMANCE MEASURES

TYPES AND CATEGORIES OF MEASURES

Driver performance measures can be categorized as follows:

1. Behavioral measures: measuring what behavior and what choices the driver exhibits (i.e., what did he/she do) by recording his or her eye movements, body movements, sequences in performing different movements and tasks, decisions made, etc.
2. Physical measures: distance, speed, acceleration, and time that can be measured with physical instruments
3. Subjective measures: based on judgments of the driver (or of an observing experimenter), for example, ratings, preferences, judgments, thresholds of perception, detection, equivalency, etc.
4. Physiological measures: physiological state of the driver based on changes in heart rate, sweat rate, oxygen intake, galvanic skin resistance, electrical activities in different skeletal muscles (electromyograph) or the heart (electrocardiograph), etc.
5. Accident-based safety performance measures: number of accidents of given type (rear end, head on, ran off the road), accident rate (number of accidents per kilometer traveled), and accident severity (damage, injury level)
6. Equivalency-based measures: comparisons of performance or driver judgments under two different conditions to determine if they are equal or different
7. Monitory measures: measures based on costs such as trip costs, energy consumption, costs and benefits related to the trip, etc.

The ultimate measure of driver performance from the safety viewpoint is based on the occurrences of accidents, such as the number of accidents of different types (e.g., ran off the road, front-end, rear-end, or side collision) or their accident rates (e.g., number of accidents of a given type per 100,000 km or miles driven). However, crashes are rare events. Even large-scale field operational tests or naturalistic driving studies are unlikely to involve many crashes. Thus, near-miss events (or near accidents) are also used in such real-world studies. The near-miss events have been shown to be a valid and useful means to understand the relative risks of various in-vehicle activities and driver behaviors (Guo et al., 2010).

SOME MEASURES USED IN THE LITERATURE

Some commonly used measures of driver performance used in the studies reported in the literature are based on statistics such as mean, median, standard deviation, percentages, percentiles of the data collected from the following:

1. Velocity
2. Lane position
3. Lane departures (lane violations, lane exceedences)
4. Lane changes
5. Time to lane crossing (i.e., time available for a driver until his or her vehicle will cross over into an adjacent lane or leave the roadway)
6. Steering wheel movements (reversals, rates)
7. Car following headway (following distance, gap between two vehicles measured in distance or time–distance)
8. Acceleration or deceleration (lateral and longitudinal)
9. Total time spent in looking at a given location or a display
10. Glance durations while viewing given objects (e.g., speedometer, radio, mirror, sign).
11. Number of glances made to use a device or to complete a given task
12. Eye fixation durations while viewing a given object
13. Number of eye fixations made on a given object
14. Percentage of time eyes were closed
15. Blink (or eye closure) rate
16. Detection rates of targets or events
17. Detection distance (on road or roadside objects)
18. Hand involvement time (e.g., time spent by the driver's hand away from the steering wheel to perform a task)
19. Eye involvement time (e.g., total time spent away from the forward road scene)
20. Driver errors (e.g., lane intrusions, slowed down, excessive speed, misread a display, unable to read a display [legibility errors], operated a wrong control, turned control in opposite direction, forgot to use a display or a control [omission or forgetting errors], took wrong turn)
21. Traffic tickets received in a given period
22. Task completion time
23. Reaction time to a given signal
24. Brake reaction time
25. Accelerator release time
26. Accelerator to brake pedal transition time
27. Time to collision (time available before a possible collision)
28. Accident involvement (e.g., ran off the road, collisions with other vehicle(s), fixed objects, pedestrians, or animals). Accident frequency and rates (e.g., number of accidents per 100,000 km or miles of travel)

For further definitions and discussions on differences between many of the above measures used by different researchers, an interested reader should refer to Savino (2009).

RANGE OF DRIVING PERFORMANCE MEASURES

The range of driver performance measures that can be used extends from the measurement of some early events, actions, or steps in a task to the measurement of final outcomes. For example, the range of events that can be measured in a target detection task while driving can involve measurements of locations and durations of eye fixations (eye search patterns), target detection response (correct detection or failure), response time to detect the target, detection distance, lane position variability while searching the target, steering wheel position during search for the target, erratic or evasive maneuver as a result of late or no detection, and accident (if it occurs) resulting from nondetection of the target. In addition, behavioral measures can provide information on how, what, and when the driver performed certain predefined steps. Measures such as total time spent, types and numbers of errors committed, and percentage of times the task was completed in an allocated time all provide information on how well a given task was completed.

The physiological measures can provide information on the physiological state of the driver (i.e., how the human body is responding while performing the task) by measuring variables such as heart rate, electromyographic potentials, sweat rate, brain waves, etc. Subjective measures are very useful where the subject can be asked to describe problems encountered during a task and provide ratings using scales developed to provide impressions of the subject on task-related characteristics such as level of difficulty or ease, magnitude of spatial dimensions, workload, comfort, etc. (also see Chapter 14 on driver workload measurements).

Thus, in determining what measures to select for a given study, it is important to assure that the researcher can obtain useful and valid information. The skill in selecting performance measures depends on the researcher's knowledge of research literature in the problem area, depth of the researcher's human factors research experience, data sensing and recording equipment availability, time and resources available, and the researcher's experience in statistical data analysis.

The selection of the dependent variables, whether they are behavioral or performance based, will depend on the problem and the driver's tasks associated with the issues in the problem. The behavioral variables are generally based on observations of driver movements and actions related to the task being performed. The performance measures, on the other hand, measure how well the driver performed the task. Some commonly used driver behavioral measures in operating in-vehicle devices are number of glances made away from the road, glance durations, percentage of time spent on the task, sequence of button pushes, etc. Some examples of performance measures are variability in lane position, variation in speed, percentage of tasks completed correctly, number of errors made during task completion, etc.

In experimental evaluations, the dependent variables are generally related to (and affected by) characteristics of drivers (e.g., young vs. old, males vs. females, experienced vs. inexperienced, familiarity of the driver, and country of origin or nationality of the driver), characteristics of the driving environment (e.g., day vs. night, dry vs. wet road, traffic speed, traffic density, and straight vs. curved road), and the characteristics of the vehicle design (e.g., analog vs. digital display, locations and types of controls, different controls and displays layouts, vehicle features or design configurations, and vehicles made by different manufacturers in a given vehicle segment).

SOME STUDIES ILLUSTRATING DRIVER BEHAVIORAL AND PERFORMANCE MEASUREMENTS

STANDARD DEVIATION OF LATERAL POSITION

The standard deviation of lateral position provides quantitative information on the variability in maintaining lane position while driving. The standard deviation is computed over a number of measurements (samples) of lateral position data, usually sampled at a preset time or distance interval (selected by the researcher) collected during driving on a test road section. Larger values of the standard deviation suggest that the driver had difficulty in driving within the left and right markings defining the driving lane, and any intrusions in the adjacent lanes would mean that the driver could have an accident with a vehicle in an adjacent lane or could have ran off the road. The Manual of Uniform Traffic Control Devices (U.S. Department of Transportation, 2003) requires that lane width delineated by lane line pavement markings should not be less than 3 m (10 ft). The lane width on the interstate highways is 3.66 m (12 ft; American Association of State Highway Officials, 2005; Fitzpatrick et al., 2000).

Green et al. (2004) found that the most commonly reported measure of driving performance was the standard deviation of lane position. They examined the data on standard deviation of lane position reported in 36 studies and found that the standard deviation values ranged between 0.05 and 0.6 m, with the mean value of the standard deviation as 0.24 m for studies conducted on roads. The mean values of the standard deviation of lane position in studies conducted in the driving simulators and test tracks were 0.30 and 0.22 m, respectively. They also found that the standard deviation of lane position increased slightly (0.002 m/year) with the driver's age.

Lambert et al. (2005) conducted a study measuring driving performance using a fixed-base driving simulator and found that the standard deviation of lane position increased by 40%–100% while performing common driver-induced distraction tasks such as reading a message from a text pager, identifying cross streets on a map, and reading a step in written directions, as compared with the average standard deviation of 0.3 m when the drivers were not performing any other tasks other than just driving on the same road. To perform these more demanding distracting tasks, the drivers made more than three eye glances away from the forward road scene.

STANDARD DEVIATION OF STEERING WHEEL ANGLE

Many researchers have measured steering wheel angle and used the standard deviation of the steering wheel angle as a measure to study the change in the driver's activity during test situations. Green et al. (2004) found that the standard deviation of steering wheel angle was one of the most commonly reported measures of driving performance. They examined the data on standard deviation of steering wheel angle reported in seven studies and found that the mean value of the standard deviation was 1.59 degrees. However, it should be noted that the variability of steering wheel angle may be elevated, yet there is no corresponding effect on lane position. This arises because lane position results from two time integrations of the steering inputs. Steering variability most directly reflects the level of effort the driver is putting into the steering task.

STANDARD DEVIATION OF VELOCITY

The standard deviation of forward velocity is a measure of a driver's ability to drive at a constant speed. The speed changes occur due to variables related to the driver (e.g., attention and distractions) and changes in the characteristics of the roadways, traffic, weather, etc. Driving with low variations in speed is also related to safety. Traffic engineers have also found that an increase in standard deviation of vehicle velocity within a traffic stream (i.e., variations of velocities between vehicles in a traffic stream) can lead to increases in accident rates.

VEHICLE SPEED

The amount of information that the driver processes to maintain his or her lateral control increases with an increase in the vehicle speed. Fitzpatrick et al. (2000) have shown that driving speed is affected by lane width. They found that the speed increased with an increase in lane width and in the presence of a median (divided roadway).

Total Task Time, Glance Durations, and Number of Glances

Total time spent by the drivers in performing a given task, total eyes-off-the-road time, number of glances made, and durations of individual glances provide information on how the drivers performed the task. Longer total eyes-off-the-road times are indications of higher complexity in the tasks. Total task time, on the other hand, is more variable and reflects the demands of the driving environment as well as the demands of the task being evaluated (Jahn et al., 2008).

Table 13.1 provides data on total task times, mean glance durations, and mean total number of glances from two reports (Green and Shah, 2004; Rockwell et al., 1973). All the tasks included in the table were visual in nature. The speedometer reading and rearview mirror viewing tasks only involved the recording of driver eye and head movements. The radio and navigation system tasks involved hand and finger movements in operating the controls as well as eye movements needed to view the displays associated with devices. The total task time was defined as the time interval between the initiation of the driver's first response to perform the task and the end of the last response. The first response in a control activation task begins with the earlier of the two events to begin the task, namely, initiation of the hand movement from its prior position (most likely from the steering wheel) or when the driver's eyes begin to turn to view the display. The last response will be the later of the two events, namely, when the driver's hand involved in operating the control reached back to the steering wheel or when the eyes moved back to the road from the display. Different researchers have defined the total task time differently, depending upon their ability to measure hand, finger, eye, and head movements.

Tasks such as tuning a radio or entering a street address or destination into a device involve a number of steps with a sequence of control activations. These tasks, thus, involve a larger number of glances and longer total task completion times. The task completion times are also significantly affected by the driver's age (Green and Shah, 2004). The older drivers may take twice the time required by the younger drivers. Green and Shah (2004) found that while dialing a telephone number the mean time to dial a digit (MTTDD) significantly increased with age, and it could be predicted by the following equation: $MTTDD = 0.55 + 0.039$ (age), where the time and the driver's age were measured in seconds and years, respectively.

In another study, Jackson et al. (2002) used the IVIS DEMAnD model (In-Vehicle Information System Design Evaluation and Model of Attention Demand) developed at the Virginia Polytechnic University's Transportation Research Center under the sponsorship of the Federal Highway

TABLE 13.1
Total Task Time, Glance Durations, and Number of Glances in Performing In-Vehicle Visual Tasks

No.	Task Description	Mean Total Task Time (s)	Mean Glance Duration (s)	Mean Total Number of Glances	Reference
1	Speedometer reading during freeway merging	NA	0.41–0.68	0.7–1.7	Rockwell et al. (1973)
2	Inside rearview mirror viewing during freeway merging	NA	0.58–0.68	1.2–2.3	Rockwell et al. (1973)
3	Outside rearview mirror viewing during freeway merging	NA	0.52–1.08	0.4–5.6	Rockwell et al. (1973)
4	Dialing a phone number (10 or 11 digits)	9.3–42.0	1.23–3.2	4.7–12.8	Green and Shah (2004)
5	Tuning a radio	7.0–27.5	0.67–2.87	2.0–15.0	Green and Shah (2004)
6	Entering street address	34.3–91.94	1.00–1.40	19.0–34.5	Green and Shah (2004)
7	Entering destination (other than street address)	13.4–159.0	1.05–2.75	4.0–33.0	Green and Shah (2004)

Administration (Hankey et al., 2001). The model was exercised to evaluate nine different tasks covering a range of attentional demands from a simple task such as glancing into a side-view mirror to operating a complex navigation system. The nine tasks were evaluated for three different vehicle configurations of interior instrument panel layouts (a center stack mounted LCD screen on the top, middle, and low locations) and two extreme levels of driver–roadway combinations (low: young driver [aged 18–30 years] under low traffic density and low roadway complexity; high: older driver [aged 60+ years] under high traffic density and high roadway complexity). The predicted values of number of glances and total task times are summarized in Table 13.2.

Driver Errors

During drive tests, drivers can be asked to use a number or controls and displays, and an experimenter seated in the front passenger seat or a video camera located behind the driver can record errors made by the driver. The error data can be analyzed by creating measures such as error frequencies or error rates observed while driving under different conditions.

Errors (or difficulties encountered by the drivers) can be described by the evaluator or research in a variety of ways:

1. Errors in locating, seeing, and reading controls or displays, made two or more short glances or one long look, looked at wrong place, focused [or squinted] to see, and leaned to see)
2. Errors in reaching and grasping (e.g., reached wrong place, leaned to operate, awkward grasp or orientation, missed the control, accidental activation of a control, and high effort in operating the control)
3. Errors in operating a control (e.g., moved in wrong direction, operated wrong control, exhibited trial and error during operation, repeated attempts made to use a control, overshot

---

TABLE 13.2
Number of Glances and Total Task Times for Nine Different Tasks under Low and High Levels of Driver–Roadway Combinations Obtained from the IVIS Demand Model

No.	In-Vehicle Task	Number of Glances		Total Task Time (s)	
		Low	High	Low	High
1	Check the driver's side mirror and locate an object present in the mirror field	1	2	1.1	2.5
2	Turn on the in-dash radio, select FM band, tune to specified frequency, and adjust radio sound volume	3	7	7.2	18
3	Locate and remove a CD from the center console, remove a CD from its case, orient, and insert in the player	3	8	6.5	21
4	Turn on a cellular phone, dial a seven-digit number, and carry on a simple conversation with nine questions	3	13	69	130.2
5	Search a specified travel route from a simple display of a in-dash navigation system	2	7	4.9	19
6	Identify required destination and current location and select a desired route from the displayed list	3	10	10.3	25.5
7	Identify required destination and current location, evaluate route, and select a desired route	16	43	35.4	91.5
8	Listen to a variety of optional spoken routes and respond verbally when the desired route is mentioned	0	0	29.9	49.4
9	Verbally define desired location through a completely voice/speech technology interface	0	0	36.9	83.5

the control setting, needed to verify setting, explanation required from the experimenter to use the control or display)

4. Vehicle behavior (e.g., changed speed [slowed down or increased speed by 5 mph or more], heading change, deviated in an adjacent lane, abrupt heading correction)

SOME DRIVING PERFORMANCE MEASUREMENT APPLICATIONS

Table 13.3 presents several driving performance measurement applications and possible measures that can be used in evaluating the problems.

TABLE 13.3
Description of the Driver Performance Measurement Applications and Possible Performance Measures

No.	Application Problem	Driver Performance Measures
1	Study on how drivers learn to drive	Standard deviation of lane position; mean velocity; standard deviation of velocity; eye fixations and glances made in different areas in the road scene, mirrors, and in-vehicle devices
2	Determine effectiveness of traffic control devices at freeway work zones	Standard deviation of lane position, mean velocity, standard deviation of velocity, eye fixations, and glances made in different areas in the road scene
3	Determine "good" car radio design	Total time spent in performing different radio tasks (e.g., turning in the radio, changing radio stations, changing bands and tuning a station, changing CDs, etc.), standard deviation of lane position, mean velocity, standard deviation of velocity, eye fixations, and glances made on the radio
4	Determine acceptable pedal layout	Percentage of driver ratings that the pedals are located "just right" using direction magnitude scales and percentage of drivers' ratings that the pedal locations are acceptable
5	Determine required field of view from a vehicle	Direction magnitude and acceptance ratings on a number of items, for example, pillar obscurations, pillar locations, down angle over the hood, up angle to view traffic signals, mirror locations, mirror width, mirror height, backlite, etc. (see Chapter 6 for more details)
6	Determine acceptable headlamp beam pattern	Seeing distances to pedestrian targets and lane lines, legibility distances to signs, discomfort glare ratings in opposed driving situations, and ratings on perception of beam pattern on pavements (Jack et al., 1995).
7	Determine acceptable navigation system	Time to perform different tasks (e.g., destination entry, determining vehicle location); standard deviation of lane position; mean velocity; standard deviation of velocity; glances durations made in different areas in the road scene, mirrors, and navigation screens; percentage of destinations correctly reached; and NASA TLX Task Load Ratings (see Chapter 14).
8	Determine effective night vision system	Glance durations and percentage of time spent on the night vision screen and forward road scene, percentage of roadway targets detected, and NASA TLX Workload Ratings
9	Determine acceptable adaptive cruise control	Standard deviation of headway between lead vehicle and subject vehicle, standard deviation of relative velocity between lead vehicle and subject vehicle, and NASA TLX Workload Ratings

CONCLUDING REMARKS

Driver behavioral and performance measurements are used to evaluate various vehicle designs and features. The concepts introduced in this chapter allow ergonomics engineers to study problems related to driver workload and vehicle evaluation issues. Chapter 14 covers methods used in the measurement of driver workload, resulting driver distractions, and their effects on driver behavior and performance. Chapter 15 covers various evaluation methods used to determine effects of ergonomic changes and improvements in vehicle designs. The evaluations also allow for the comparison of different vehicle designs in terms of how they affect the driver's behavior and performance.

REFERENCES

American Association of State Highway Officials. 2005. *A Policy on Design Standards: Interstate System.* 5th ed.

Fitzpatrick, K., L. Elefteriadou, D. Harwood, J. Collins, J. McFadden, I. B. Anderson, R.A. Krammes, N. Irizarry, K. Parma, K. Bauer, and K. Passetti. 2000. Speed prediction for two-lane rural highways. Report FHWA-RD-99-171. Federal Highway Administration, Washington, DC.

Green, P., B. Cullinane, B. Zylstra, and D. Smith. 2004. Typical values for driving performance with emphasis on the standard deviation of lane position: A summary of the literature. A report on safety vehicles using adaptive interface technology (SAVE-IT, Task 3A). The University of Michigan Transportation Research Institute, Ann Arbor, MI.

Green, P., and R. Shah. 2004. Task times and glance measures of the use of telematics: A tabular summary of the Literature. A report on safety vehicles using adaptive interface technology (SAVE-IT, Task 6). The University of Michigan Transportation Research Institute, Ann Arbor, MI.

Guo, G., S. Klauer, M.T. McGill, and T. Dingus. 2010. Evaluating the relationship between crashes and near-crashes: Can near-crashes serve as a surrogate safety metric for crashes? Report DOT-HS-811-382. National Highway Traffic Safety Administration, Washington, DC.

Hankey, J. M., T. A. Dingus, R. J. Hanowski, W. W. Wierwille, and C. Anderws. 2001. In-vehicle information systems behavioral model and design support: Final report. Report FHWA-RD-00-135 sponsored by the Turner-Fairbank Highway Research Center of the Federal Highway Administration, Virginia Tech Transportation Institute, Blacksburg, VA.

Jack, D. D., S. M. O'Day, and V. D. Bhise. 1995. Headlight beam pattern evaluation: Customer to engineer to customer—A continuation. SAE Paper 950592. Presented at the 1995 SAE International Congress, Detroit, MI.

Jackson, D., J. Murphy, and V. D. Bhise. 2002. An evaluation of the IVIS-DEMAnD driver attention demand model. Paper presented at the 2002 Annual Congress of the Society of Automotive Engineers Inc., Detroit, MI.

Jahn, G., J. Krems, and C. Gelau. 2008. Skill acquisition while operating in-vehicle information systems: Interface design determines the level of safety-relevant distractions. *Human Factors*, 51(2), 136–151.

Lambert, S., S. Rollins, and V. Bhise. 2005. Effects of common driver induced distraction task on driver performance and glance behavior. Paper presented at the 2005 Annual Meeting of the Transportation Research Board, Washington, DC.

Rockwell, T. H., V. D. Bhise, and Z. A. Nemeth. 1973. Development of a computer based tool for evaluating visual field requirements of vehicles in merging and intersection situations. Vehicle Research Institute Report. Society of Automotive Engineers Inc., New York.

Savino, M. R. 2009. Standardized names and definitions for driving performance measures. MS Thesis, Tufts University.

Tarrants, W. E. 1980. *The Measurement of Safety Performance.* New York: Garland STPM Press.

Tijerina, L., E. Parmer, and M. J. Goodman. 1998. Workload assessment of route guidance system destination entry while driving: A test track study. *Proceedings of the 5th ITS World Congress*, Seoul, Korea.

U.S. Department of Transportation, 2003. *Manual of Uniform Traffic Control Devices for Streets and Highways.* Prepared by the Federal Highway Administration, Washington, DC.

14 Driver Workload Measurement

INTRODUCTION

Driver workload measurement is involved in developing and applying methods to measure the total amount of physical and mental resources and effort due to the tasks that a driver performs under any given situation. The area is of particular importance as the technological advances are generating new features to enable the drivers to access and use more information from various carried-in and in-vehicle devices while driving. Answering the question of what tasks can a driver safely perform simultaneously in a given situation, thus, assumes that we are able to measure the driver's total workload and compare it with the driver's capabilities and determine if the driver has sufficient spare capacity left for any emergencies or is overloaded.

The problem becomes more complex because both the driver's total workload and the driver's capabilities (or capacities) to perform the driving and nondriving tasks are not constant during driving. The total number of tasks that the driver needs to perform at a given instant can include (a) driving-related tasks and responding to changing situations (e.g., variations in traffic, roadway, weather) that are outside the driver's control and must be performed under certain time pressure (these tasks are generally involuntary in nature as they cannot be performed under the driver's discretion) and (b) other nondriving-related (voluntary) tasks that can be performed more or less at the driver's discretion.

The objectives of this chapter are to review various approaches and methods available to measure and evaluate the driver workload.

DRIVER TASKS AND WORKLOAD ASSESSMENT

The driver's total workload includes all driving and nondriving tasks:

1. Driving tasks: These include monitoring the roadway and making lateral and longitudinal control actions by using primary controls (i.e., steering and pedals) and manipulating other safety-related driving controls (e.g., defrosting windshield) and displays. Also included as a part of the driving tasks are responding to the demands from the roadway (e.g., curves, lane drops, merges), executing maneuvers demanded by traffic or other vehicles (e.g., changing lanes, passing), and responding to other sudden demands (e.g., avoiding colliding with a crossing pedestrian).
2. Nondriving tasks: These include (mostly voluntary) reading displays and operating controls and use of secondary interfaces (e.g., climate controls, entertainment devices). It also includes more discretionary activities like talking with other passengers, reading maps, using cell phones, reading notes, drinking beverages, eating, attending to other passenger needs, looking at billboards, pedestrians, etc.

The above tasks require mental and physical work. The mental work involves information acquisition (involving sensing, detecting, recognizing) and processing information (searching, selecting, and integrating sensed information from different modalities; analyzing; retrieving/storing information in human memory systems; and decision making) and executing and making control actions. The physical work includes generation of muscular forces to produce coordinated movements of different body parts (e.g., head, hand, arm, foot, leg, and torso) with needed speeds and accuracies.

The workload assessment, thus, involves answering questions such as how busy is the driver? How many tasks can the driver handle safely? Would the driver be overloaded by this task under normal driving conditions? The driver has limited capabilities to perform tasks. If the demands of the concurrent tasks are greater than the driver's capabilities, then (a) the driver may perform the tasks but experience higher stress, (b) the driver may make errors, (c) the driver may slow down, (d) the driver may not perform some tasks or parts of some tasks, or (e) some combination of the above depending on priorities and capabilities of the driver.

PRESENT SITUATION IN THE INDUSTRY

Despite of a considerable amount of research reported in the literature in this area and because of the complexities associated with the combination of different mental and physical tasks associated while driving, currently multiple methods are used in the automotive industry to decide on what new in-vehicle devices and features are safe enough to be incorporated into a production vehicle. One method that has been adopted by some automakers to limit visual distraction caused by in-vehicle devices with visual–manual interfaces is the occlusion technique (Alliance of Automobile Manufacturers, 2006). However, automobile manufacturers often resort to use of multiple approaches and methods to evaluate the new devices. The positive and negative aspects of the information gathered in these evaluations are reviewed and discussed with different levels of subject matter, management, and legal experts to determine if a product feature or a device should be incorporated into a production vehicle.

CONCEPTS UNDERLYING MENTAL WORKLOAD

The concept of mental workload has been extensively studied and is being constantly researched in the field of cognitive psychology. A number of researchers have used the concept to explain how the human operator processes information to perform complex tasks and developed models to explain human performance and to measure workload (International Standards Organization [ISO], 2008; Meshkati et al., 1992; Tijerina et al., 2000; Society of Automotive Engineers Inc. [SAE], 2009a and 2009b). Realizing that the area of mental models and workload measurements are evolving, the following brief statements will summarize approaches and concepts used in understanding the concept of metal workload.

1. Humans have limited information-processing information-processing capacity. Simple models of information processing have shown that reaction times can be used to measure the driver's information processing capacity (see Chapter 4). The amount of processing capacity of the driver does not remain constant during driving due to changes in number of factors such as attention, distraction, fatigue, roadway and traffic.
2. Humans have multiple resources to process information. These include different input modalities such as vision and audition and different output modes such as manual output (i.e., hand and foot manipulations) and vocal output (i.e., voice utterances). In addition, central resources use verbal as well as spatial codes for sensed information and response selection. The central processor can act as a serial, parallel, or hybrid processor depending on the operator characteristics such as practice, experience, stress, attention level and age. Both ambient (i.e., peripheral) and focal (i.e., foveal) vision are used (Wickens, 1992).
3. Mental workload is the specification of the amount of information processing capacity that is used for task performance. Demand is determined by the goal that has to be attained by means of task performance.
4. Workload can be defined as that portion of the driver's limited capacity that is actually required to perform a particular task.
5. Workload is not only task specific but it is also person specific and situation specific. Thus, each individual would have different workload while performing the same set of tasks.

Also, the same person will have different workload associated with the same task if it is performed under different conditions.

6. Workload is not an inherent property, but rather, it emerges from the interaction between the requirements of a task, the circumstances under which it is performed, and the skills, behaviors, and the perception of the operator (Hart and Staveland, 1988).

7. The complexity of a task increases with an increase in the number of processing operations that are required to perform the task.

8. The difficulty of a task is related to the processing effort (e.g., amount of resources consciously allocated) that is required by the individual for task performance.

9. Spare capacity can be defined as the additional or excess information processing capacity available at any given instant. Under most driving situations, drivers have considerable spare capacity available. However, addition of tasks can reduce or even eliminate the spare capacity and leave the driver with insufficient capacity to perform the required amount of information processing, especially to deal with sudden emergencies.

10. The term "visual spare capacity" is used to refer only to the unused capacity of the visual resource available to process visual information acquired through the driver's peripheral and foveal vision. An occlusion task that involves controlling the "open" durations (i.e., when vision is unobstructed) and "closed" durations (when visual information is not available due to occlusion) is one of the principal methods used to measure visual spare capacity (see Chapter 4).

11. Many factors affect the driver's workload. The factors can be categorized as (a) driver state affecting factors, for example monotony, fatigue, sedatic drugs, and alcohol; (b) driver trait factors, for example experience, skills, age, and strategy; (c) environmental factors, for example road environmental demands, traffic demands, and weather and (d) vehicle factors, for example driver–vehicle interface, automation, and feedback. Thus, performance, effort, and spare capacity may or may not be related because of variations in the driver's skills, motivation, attention, capabilities, etc.

12. Most driver failures occur due to information-processing failures, that is, inability of the driver to make the right decisions at the right time and right place. The information failures can occur due to situations such as inadequate information gathering, expectancy violations, faulty decision making, or making an inappropriate response.

METHODS TO MEASURE DRIVER WORKLOAD

1. Driver performance measurements: Various driver performance measurement methods and measures can be used to determine the effect of the driver workload on driver performance in performing different tasks (see Chapter 13). The changes in levels of performance measures obtained while the driver is simply driving (i.e., baseline driving tasks) versus when the driver is asked to perform tasks in addition to driving (i.e., dual task: driving and dialing a cell phone) have been used to assess the effects of driver workload. Table 14.1 presents results from four studies showing worsening of driver performance by a factor of 1.12–1.88 when the drivers were asked to perform different distracting secondary tasks in addition to primary driving task.

2. Physiological measurements: These measures are based on the assumption that workload will affect bodily functions. They are based on measuring effects of arousal, excitement, stress/tenseness, thought processes, use of body movements through muscle activations, etc., that are caused during performance of the tasks.

   A number of physiological measurements have been used to determine the effects of the workload on the human operator. Some examples of the physiological measurements are heart rate, respiration rate, brain's spontaneous electrical activity from electroencephalograms and evoked potential recordings, electrical activity of the heart from

TABLE 14.1
Illustrative Results from Four Performance-Based Studies

| No. | Task Description | | Measure | Baseline: Driving Only | Driving and Secondary Task | Ratio of Dual Task to Single Task (%) | Reference |
	Primary Task	Secondary Task		Single task (S)	Dual task (D)	(D/S)%	
1	Car following on a freeway	Exchanging text messages	Standard deviation of following distance	11.9 m	17.9 m	1.5	Drews et al. (2009)
			Lane crossings per kilometer	0.26	0.49	1.88	
2	Driving on a simulated two-lane roadway at about 50 mph	Reading cross streets on a map	Standard deviation of lateral position	0.27 m	0.41 m	1.5	Lambert et al. (2005)
		Read a short message from a text pager	Standard deviation of lateral position	0.34 m	0.47 m	1.41	
3	Driving on a simulated roadway at 60 kph	Sending text message using cell phone	Mean time headway	4.3 s	6.4 s	1.49	Hosking et al. (2009)
			Standard deviation of lateral position	0.2 m	0.29 m	1.45	
4	Following a lead vehicle while braking from 65 mph to 30–45 mph in a fixed-base driving simulator	Naturalistic conversing on a hands-free cell phone with a research confederate after 4 days of practice	Collision frequency (vehicle contacted objects in the environment)	7	12	1.71	Cooper and Strayer (2008)
			Brake reaction time	1.00 s	1.20 s	1.2	
			Following distance	22.5 m	25.1 m	1.12	

electrocardiograms, electrical activity of muscles from electromyograms, electrical activities of the eye muscles from electro-occulargrams, galvanic skin response, body and skin temperatures, sweat rate, pupil size, and eye blink rate.

The variations in the body functions due to extensive physical biomechanical workload can be measured by the use of heart rate, respiration rate, oxygen intake, and electromyograms. These measures related to muscular activities are relatively easier to measure and interpret as compared with physiological variations due to changes in the mental workload during driving. Brookhuis and Waard (2001) have reported studies that showed that the drivers' heart rate increased under higher stress and workload, for example, driving through traffic circles increased heart rate as compared with that through straight roads, and awaiting a traffic light change increased heart rate variability. Recently, Mehler et al. (2009) examined the sensitivity of heart rate, skin conductance, and respiration rate as measures of mental workload in simulated driving environment using a sample of 121 young adults. Their results showed that as the mental workload imposed by the n-back task (recalling a one-digit number from zero, one, or two digits back when one-digit numbers are presented at a constant rate) increased the heart rate, the skin conductance, and the respiration rate. Verwey and Zaidel (2000) found that the eye blink rate was related to drowsiness, suggesting that the frequency of eye closures exceeding 1 s is an indicator of drowsiness.

The reliability of physiological measures in measuring the driver's mental workload is poor because of large variations among individuals and many factors (e.g., anxiety, stress) that affect these measures. The physiological measures are rarely used during the vehicle development process because (a) the links between the physiological measures and real-world performance are not clearly understood, (b) the physiological measures are difficult to obtain during the driving environment, and (c) the high expense and time associated in collecting large amounts of data and analyses requiring more sophisticated techniques.

3. Subjective assessments: Subjective ratings on the level of difficulty, stress, discomfort, mental workload, physical workload, etc., provided by subjects during performance of different tasks are commonly used as measures to assess the workload (Tsang et al., 1996; Wierwille and Eggemeier, 1993). Three well-developed subjective workload measurement techniques used in this field are (1) the NASA-Task Load Index (TLX), (2) the Subjective Workload Assessment Technique (SWAT), and (3) the Workload Profile (WP); Rubia et al., 2004).

The NASA-TLX is a multidimensional rating procedure that derives an overall workload score based on a weighted average of ratings on six subscales (Hart and Staveland, 1988). These subscales include Mental Demand, Physical Demand, Temporal Demand, Own Performance, Effort, and Frustration. The method has been used to assess workload in various human–machine environments such as aircraft cockpits, workstations, process control environments, and various actual driving as well as in simulated driving situations (Bhise and Bhardwaj, 2008).

The subscales can be specified by 5-point, 10-point, or 100-point interval scales. The standardized descriptors (questions) used for each subscale category and adjectives used to define their end points are presented in Table 14.2. The ratings on individual subscales can be used as evaluation scores, or an overall workload score can be obtained from a weighted sum of the ratings on the six scales. The weightings of the subscales can be obtained after the subjects have performed all the tasks. A paired comparison method (see Chapter 15) can be used to obtain the weightings based on the importance of the subscale categories associated with the tasks.

The SWAT involves asking the operator to rate workload using three scales, namely, Time Load, Mental Effort Load, and Psychological Stress Load. Each scale has three levels: Low, Medium, and High. The descriptors used to define the three levels of each of the three scales are presented in Table 14.3. The method uses conjoint measurements and scaling techniques to develop a single interval rating scale (Reid and Nygren, 1988).

TABLE 14.2

Description of the Six Subscales Used in the Measurements of the NASA-Task Load Index

No.	Subscale (Workload Attribute)	Adjectives Used to Describe the Low and High End Points of the Scale	Questions Used to Describe the Scale Attribute
1	Mental demand	Low, high	How much mental and perceptual activity was required (e.g., thinking, deciding, calculating, remembering, looking, searching)? Was the task easy or demanding, simple or complex, exacting or forgiving?
2	Physical demand	Low, high	How much physical activity was required (e.g., pushing, pulling, turning, controlling, activating)? Was the task easy or demanding, slow or brisk, slack or strenuous, restful or laborious?
3	Temporal demand	Low, high	How much time pressure did you feel due to the rate or pace at which the tasks of task elements occurred? Was the pace slow and leisurely or rapid and frantic?
4	Performance	Good, poor	How successful do you think you were in accomplishing the goals of the task set by the experimenter (or yourself)? How satisfied were you with your performance in accomplishing these goals?
5	Effort	Low, high	How hard did you have to work (mentally and physically) to accomplish your level of performance?
6	Frustration level	Low, high	How insecure, discouraged, irritated, stressed, and annoyed versus secure, gratified, content, relaxed, and complacent did you feel during the task?

TABLE 14.3

Descriptors Used to Define Three Levels of the Time Load, Mental Effort Load, and Psychological Stress Load Scales Used in the Subjective Workload Assessment Technique

No.	Scale	Level	Descriptors
1	Time load	Low	Often have spare time. Interruptions or overlap among activities occur infrequently or not at all.
		Medium	Occasionally have spare time. Interruptions or overlap among activities occur infrequently.
		High	Almost never have spare time. Interruptions or overlap among activities are very frequent or occur all the time.
2	Mental effort load	Low	Very little conscious mental effort or concentration required. Activity is almost automatic, requiring little or no attention.
		Medium	Moderate conscious mental effort or concentration required. Complexity of activity is moderately high due to uncertainty, unpredictability, or unfamiliarity. Considerable attention required.
		High	Extensive mental effort and concentration are necessary. Very complex activity requiring total attention.
3	Psychological stress load	Low	Little confusion, risk, frustration, or anxiety exists and can be easily accommodated.
		Medium	Moderate stress due to confusion, frustration, or anxiety noticeably adds to workload. Significant compensation is required to maintain adequate performance.
		High	High to very intense stress due to confusion, frustration, or anxiety. High extreme determination and self-control required.

The WP method is based on the multiple resource model (Wickens, 1987). It considers the following eight workload dimensions as attentional resources: (1) perceptual/central processing, (2) response selection and execution, (3) spatial processing, (4) verbal processing, (5) visual processing, (6) auditory processing, (7) manual output, and (8) speech output. The subjects are asked to provide proportions for each of the eight workload dimensions used in each task (in a random order) after they have experienced all the tasks. Thus, each task is evaluated by providing eight ratings, each between 0 and 1, to represent the proportion of each attentional resource used in the task. Thus, a rating of 0 means that the task did not require the dimension and 1 means that the task required maximum attention. The ratings on these eight dimensions of each task are later summed to obtain an overall workload rating for the task.

The subjective methods are commonly used in vehicle development because they are easier to obtain (require no instrumentation) and have high "face validity" as the voice of the customer. The disadvantages of the subjective methods are that the rater may find it difficult to understand many issues associated with comparing different products and situations and the agreement between different raters may not be unanimous. Chapter 15 provides additional information on subjective methods used in the industry.

4. Secondary task performance measurement: This approach uses an artificially added task called a secondary task while performing a primary task and assumes that an upper limit exists on the capacity of the driver to gather and process information. The performance in the secondary task is used to measure the driver's workload on the assumption that adding the secondary task on top of the primary driving task (e.g., lateral and longitudinal control of the vehicle) will increase the driver's total workload; if the secondary task is sufficiently difficult, the driver would reach or exceed his or her overall capacity to perform both the tasks. By carefully controlling the driver's priorities through instructions (e.g., the driver must maintain constant 100 km/h speed and keep the vehicle in the lane all the time), the driver will be asked to maintain performance in the primary task. Thus, the level of performance in the secondary task will indicate the amount of workload or capacity taken up by the primary task.

Many different secondary tasks have been used in the literature. Some examples of secondary tasks are peripheral detection tasks, arithmetic addition tasks, repetitive tapping tasks, time estimation, random number generation, choice reaction time tasks, critical tracking tasks, visual search tasks, and memory search tasks. For example, Olsson and Burns (2000) measured reaction time to detect peripheral targets and hit rate (proportion of targets correctly detected) while driving a car under a baseline driving condition and in other conditions involving the baseline driving and performing radio station tuning and CD tasks (e.g., turn on the CD mode and select track). They found that when the drivers were asked to perform these additional radio and CD tasks, their peripheral detection performance, measured by both the reaction time and hit rate, were degraded as compared with their performance in the baseline driving condition.

The use of the secondary task as a method to measure driver workload has a number of shortcomings. If the introduction of the secondary task modifies or interferes the driver's primary task, then the driver may be forced to change his or her strategy and thus may distort the load imposed by the primary task. The interference in the primary task is greater when the tasks share the same response resources than where the responses occupy different resources. Further, it is difficult for some drivers to maintain the same level of attention and priority in performing the primary task when the secondary task is introduced.

5. ISO Lane Change Test (LCT): The lane change task proposed by ISO (2008) is a simple laboratory dynamic dual-task method that quantitatively measures performance degradations in a primary driving task while a secondary task is being performed. The primary task involves driving on a 3-km length straight three-lane roadway at 60 km/h constant speed and making a series of lane changes as indicated on pairs of signs located on the road shoulders. The average distance between the pairs of signs is 150 m. Thus, the driver

is forced to make 20 lane changes during the 3-km drive. The driver sits in front of a steering wheel mounted on a table and views the driving scene in a video monitor located on the table (see Figure 14.1). A number of secondary tasks can be used for the driver to perform while performing the above-described primary lane change task. The entire procedure including its software, test track, sign configuration, laboratory setup, data collection, primary and secondary task performance measures, and analysis procedure is standardized and specified in the ISO document.

The primary task performance measure is the average deviation in the path of the vehicle, called the mean deviation (MDEV). It is calculated with respect to a reference path trajectory. Two methods are provided for selecting the reference trajectory. The first method, called the adaptive model, method uses the same subject's trajectory obtained at the end of the practice session under the baseline condition (which involves performing the primary driving task with the lane changes only, i.e., without a secondary task) as the reference path trajectory. The second method, called the normative model, uses the reference trajectory as a "nominal trajectory" specified under an idealistic set of assumptions (to simulate an ideal driver who changes lane with 0.6-s reaction time and drives exactly at the center of the driving lane), and it is the same of all subjects. The use of the normative model method is optional.

The LCT procedure has been used by a number of researchers to evaluate its repeatability, consistency, and usefulness. The method was used by Jothi (2009), and the study is discussed later in this chapter.

6. IVIS DEMAnD Model: A computer model known as the In-Vehicle Information System Design Evaluation and Model of Attention Demand (called the IVIS DEMAnD Model) can be used to predict the level of driver distraction associated with an in-vehicle device. The model was developed by Hankey et al. (2001) to allow automotive designers and safety engineers to predict driver workload levels on either an existing or a proposed vehicle information system. Using this model, an engineer is able to enter dimensional data for a given vehicle, along with the locations of controls and displays for the system being analyzed, and the model uses empirical data from a database of human factors to predict the driver workload associated with the system or a task completed on that system.

Jackson and Bhise (2002) exercised the model under nine different driver attention task levels, ranging from a simple task, such as glancing into a side-view mirror, to a very complex task, such as operating an in-vehicle navigation system. The nine driver tasks were evaluated using three different vehicle configurations and two levels of driver–roadway complexity. In addition, a real-world study was conducted to measure the drivers' visual performance using four of the above nine tasks for comparison with model predictions. The model predicted values of maximum number of glances, total task performance time, and a model-rating feature called the "figure of demand" were compared for each of the nine tasks. The results showed the following: (1) Driver task performance behavior was influenced substantially more by the differences in the level of driver/roadway/traffic combinations than by the differences in the test vehicle configurations. (2) The driver performance data collected under the actual driving conditions compared very well with the model predictions. (3) Overall, the IVIS DEMAnD Model was found to be a good early attempt at modeling the effect on the driver performance using the in-vehicle information systems.

7. SAE J2364 and J2365 recommended practices: The SAE published two recommended practices, namely, SAE J2364 and SAE J2365, to measure driver workload for the navigation and route guidance system tasks (SAE, 2009a, 2009b).

The SAE J2364 practice presents a laboratory-based procedure to measure total time taken by subjects to perform a given task related to a navigation and route guidance system. The method, thus, assumes that a working model of the in-vehicle navigation and route guidance system is available for the laboratory test. Two methods of obtaining total time can be used. The first method requires the driver to do the entire task uninterrupted, whereas the second method allows the use of an occlusion device. The subjects for the test should be licensed older drivers (aged 45–65 years) who are initially not familiar with the device. The total time taken to complete the task is measured after five practice trials. The SAE practice recommends using 10 subjects and measuring total time three times (three trials after the first five practice trials) for each subject. The task is performed under a static or non-driving situation in a laboratory bench test type situation. For the occlusion method, the total time is obtained by summing all the glances through the occlusion device made to the device or task being evaluated. The task is considered to be acceptable if it can be completed in less than 15 s under uninterrupted method and 20 s under the occlusion method. This method is known in the industry as the SAE's 15-second rule.

The SAE J2365 practice presents an analytical method to calculate the time to complete a given in-vehicle navigation and route guidance task. The advantage of this method is that it does not require a working version of the device and it can be used in an early design phase. The calculation method is based on the Goals, Operators, Methods and Selection Rules Model described by Card et al. (1980, 1983). The method essentially involves (a) breaking down the task into a series of simple steps; (b) applying predetermined times by considering appropriate mental operations, key strokes, age multipliers, and other operators in each step; (c) summing the assigned times for each subgoal (e.g., move hand to the device, select a city) and goal (e.g., enter a destination using the street address method) in each step; and (d) summing the completion times over all steps to obtain an estimate of the total time required. This method is especially useful in the early design phase in which a number of alternate design concepts and their operational features can be evaluated by comparing the total time estimates to perform each given task. The acceptability of the design can be also judged by comparing the total time with a preselected requirement such as the 15-second rule. This method does not consider voice-activated controls, voice outputs, and communication between the driver and others, or passenger operation.

SOME STUDIES ILLUSTRATING DRIVER WORKLOAD MEASUREMENTS

The most common approach used by many researchers involves asking subjects to drive under a baseline driving situation (when the subjects are only driving and not performing any other secondary tasks) and comparing their performance in the baseline situation with the performance obtained while performing dual tasks (i.e., performing the baseline task and additional secondary task). Various secondary tasks have been used by different researchers while performing driving tasks under different baseline situations (e.g., open road driving, car following, lane changing). Five different studies reported in the literature are presented below.

Destination Entry in Navigation Systems

Tijerina et al. (1999) evaluated four commercially available navigation systems in a test car. Sixteen test participants (8 males and 8 females; 8 below age 35 years and above age 55 years) drove a 1993 Toyota Camry with microDAS (data acquisition system) that captured travel speed, lane position, and lane exceedences, as well as video of road scene and eye glance behavior at a sampling rate of 30 Hz. The participants drove the vehicle on a 12-km (7.5-mile) multilane oval track at speeds between 72 and 96 km/h (45–60 mph). The participants were asked to enter point-of-interest destinations in each of the four systems (tested in random order) along with the other two tasks that included (a) dialing an unfamiliar 10-digit phone number in a handheld cell phone and (b) entering a radio station manually in a modern radio.

The response measures used in the study included (a) total time to complete the task, (b) mean glance frequency to the device, (c) average glance duration, (d) average total eyes-off-the-road time, and (e) number of lane exceedances during each task trial.

The results showed the following: (a) The average total time spent in destination entry was 68 s for the younger drivers and 118 s for the older drivers, whereas the cell phone 10-digit entry and the radio tuning tasks took on average about 25 and 22 s, respectively. (b) The mean number of glances made by the subjects in the visual–manual destination entry tasks ranged from 22 to 33, whereas the glances made during audio entry of the destinations, cell phone 10-digit dialing and radio tuning took 4, 8, and 6 glances, respectively. (c) The average glance duration during visual–manual entry tasks in all the devices ranges between about 2.5 and 3.2 s. (d) The total eyes-off-the-road time ranged between 60 and 90 s for the visual–manual destination entry tasks, and the cell phone 10-digit dialing and radio tuning tasks took on average 17 and 16 s, respectively. (e) The average number of lane exceedances per trial during the destination entry tasks ranged from 0 (for a voice-activated audio navigation system) to 0.9, whereas the cell phone dialing and radio tuning tasks had 0.06 and 0.2 lane exceedances per trial, respectively.

HANDHELD VERSUS VOICE INTERFACES FOR CELL PHONES AND MP3 PLAYERS

Owens et al. (2010) conducted an on-road study to compare driver performance in using handheld versus in-car voice-control interfaces. They asked 21 participants to drive a test vehicle on straight sections of a divided secondary road with a 105-km/h (65-mph) speed limit and asked the drivers to perform the following tasks: (1) baseline: only driving at the beginning and end of the study, (2) dial a contact person's number, (3) converse on the phone with an experimenter on a predetermined topic of interest, and (4) play a music track. Tasks 2–4 were performed by using two types of equipment: (1) a personal handheld cell phone and a personal MP3 player and (2) the SYNC voice control interface in a 2010 Mercury Mariner (midsize SUV).

They used an onboard data acquisition system to measure steering wheel position, and video cameras were used to measure number of glances on the user interface, maximum glance duration, and task duration. After each task, the participants were also asked to provide a rating on mental demand using a 1 to 7 scale (1 = low demand, 7 = high demand).

The results of the study showed the following:

1. The task duration times were significantly shorter while using the voice interface as compared with the handheld devices. For dialing a 10-digit phone number, the voice interface took about 10 s as compared with about 15–20 s using the handheld cell phone. The track play tasks took about 10 s with the voice interface as compared with about 35–40 s using the handheld MP3 players.
2. The steering variance in the baseline condition was about 1.0 deg^2. The variance increased while using both the phone and MP3 handheld devices from about 1.2 to 1.5 deg^2 for the younger drivers to 2.0 deg^2 for the older drivers. The steering variance during the use of voice interface did not increase from the baseline.
3. The number of glances made on the handheld devices was on average about 8–9 for the dialing and 15–20 for the track playing tasks. With the voice interface, the average number of glances for the two types of tasks ranged between 3 and 4 glances.
4. The mental demand ratings for the handheld devices ranged from about 3 to 5.5 as compared with 2–3 for the voice interface. The mental demand ratings for the older drivers were always higher by 1.5–2 points as compared with the younger drivers. The mental demand ratings for the baseline condition were on average 1.5 and 2.0 points for the younger and older drivers, respectively.

These results indicate that a voice interface offers advantages of a visual–manual or handheld interface for command and data entry, contrary to the view that "there is no difference between hands-free (voice) and hand-held."

TEXT MESSAGING DURING SIMULATED DRIVING

Drews et al. (2009) evaluated the effects of text messaging while driving a fixed-base driving simulator, with three screens providing approximately 180 forward degree field of view. They compared performance of 40 participants while driving (single task) and driving and text messaging (dual task). The participants drove in the right lane of a 51-km (32-mile) simulated multilane rural and urban beltway. During each trial, the participants followed a pace car that was programmed to brake at 42 randomly selected intervals and would continue to decelerate until the participant depressed the brake pedal, at which point the pace car would begin to accelerate to the normal freeway speed. During the dual-task condition, the participants used their own cellular phones and received and composed text messages.

The driving performance was measured by using (a) brake onset time (reaction time of the following driver), (b) following distance, (c) standard deviation of the following distance, (d) minimal following distance, (e) lane crossings per kilometer, (f) lane reversals per kilometer, and (g) gross lateral displacement.

Their results showed the following:

1. Mean brake onset time increased from 0.88 s in the baseline driving to 1.08 s in the dual-task driving.
2. The average following distance increased from 29.1 m in the baseline driving to 34.3 m in the dual-task driving.
3. The minimal following distance decreased from 9.0 m in the baseline driving to 6.8 m in the dual-task driving.
4. The lane crossing frequency, lane reversal frequency, and the gross lateral displacement increased in the dual-task driving as compared with the baseline driving by 88%, 26%, and 26%, respectively. The authors, thus, concluded that the text messaging during driving has a negative impact on the simulated driving.

COMPARISON OF DRIVER BEHAVIOR AND PERFORMANCE IN TWO DRIVING SIMULATORS

Bhise and Bhardwaj (2008) conducted a study to compare driving behavior and performance of drivers in two different fixed-base driving simulators (namely, FAAC [produced by FAAC, Inc., Ann Arbor, MI] and STI [produced by Systems Technology, Inc., Hawthorne, CA]) while performing the same set of distracting tasks under geometrically similar freeway and traffic conditions. The FAAC simulator had a wider three-screen road view with steering feedback as compared with the STI simulator, which had a single screen and a narrower road view and had no steering feedback. Twenty-four subjects (12 younger and 12 mature) drove each simulator on a freeway-type roadway with geometrically similar characteristics and were asked to perform a set of nine different tasks involving different distracting elements.

The nine tasks were as follows:

1. Collect Ontario map from the map compartment (on the lower part of the driver's door).
2. Answer the cell phone (after the subject's own cell phone placed in the center console rang).
3. Collect 65 cents from the coin holder located in the center console.
4. Switch (radio) to FM and tune to preset 4.
5. Collect two yellow and two red candies (placed in the console).
6. Sip the beverage (water bottle) from the cup holder (located in the front part of the center console).
7. Search for the keys located in the center console (inside compartment with a hinged lid).
8. Check if there is voice mail on the cell phone.
9. Replace the CD in the CD player (remove a CD and insert a different CD from a CD case placed on the center console).

The following performance measures were obtained:

1. Number of glances made away from the forward scene to complete the task
2. Longest glance duration
3. Total task completion time
4. Number of lane deviations during task completion
5. Maximum lane deviation
6. Vehicle speed
7. Number of crashes
8. Average mental demand rating
9. Average physical demand rating
10. Average temporal demand rating
11. Average performance demand rating
12. Average effort rating
13. Average frustration rating

The results showed that driver behavioral measures, such as number of glances made in performing a task, total task completion time, and the NASA-TLX workload ratings (on 10-point scales) differed due to the differences in the tasks. However, the behavioral measures and the NASA-TLX ratings showed remarkably similar behavior in the two simulators. Overall, tasks 4 (switch (radio) to FM and tune to preset 4) and 7 (search for the keys located in the center console) were the least demanding; on the other hand, task 3 (collect 65 cents from the coin holder located in the center console) and task 8 (check if there is voice mail on the cell phone) were the most demanding. On the other hand, the drivers' driving performance, measured by maximum lane deviation, average speed, and number of accidents, was significantly different in the two simulators. The driver performance was significantly better while driving the FAAC simulator than the STI simulator. The results, thus, showed that while the demand placed on the drivers due to the distracting tasks produced similar glance behavior and task loadings in the two simulators, the narrower road view and lack of steering feedback in the STI simulator produced substantially degraded driving performance than the performance observed in the FAAC simulator.

APPLICATIONS OF THE ISO LCT

The LCT (Lane Change Task) proposed by the ISO subcommittee (ISO/TC2/SC22) (ISO, 2008) was conducted in 2008 at different sites in different countries to test the calibration and replication capability of the LCT method. One of the test sites was the Vehicle Ergonomics Laboratory at the University of Michigan-Dearborn campus. The study was conducted by Jothi (2009) under the guidance of the Alliance of Automotive Manufacturers members and Professors Bhise and Rodrick at the university.

The test setup is presented in Figure 14.1. The primary task of the driver during the entire testing was to drive the lane change course at a constant 60 km/h and make quick lane changes as indicated by each pair of signs. The driver's view of the lane change course presented in the screen is shown in Figure 14.2.

Twenty-four subjects participated in the University of Michigan-Dearborn study. After extensive familiarization with the driving simulator and the lane change procedure, each subject was asked to drive the 3-km route 10 times and perform the lane change task. The first and last (tenth) runs involved driving only with the lane changes (single task), and thus they were called baseline 1 and baseline 2, respectively. The middle eight runs were randomly assigned to the dual tasks involving four types of secondary tasks. Each secondary task had two levels—easy and hard. The two levels of tasks for each type were conducted sequentially (easy level first and hard next) in separate runs. The secondary tasks were as follows: (1) critical tracking task (easy and hard), (2) visual search task (easy and hard levels), (3) Sternberg memory task (easy and hard levels), and (4) nomadic task with

FIGURE 14.1 The laboratory setup of the LCT test showing the driver's screen on the left and the secondary task screen and controls on the right-side table.

FIGURE 14.2 The driver's view of the LCT screen showing a series of side mounted signs indicating driving lane.

a TomTom navigation system (easy and hard levels). During the dual-task runs, the subjects were asked to perform the selected secondary task (CCT) continuously, in repeated trials, throughout the entire run while performing the primary task of driving and constantly changing the lanes.

The critical tracking task (CTT) involved stabilizing an increasingly unstable vertically moving bar within a marked interval displayed in the right-hand screen. The subject controlled the bar by using two up and down arrow keys of a keypad placed on the right-hand side of the table (see Figure 14.1). The stability parameters were changed to create easy and hard tracking difficulty levels.

The visual search task (called the surrogate reference task [SURT]) presented in the right-hand screen involved detecting a larger diameter target ring among many scattered smaller background rings with identical diameter. The number of background rings and the difference between the diameter of the target and the background rings were varied to create easy and difficult tasks. The easy task had a larger diameter target ring and less number of background rings as compared with the difficult level.

The Sternberg task (called the cognitive task [COTA]) involved the subject to first listen to set of three (for easy level) or six (for hard level) randomly selected single-digit numbers and then was presented another probe digit. The subject had to determine if the probe digit was included in the set of digits presented earlier and respond by pushing "yes" or "no" keys placed on the right-side table.

The nomadic task involved the subject selecting a required screen of a TomTom navigational system placed on the right-side table and adjusting the sound volume to the required level provided by the experimenter for the easy level. For the hard level, the subject was given a destination address on a 76 × 127 mm (3″ × 5″) card and was asked to enter the city name, street name, and the street number. The subjects were asked to perform the nomadic tasks throughout the entire length of the runs while driving and lane changing.

The data collection software was programmed to store data for the primary and secondary tasks for all the runs. The stored data were analyzed to obtain mean difference in lateral deviation trajectory (MDEV) using both the adaptive and the normative models for each run of each subject. The adaptive model used the subject's qualifying lateral position profile obtained at the end of the practice runs. The normative model used the basic lateral profile obtained by assuming that an ideal driver makes all the lane changes with 0.6-s response time and maintains the vehicle in the center of the lane at all times except during the lane changes. The mean values of the lateral deviations obtained from the data over all the 24 subjects for the 10 runs are shown in Figure 14.3.

From Figure 14.3, the following observations can be made:

1. The MDEV values obtained from the normative model (labeled as N) were larger than the mean MDEV values obtained from the adaptive model (labeled as A) for any given run condition. This is expected because in the computation of differences in the lateral positions, the adaptive model uses the subject's actual lateral profile obtained while performing

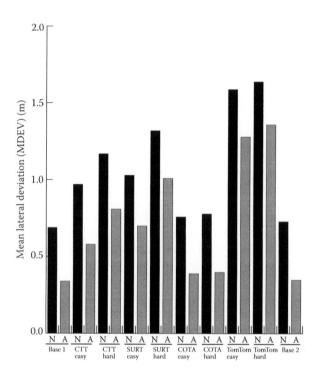

FIGURE 14.3 Mean lateral deviations (MDEV) values obtained by using the normative (N) and the adaptive (A) models for the baseline (called base 1 and base 2) and dual-task conditions.

the primary task only during the qualifying run after the practice trials as the reference profile. On the other hand, the normative model uses the ideal lateral profile as the reference to compute the differences in lateral position.

2. The baseline values (labeled as base 1 and base 2) of MDEV were about 0.7 and 0.35 m, respectively, for the normative and the adaptive models.

3. The MDEV values of all other runs with the dual tasks were larger than the corresponding baseline values obtained under the primary (single) task in baselines 1 and 2.

4. The mean MDEV value for the easy level of any given secondary task was smaller than the mean MDEV value of the hard level of the secondary task. Thus, the MDEV measure was found to be sensitive to differences in the levels of the secondary tasks.

5. The magnitude of the difference in the mean MDEV values in any given dual-task run and the MDEV value in the corresponding mean MDEV value of the baseline, thus, indicates (or measures) the amount of additional workload imposed by the secondary task.

CONCLUDING COMMENTS

The above-described five studies show that currently many methods and performance measures are used to determine the effects of addition of different in-vehicle tasks to the baseline driving situations. The NASA-TLX, SWAT, WP, SAE's total task time, and ISO LCT are examples of methods, each of which has the capability of providing a single overall measure of driver workload. However, criterion limits are presently not established to determine acceptable and unacceptable levels of driver workload, and each method has some major shortcomings. The subjective methods cannot be used without the availability of actual prototype hardware, and extensive time is required to conduct evaluations by using actual subjects. The LCT test procedure lacks the validity because the drivers in the real-world driving do not change lanes continuously at the high-frequency rate built into the procedure, and most drivers under most driving situations can voluntarily decide on when and how quickly to change a lane. SAE J2364 and 2365 test procedures are not based on actual driving. Another currently used occlusion method adopted by the U.S. (Alliance of Automobile Manufacturers, 2006) is only applicable to evaluation of tasks involving visual–manual interfaces.

While many of the above-discussed methods provide useful information, none of them can be used alone to decide on acceptability of any new in-vehicle feature. The industry would prefer to use methods that are objective, less time-consuming, repeatable, and precise. Thus, until better methods are developed, the decision makers will continue to rely on using combinations of many of the existing workload measurement methods under various laboratory, driving simulators, and field studies and supplement their findings with additional information obtained from other sources, such as ergonomics experts, customers, and benchmarking prototypes with other available in-vehicle devices.

REFERENCES

Alliance of Automobile Manufacturers. 2006. Statement of principles, criteria and verification procedures on driver interactions with advanced in-vehicle information and communication systems. Southfield, MI: Alliance of Automobile Manufacturers Driver Focus-Telematics Working Group.

Bhise, V. D., and S. Bhardwaj. 2008. Comparison of driver behavior and performance in two driving simulators. SAE Paper 2008-01-0562. Presented at the SAE World Congress, Detroit, MI.

Brookhuis, K. A., and D. Waard. 2001. Assessment of driver's workload: Performance and subjective and physiological indexes. In *Stress, Workload and Fatigue*, ed., P. A. Hancock, and P. A. Desmond. Mahwah, NJ: Lawrence Erlbaum Associates, Inc.

Card, S. K., T. P. Morgan, and A. Newell. 1980. The keystroke-level model for user performance time with interactive systems. *Communications of the ACM*, 23(7), 396–410.

Card, S. K., T. P. Morgan, and A. Newell. 1983. *The Psychology of Human–Computer Interaction*. Hillsdale, NJ: Lawrence Erlbaum Associates.

Cooper, J. M., and D. L. Strayer. 2008. Effects of simulator practice and real-world experience on cell phone related driver distraction. *Human Factors*, 50(6), 893–902.

Drews, F., H. Yazdani, C. N. Godfrey, J. M. Cooper, and D. L. Strayer. 2009. Text messaging during simulated driving. *Human Factors*, 51(5).

Hankey, J. M., T. A. Dingus, R. J. Hanowski, W. W. Wierwille, and C. Anderws. 2001. In-vehicle information systems behavioral model and design support: Final report. Report FHWA-RD-00-135. Sponsored by the Turner-Fairbank Highway Research Center of the Federal Highway Administration, Virginia Tech Transportation Institute, Blacksburg, VA.

Hart, S. G., and L. E. Staveland. 1988. Development of NASA-TLX (Task Load Index): Results of empirical and theoretical research. In *Human Mental Workload*, ed. P. A. Hancock and N. Meshkati, 139–183. Amsterdam: North Holland.

Hitt, J. M. II, J. P. Kring, E. Daskarolis, C. Morris, and M. Mouloua. 1999. Assessing mental workload with subjective measures: An analytical review of the NASA-TLX index since its inception. In *Proceedings of the 43rd Annual Meeting of the Human Factors and Ergonomics Society*, Santa Monica, CA.

Hosking, S. G., K. L. Young, and M. A. Regan. 2009. The effects of text messaging on young drivers. *Human Factors*, 51, 582–592.

International Standards Organization (ISO). 2008. Road vehicles: Ergonomic aspects of transportation and control systems—Simulate lane change test to assess in-vehicle secondary task demand. Draft of ISO/DIS 26022. Prepared by the Technical Committee ISO/TC22, Road Vehicles, Subcommittee SC13, Ergonomics.

Jackson, D., and V. D. Bhise. 2002. An evaluation of the IVIS-DEMAND driver Attention Model. SAE Paper 2002-01-0092. Paper presented at the SAE International Congress, Detroit, MI.

Jothi, V. 2009. Applications of the ISO lane change procedure. Master's thesis. University of Michigan-Dearborn, Dearborn, MI.

Lambert, S., S. Rollins, and V. D. Bhise. 2005. Effects of driver induced distraction tasks on driver performance and glance behavior. In *Proceedings of the Annual Meeting of the Transportation Research Board*, Washington, DC.

Mehler, B. R., B. Reimer, J. F. Coughlin, and J. A. Dusek. 2009. Impact of incremental increases in cognitive workload on physiological arousal and performance in young adult drivers. Transportation Research Record. *Journal of the Transportation Research Board*, 2138.

Meshkati, N., P. Hancock, and M. Rahimi. 1992. Techniques in mental workload assessment. In *Evaluation of Human Work, A Practical Ergonomics Methodology*, ed. J. Wilson and E. Corlett. 605–627. London: Taylor & Francis.

Olsson, S., and P. C. Burns. 2000. Measuring driver visual distraction with a peripheral detection task. Department of Education and Psychology, Linkoping University, Sweden.

Owens, J. M., S. B. McLaughlin, and J. Sudweeks. 2010. On-road comparison of driving performance measures when using handheld and voice-control interfaces for cell phones and MP3 players. SAE Paper 2010-01-1036. Presented at the 2010 SAE World Congress, Detroit, MI.

Reid, G. B. and T. E. Nygren. 1988. The subjective workload assessment technique: A scaling procedure for measuring mental workload. In *Human Mental Workload*, ed. P. A. Hancock and N. Meshkati, 139–183. Amsterdam: North Holland.

Rubia, S., E. Diaz, J. Martin, and J. M. Puente. 2004. Evaluation of subjective mental workload: A comparison of SWAT, NASA-TLX, and workload profile methods. *Applied Psychology: An International Review*, 53(1), 61–86.

Society of Automotive Engineers Inc. 2009a. SAE J2364 Recommended practice: Navigation and route guidance function accessibility while driving. In *SAE Handbook*. Warrendale, PA: Society of Automotive Engineers.

Society of Automotive Engineers Inc. 2009b. SAE J2365 Recommended practice: Calculation of the time to complete in-vehicle navigation and route guidance tasks. In *SAE Handbook*. Warrendale, PA: Society of Automotive Engineers Inc.

Tijerina, L, E. Parmer, and M. J. Goodman. 2000. Driver distraction with wireless telecommunications and route guidance systems. Report DOT HS809-069. National Highway Traffic Safety Administration, Washington, DC.

Tijerina, L, E. Parmer, and M. J. Goodman. 1999. Driver workload assessment of route guidance system destination entry while driving: A test track study. Transportation Research Center, East Liberty, OH.

Tsang, P. S., and V. L. Velazquez. 1996. Diagnosticity and multidimensional subjective workload ratings. *Ergonomics*, 39(3), 358–381.

Verwey, W. B., and D. M. Zaidel. 2000. Predicting drowsiness accidents from personal attributes, eye blinks, and ongoing driving behaviour. *Personality and Individual Differences*, 28(1), 123–142.

Wickens, C. D. 1987. Information processing, decision making, and cognition. In *Cognitive Engineering in the Design of Human: Computer Interaction and Expert Systems*, ed. G. Salvendy. Amsterdam: Elsevier.

Wickens, C. D. 1992. *Engineering Psychology and Human Performance*. New York: Harper Collins.

Wierwille, W. W., and F. T. Eggemeier. 1993. Recommendation for mental workload measurement in a test and evaluation environment. *Human Factors*, 35, 263–281.

15 Vehicle Evaluation Methods

OVERVIEW ON EVALUATION ISSUES

An automotive product is used by a number of users in a number of different usages. To assure that the vehicle being designed will meet the needs of its customers, the ergonomic engineers must conduct evaluations of all ergonomic vehicle features under all possible usages. A usage can be defined in terms of each task that needs to be performed by a user to meet a certain objective. A task may have many steps or subtasks. For example, the task of getting into a vehicle would involve a user to perform a series of subtasks such as (a) unlocking the door, (b) opening the door, (c) entering the vehicle and sitting in the driver's seat, and (d) closing the door. The ergonomic evaluations are conducted for a number of purposes such as (a) to determine if the users will be able to use the vehicle or its features, (b) to determine if the vehicle has any unacceptable features that will generate customer complaints after its introduction, (c) to compare the user preferences for a vehicle or its features with other vehicles, and (d) to determine if the product will be perceived by the users to be the best in the industry. The purpose of this chapter is to review methods that are useful in ergonomic evaluations of vehicles.

The evaluations can be conducted by collecting data in a number of situations. Some examples of data collection situations are given below:

1. A product (vehicle or one or more of its systems, chunks [portion of the vehicle], or features) is shown to a user, and the user's responses (e.g., facial expressions, verbal comments) are noted (or recorded). (This situation occurs when a concept vehicle is displayed in an auto show.)
2. A product is shown to a user, and then responses to questions asked by an interviewer are recorded. (This situation occurs in a market research clinic.)
3. A customer is asked to use a product, and then responses to a number of questions asked in a questionnaire or asked by an interviewer are recorded. (This situation can occur in a drive evaluation.)
4. A user is asked to use a number of products, and the user's performance in completing a set of tasks on each of the products is measured. (This situation can occur in a performance measurement study using a set of vehicles [or alternate designs of a vehicle system] in test drives.)
5. A user is asked to use a number of products and then asked to rate the products based on a number of criteria (e.g., preference, usability, accommodation, effort). (This situation occurs in field evaluations using a number of vehicles—the manufacturer's test vehicle and other competitive vehicles.)
6. A sample of drivers are provided with instrumented vehicles that record vehicle outputs and video data of driver behavior and performance as the participants drive where they wish, as they wish, for weeks or months each. This is probably the only valid method to discover what drivers actually do over time in the real world. (This situation occurs in naturalistic driving behavior measurement studies.)

The above examples illustrate that an ergonomics engineer can evaluate a vehicle or its features by using a number of data collection methods and measurements.

ERGONOMIC EVALUATIONS DURING VEHICLE DEVELOPMENT

During the entire vehicle development process, a number of evaluations are conducted to assure that the vehicle being designed will meet the needs of the customers. The design issues and ergonomic considerations covered in all the chapters in this book need to be systematically evaluated to assure that all design requirements are considered and appropriate evaluation methods are used. The results of the evaluations are generally reviewed in the vehicle development process at different milestones with various design and management teams.

Table 15.1 provides a summary of ergonomic evaluations and the evaluation methods used in the entire vehicle development process. The systems engineering model provided in Figure 11.1 is used here to provide reference to the timings of different events in the vehicle development process (see bottom part of Table 15.1). The left two columns in the table present the order and brief description of the general areas for evaluation needs. The middle columns provide types of evaluation methods used during the vehicle development process. The timings of the evaluations are indicated by two- or three-letter codes that refer to the part of the systems engineering V model presented in the lower portion of the table. The right-hand column presents some details of the evaluation methods, requirements, and issues to be addressed in each step. The second last column provides chapter number where these areas are covered in this book.

EVALUATION METHODS

Table 15.2 provides a summary of methods categorized by combinations of types of data collection methods and types of measurements.

The left-hand column of Table 15.2 shows that the data can be collected by using methods of observation, communication, and experimentation. In the observation method, it is assumed that a subject performing a task can only be observed by an experimenter, observations reported by the subjects can be recorded, or the data can be recorded (e.g., by using a camera) for later observations by an experimenter. In the communication method, the subject (or the experimenter) can be asked to report about the problems experienced during performance of a task or asked to provide ratings on his or her impressions about the task. In the experimentation method, the test situations are designed by deliberate changes in combinations of certain independent variables, and the responses are obtained by using combinations of methods of observation and/or communication. For further information on many available methods of data collection and their advantages and disadvantages, the reader should refer to Chapanis (1959) and Zikmund and Babin (2009).

The types of measurements can be categorized as objective or subjective as shown in Table 15.2. The objective measures can be defined here as measurements that are not affected by the subject performing the tasks or by the experimenter observing or recording the subject's performance. The objective measures are generally obtained by use of physical instruments or by unbiased and trained experimenters. The subjective measurements are generally based on the subject's perception and experience during or after performing one or more tasks. The objective measures are generally preferred because they are more precise and unbiased. However, there are many vehicle attributes that cannot be measured without using human subjects as "measuring instruments." After the users have experienced the vehicle, they are better able to express their perceived impressions about the vehicle and its characteristics by the use of methods of communication.

The following section provides additional information on the methods of data collection relevant to ergonomic evaluations.

METHODS OF DATA COLLECTION AND ANALYSIS

OBSERVATIONAL METHODS

In observational methods, information is gathered by direct or indirect observations of subjects during their product usages to determine if the product is easy or difficult to use. An observer can

TABLE 15.1
Summary of Ergonomic Evaluations and the Evaluation Methods Used in the Vehicle Development Process

No.	Vehicle Evaluation Need	Type of Evaluation Methods and Their Application Timings[a] in the Vehicle Design Process				Chapter Number to Refer for Additional Info.	Some Details on Evaluation Methods and Issues
		Checklists and Judgments of Experts	Applications of Data-Based Requirements, Models, and Standards	Static Tests—Laboratory and Simulators	Drive Tests and Evaluations		
1	Driver positioning, primary controls, and occupant locations	SRT	LE			3, 16	CAD/CAE applications of SAE standards J1516, J1517, J4002, J4003, J4004, J287, J1050; special populations
2	Available space to locate controls and displays	LL	LE			5	Controls and display concepts and location considerations; CAD applications of SAE J941, J287, minimum reach, down angle requirement, and J1050
3	Entry/exit evaluations for body door openings, seats, steering wheel, and armrest	LE		LM		8	Historic data from customer responses; CAD applications of digital manikins; task analyses; subjective evaluation using rating scales
4	Field-of-view evaluations	LM	LM			6	Historic customer response; CAD/CAE evaluations of FMVSS 103, 104, 111, SAE J1050
5	Instrument panel, door trim panel, and console		LM			5	Location requirements in CAD, SAE J1138, J1139, FMVSS 101
6	Interior buck evaluation	LE		LM		2,3	SAE J1826/J4002; CAD/CAE manikin-based evaluations; subjective ratings
7	Sunlight reflections, veiling glare, and legibility evaluations		LM	LM	RM	12	CAD ray tracing analyses of reflections; veiling glare analysis; subjective assessments of sunlight reflections (outdoors and in daylight simulation rooms)
8	Controls and displays operability	LE	LM	LM	RM	4, 5, 14	Driver workload evaluations in simulators and field tests with prototypes
9	Interior lighting and graphics evaluations		LM	RM	RL	12	Visibility and legibility models; subjective assessments of interior lighting and lighted components for appearance, harmony, and legibility
10	Vehicle lighting	LE	LL	RM	RM	7,12	Technology feasibility; visibility prediction models; FMVSS108 and SAE lighting standards; night drives to evaluate perceptions of beam patterns; subjective assessments of lighted appearance of signal lamps

continued

TABLE 15.1 (Continued)
Summary of Ergonomic Evaluations and the Evaluation Methods Used in the Vehicle Development Process

No.	Vehicle Evaluation Need	Type of Evaluation Methods and Their Application Timings[a] in the Vehicle Design Process				Chapter Number to Refer for Additional Info.	Some Details on Evaluation Methods and Issues
		Checklists and Judgments of Experts	Applications of Data-Based Requirements, Models, and Standards	Static Tests— Laboratory and Simulators	Drive Tests and Evaluations		
11	Engine service evaluations	LE	LM		RM	9	Task analyses; evaluation of labels and hand clearances
12	Trunk space and cargo loading/ unloading	LE	SRT		RM	9	SAE J1100 luggage/cargo volumes; biomechanical models and task analyses
13	Craftsmanship	LL		RL	RL	10	Evaluation of pleasing perceptions—fits/gaps, materials, textures and color harmony, etc.
14	Final drive evaluations (entire vehicle)				RL	14, 15, 16	Checklists, subjective evaluations by experts, customers, and management personnel

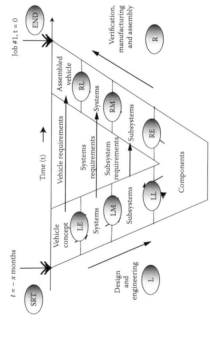

a　Timing within the systems engineering V model (see figure above).
SRT = start of the program; L = left (throughout left side of V); LE = left side early; LM = left side middle; LL = left side late; R = right (throughout right side); RE = right side early; RM = right side middle; RL = right side late; END = end of the program.

TABLE 15.2
Evaluation Methods

Type of Data Collection Method	Type of Measurements	
	Objective Measurements	**Subjective Measurements**
Observation	Experimenter observed or data recorded with instruments; behavior observations (glances, durations, errors, difficulties, conflicts), and near accidents	Checklists completed by subjects based on their observations
Communication	Experimenter reported objective measures (e.g., speed, events)	Subject reported—detections, identifications; responses in checklists; responses (or ratings) using nominal, interval, and ratio scales; problems, difficulties, and errors during operation of equipment
Experimentation	Measurements with instrumentation: performance measurements, behavioral measurements	Obtained from subjects: ratings, behavioral measurements (difficulties, errors, etc.)

directly observe or a video camera can be set up, and its recordings can be played back at a later time. The observer needs to be trained to identify and classify different types of predetermined behaviors, events, problems, or errors that a subject commits during the observation period. The observer can also record durations of different types of events, number of attempts made to perform an operation, number and sequence of controls used, number of glances made, etc. Some events such as accidents are rare, and they cannot be measured through direct observations due to excessive amount of direct observation time needed until sufficient accident data are collected. However, information about such events can be obtained through reports of near accidents (i.e., situations where accidents almost occurred but were averted) and indirect observations (e.g., through witnesses or from material evidence) gathered after such events. Therefore, the information gathered through indirect observations may not be very reliable due to a number of reasons (e.g., witness may be guessing or even deliberately falsifying or objects associated with the event of interest may be displaced or removed).

COMMUNICATION METHODS

The communication methods involve asking the user or the customer to provide information about his or her impressions and experiences with the product. The most common technique involves a personal interview where an interviewer asks a user a series of questions. The questions can be asked prior to usage of the product, during usage, or after usage. The user can be asked questions that will require the user to (a) describe the product or the impressions about the product and its attributes (e.g., usability), (b) describe the problems experienced while using the product (e.g., difficult to read a label), (c) categorize the product using a nominal scale (e.g., acceptable or unacceptable; comfortable or uncomfortable; liked or disliked), (d) rate the product on one or more scales describing its characteristics and/or overall impressions (e.g., workload ratings, comfort ratings, difficulty ratings), or (e) compare the products presented in pairs based on a given attribute (e.g., ease of use, comfort, quality feel during operation of a control).

Some commonly used communication methods in product evaluations include (1) rating scales: using numeric scales and scales with adjectives (e.g., acceptance ratings and semantic differential scales); (2) paired-comparison-based scales (e.g., using Thurstone's method of paired comparisons and analytical hierarchical method) described later in this chapter.

In addition, many tools used in fields such as industrial engineering, quality engineering and design for Six Sigma, and safety engineering can be used. Some examples of such tools are process charts, task analysis, arrow diagrams, interface diagrams, matrix diagrams, quality function deployment, Pugh analysis, failure mode and effects analysis, and fault tree analysis. The above-mentioned tools rely heavily on the information obtained through the methods of communication from the users/customers and members of the multifunctional design teams. Additional information on many of these tools can be obtained from Besterfield et al. (2003), Creveling et al. (2003), and Yang and El-Haik (2003).

EXPERIMENTAL METHODS

The purpose of experimental research is to allow the investigator to control the research situation (e.g., selecting a vehicle design, test condition) so that causal relationships between the response variable and independent variables (that define the vehicle characteristics, e.g., interface configuration, type of control, type of display, operating forces) may be evaluated. An experiment includes a series of controlled observations (or measurements of response variables) undertaken in artificial (test) situations with deliberate manipulations of combinations of independent variables to answer one or more hypotheses related to the effect of (or differences due to) the independent variables. Thus, in an experiment, one or more variables (called independent variables) are manipulated, and their effect on another variable (called dependent or response variable) is measured, whereas all other variables that may confound the relationship(s) are eliminated or controlled.

The importance of the experimental methods is that they help identify the best combination of independent variables and their levels to be used in designing the vehicle and thus provide the most desired effect on the users, and when the competitors' products are included in the experiment along with the manufacturer's product, the superior product can be determined. To ensure that this method provides valid information, the researcher designing the experiment needs to ensure that the experimental situation is not missing any critical factor related to the performance of the product or the task being studied. Additional information on the experimental methods can be obtained from Kolarik (1995) or other textbooks on design of experiments.

OBJECTIVE MEASURES AND DATA ANALYSIS METHODS

Depending on the task used to evaluate a product, task performance measurement capabilities, and instrumentation available, the ergonomics engineer would design an experiment and procedure to measure dependent measures. The objective measures can be based on physical measure such as time (taken or elapsed), distance (position or movements in lateral, longitudinal, or vertical directions), velocities, accelerations, events (occurrences of predefined events), and measures of user's physiological state (e.g., heart rate). The recorded data are reduced to obtain the values of the dependent measures and their statistics such as means, standard deviations, minimum, maximum, and percentages above and/or below certain preselected levels. The measured values of the dependent measures are then used for statistical analyses based on the experiment design selected for the study. Some examples of applications involving objective measures are provided in Chapters 13 and 14.

SUBJECTIVE METHODS AND DATA ANALYSIS

The subjective methods are used by the ergonomics engineers because in many situations (a) the subjects are better able to perceive characteristics and issues with the product, and thus they can be used as the measurement instruments; (b) suitable objective measures do not exist; and (c) the subjective measures are easier to obtain.

Pew (1993) has pointed out several important points regarding subjective methods. Subjective data must come from the actual user rather than the designer, the user must have an opportunity to experience the conditions to be evaluated before providing opinions, care must be taken to collect the subjective data independently for each subject, and the final test and evaluation of a system should not be based solely on subjective data.

The two most commonly used subjective measurement methods during the vehicle development process are (a) rating on a scale and (b) paired-comparison-based methods. These two methods are presented below.

RATING ON A SCALE

In this method of rating, the subject is first given instructions on the procedure involved in evaluating a given product including explanations on one or more of the product attributes and the rating scales to be used for scaling each attribute. Interval scales are used most commonly. Many different variations are possible in defining the rating scales. The interval scales can differ due to (a) how the end points of the scales are defined, (b) number of intervals used (note: odd number of intervals allow use of a midpoint), and (c) how the scale points are specified (e.g., without descriptors vs. with word descriptors or numerals).

Figure 15.1 presents eight examples of interval scales. The first four scales (a through d) are numeric scales with end points defined by descriptors (words or adjectives). The first two scales have 10 points, and their numeric values range from 0 to 10. On the other hand, the remaining scales (d through h) have clearly defined midpoints, and numbers and/or adjectives (or descriptors) are used in defining each scale marking. The use of adjectives or descriptors can help subjects in understanding the levels of the attribute associated with the scale. The use of midpoints (e.g., in scale e or f) allows the subject to choose the middle category if the subject is unable to decide if the product attribute in question falls on one or the other side of the scale. The use of a scale such as scale e also allows the subject to first decide if the product was easy or difficult to use and then select the level by using the adjectives "somewhat" or "very." Even number of intervals can also be used where the subject should be forced to decide between either side of the scale. Thus, the midpoint associated with the inability of the subject to decide between the two sides will be removed. The scales with 5 or less points are easier for the subjects to use as compared with scales with larger number of intervals. The direction magnitude scales such as scales g and h are particularly useful in evaluating vehicle dimensions. In these scales, the midpoint is defined by the words "about right," and thus, a large percentage of responses in this category helps confirm that the evaluated product dimension was designed properly. On the other hand, a skewed distribution of responses to the left or the right side on the scale will indicate a mismatch in terms of both the direction and magnitude of the problem with the dimension.

Table 15.3 illustrates how the direction magnitude and the 10-point acceptance scales together can be used to evaluate a number of interior dimensions in a vehicle package. Here a subject using this form as an evaluation form will provide two ratings for each line item. First rating will involve choosing/circling one of the three alternatives in the direction magnitude scale. And the second rating will involve a number between 1 and 10 indicating acceptance rating. The distribution of responses (obtained from evaluation data from a large number of subjects) on each direction magnitude scale provides feedback to the designer on how the dimension corresponding to the scale was perceived in terms of its magnitude, and the ratings on the acceptance scale provide the level of acceptability of the dimension. For example, if the ratings on item number 5 (gas pedal lateral location) in Table 15.3 showed that 80% of the subjects rated the gas pedal location as "too much to the left" on the direction magnitude scale and the average rating on the 10-point acceptance scale was 4.0, the designer can conclude that the gas pedal needs to be moved to the right to improve its acceptability. The author found that such use of dual scales was very helpful in fine-tuning the vehicle dimensions in the early stages of the vehicle design process.

PAIRED-COMPARISON-BASED METHODS

The method of paired comparison involves evaluating products presented in pairs. In this evaluation method, each subject is essentially asked to compare two products in each pair using a predefined

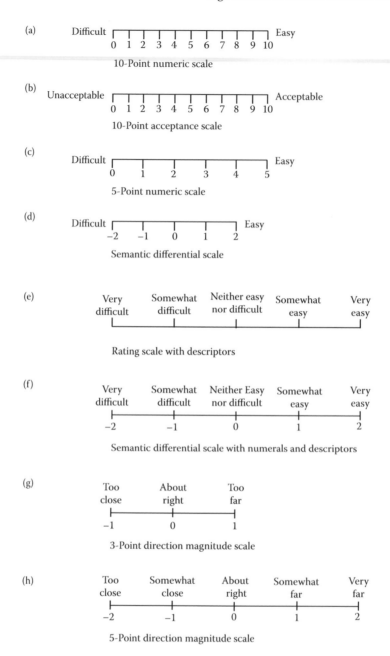

FIGURE 15.1 Examples of rating scales.

procedure and is asked to simply identify the better product in the pair on the basis of a given attribute (e.g., comfort, usability). (If the respondent says there is no difference between the two products, the instruction would be to randomly pick one of the pair. The idea is that, if there truly is no difference in that pair among the respondents, the result will average out to 50:50.) The evaluation task of the subject is, thus, easier as compared with rating on a scale. However, if n products have to be evaluated, then the subject is required to go through each of the $n(n-1)/2$ possible number of pairs and identify the better product in each pair. Thus, if five products need to be evaluated, then the number of possible pairs would be $5(5-1)/2 = 10$. The major advantage of the paired comparison approach is that it makes the subject's tasks simple and more accurate as the subject has to only compare the

TABLE 15.3
Illustration of Vehicle Package Evaluation Using Direction Magnitude and Acceptance Rating Scales

Item No.	Driver Package Consideration	Rating Using Direction Magnitude Scale			Acceptance Rating: 1 = Very Unacceptable, 10 = Very Acceptable
1	Steering wheel longitudinal (fore/aft) location	Too close	About right	Too far	_____
2	Steering wheel vertical (up/down) location	Too low	About right	Too high	_____
3	Steering wheel diameter	Too small	About right	Too large	_____
4	Gas pedal fore/aft location	Too close	About right	Too far	_____
5	Gas pedal lateral location	Too much to left	About right	Too much to right	_____
6	Lateral distance between the gas pedal and the brake pedal	Too small	About right	Too large	_____
7	Gas pedal to brake pedal liftoff	Too small	About right	Too large	_____
8	Gearshift lateral location	Too much to left	About right	Too much to right	_____
9	Gearshift location longitudinal location	Too close	About right	Too far	_____
10	Height of the top portion of the instrument panel directly in front of the driver	Too low	About right	Too high	_____
11	Height of the armrest on driver's door	Too low	About right	Too high	_____
12	Belt height (lower edge of the driver's side window)	Too low	About right	Too high	_____
13	Space above the driver's head	Too little	About right	Too generous	_____
14	Space to the left of the driver's head	Too little	About right	Too generous	_____
15	Knee space (between instrument panel and right knee with foot on the gas pedal)	Too little	About right	Too generous	_____
16	Thigh space (between the bottom of the steering wheel and the closest lower surface of the driver's thighs)	Too little	About right	Too generous	_____

two products in each trial and only identify the better product in the pair. The disadvantage of the paired comparison approach is that as the number of products (n) to be evaluated increases, the number of possible paired comparison judgments that each subject needs to make increases rapidly (proportional to the square of n), and the entire evaluation process becomes very time consuming.

We will review two commonly used methods based on the paired comparison approach, namely (a) Thurstone's method of paired comparisons and (b) the analytical hierarchical method. Thurstone's method allows us to develop scale values for each of the n products on a z scale (z is a normally distributed variable with mean equal to 0 and standard deviation equal to 1) of desirability (Thurstone, 1927), whereas the analytical hierarchical method allows us to obtain relative importance weights of each of the n products (Satty, 1980). Both the methods are simple and quick to administer and have the potential of providing more reliable evaluation results as compared with other subjective methods where a subject is asked to evaluate one product at a time.

THURSTONE'S METHOD OF PAIRED COMPARISONS

Let us assume that we have five products (or designs or issues) that need to be evaluated. The five products are named: S, W, N, P, and K. The 10 possible pairs of the product are (1) S and W, (2) S and N, (3) S and P, (4) S and K, (5) W and N, (6) W and P, (7) W and K, (8) N and P, (9) N and K, and (10) P and K. The steps to be used in the procedure are presented below.

Step 1: Select an Attribute for Evaluation of the Products

The purpose of the evaluation is to order five products along an interval scale based on a selected attribute. Let us assume that the five products are outside door handles used to open the driver's door. The five designs are assumed to differ due to the shape of their grasp areas and operating movements. The attribute selected is "ease of operation of the door handle during door opening."

Step 2: Prepare the Products for Evaluations

It is further assumed that for the evaluations, five identical doors have been built and mounted in five identical vehicle bodies. Each door is fitted with one of the five door handles with their latches and latching mechanisms. The five vehicle bodies will be positioned in the same orientations in a test area.

Step 3. Obtain Responses of Each Subject on all Pairs

It is also assumed that 80 subjects will be selected randomly from the population of the likely owners of the vehicle for the evaluation study.

Each subject will be brought in the test area separately by an experimenter. The experimenter will provide instructions to the subject and ask the subject to open and close each door within each selected pair of doors and ask the subject to select the door handle that is easier to open in each pair. The pairs of doors will be presented in a random order to each subject, and the random order will be different for each subject.

The responses of an individual subject are illustrated in Table 15.4. Each cell of the table presents "yes" or "no" depending on if the handle shown in the column was better (easier to open) than the handle shown in the row. It should be noted that only the 10 cells above the diagonal (marked by X) need to be evaluated.

Step 4. Summarize Responses of All Subjects in Terms of Proportion of Product in the Column Better than the Product in the Row

After all the subjects have provided responses, the responses are summarized as shown in Table 15.5 by assigning 1 to a "yes" response and 0 to a "no" response. Thus, the cell corresponding to the W column and S row indicates that only 1 out the 80 subjects judged the handle W to be better than handle S.

The complements of the summarized ratings in Table 15.5 are entered in the cells below the diagonal as shown in Table 15.6. For example, the complement of "1/80 responses of product W better than product S" is "79/80 responses of product S better than product W." The proportions in Table 15.6 are expressed in decimals in Table 15.7. Each cell in the matrix presented in Table 15.7 thus represents proportion p_{ij} indicating the proportion of responses in which the product in the ith column was preferred over the product in the jth row.

Step 5: Adjusting p_{ij} values

To avoid the problem of distorting the scale values (computed in next step) of the products (when p_{ij} values are very small [close to .00] or close to 1.00), the proportion values in Table 15.7 above .977 are set to .977 and the proportion values below .023 are set to .023 as shown in Table 15.8.

TABLE 15.4
Responses of an Individual Subject for the Ten Possible Product Pairs

	S	W	N	P	K
S	X	No	No	No	No
W		X	No	No	Yes
N			X	No	Yes
P				X	Yes
K					X

Note: A "yes" response indicates that the product shown in the column is better than the product in the row. A "no" response indicates that the product shown in the row was better than the product shown in the column.

TABLE 15.5
Number of Subjects Preferring Product in the Column over the Product in the Row Divided by Number of Subjects

	S	W	N	P	K
S	X	1/80	3/80	2/80	4/80
W		X	3/80	30/80	50/80
N			X	30/80	50/80
P				X	60/80
K					X

TABLE 15.6
Response Ratio Matrix with Lower Half of the Matrix Filled with Complementary Ratios

	S	W	N	P	K
S	X	1/80	3/80	2/80	4/80
W	79/80	X	3/80	30/80	50/80
N	77/80	77/80	X	30/80	50/80
P	78/80	50/80	50/80	X	60/80
K	76/80	30/80	30/80	20/80	X

Step 6: Computation of Z Values and Scale Values for the Products

In this step, the values of the proportions (p_{ij}) in each cell are converted into Z values by using the table of standardized normal distribution found in any standard statistics textbook. For example, the value of $p_{21} = .023$ is obtained by integrating the area under the standardized normal distribution

TABLE 15.7

Proportion of Preferred Responses (p_{ij})

		$i = 1$	$i = 2$	$i = 3$	$i = 4$	$i = 5$
		S	W	N	P	K
$j = 1$	S	X	0.013	0.038	0.025	0.050
$j = 2$	W	0.988	X	0.038	0.375	0.625
$j = 3$	N	0.963	0.963	X	0.375	0.625
$j = 4$	P	0.975	0.625	0.625	X	0.750
$j = 5$	K	0.950	0.375	0.375	0.250	X

TABLE 15.8

Adjusted Table of p_{ij} (if $p_{ij} > 0.977$, Then Set $p_{ij} = 0.977$; If $p_{ij} < 0.023$, Then Set $p_{ij} = 0.023$)

		$i = 1$	$i = 2$	$i = 3$	$i = 4$	$i = 5$
		S	W	N	P	K
$j = 1$	S	X	0.023	0.038	0.025	0.050
$j = 2$	W	0.977	X	0.038	0.375	0.625
$j = 3$	N	0.963	0.963	X	0.375	0.625
$j = 4$	P	0.975	0.625	0.625	X	0.750
$j = 5$	K	0.950	0.375	0.375	0.250	X

TABLE 15.9

Values of Z_{ij} Corresponding to Each p_{ij} and Computation of Scale Values (S_i)

		$i = 1$	$i = 2$	$i = 3$	$i = 4$	$i = 5$
		S	W	N	P	K
$j = 1$	S	X	−1.995	−1.780	−1.960	−1.645
$j = 2$	W	1.995	X	−1.780	−0.319	0.319
$j = 3$	N	1.780	1.780	X	−0.319	0.319
$j = 4$	P	1.960	0.319	0.319	X	0.674
$j = 5$	K	1.645	−0.319	−0.319	−0.674	X
	ΣZ_{ij}	7.381	−0.215	−3.561	−3.272	−0.333
	S_i	2.088	−0.061	−1.007	−0.925	−0.094

Note: Z_{ij} = value of NORMINV(p_{ij},0,1) function from Microsoft Excel.

curve (with mean equal to 0 and standard deviation equal to 1.0) from minus infinity to −1.995. Thus, a Z value of −1.995 provides p value of .023. The Z values can also be obtained by using a function called NORMINV by setting its parameters as (p_{ij},0,1) in Microsoft Excel. The Z values (Z_{ij}) obtained by converting all the proportion (p_{ij}) values in Table 15.8 by using the above conversion procedure are shown in the matrix on the top part of Table 15.9.

The Z values obtained in each column are summed, and the scale values for each product (S_i) are obtained by using the following formula (see last two rows of Table 15.9):

$$S_i = (\sqrt{2} / n) \sum Z_{ij}$$

where n = number of products used in paired comparisons.

The bottom row of Table 15.9 presents the scale values (S_i) for each product (note: using $n = 5$ in the above formula). It should be noted that the sum of the scale values computed from the above formula is equal to 1.0 (i.e., $\sum S_i = 1.0$).

Figure 15.2 presents a bar chart of the scale values (S_i) of the five products shown in Table 15.9. Thus, the above procedure shows that by using the Thurstone's method of paired comparisons scale values of the products are obtained. The scale values indicate the strength of the relative preference of each of the products in the set of the n products. The unit of the scale values is in number of standard deviations, and the 0 value on the scale corresponds to the point of indifference (i.e., the product with the 0 scale value is neither liked [preferred] nor disliked [not preferred]). Thus, in this example, product S is the best (most preferred) among the five products, and product N is least preferred.

ANALYTICAL HIERARCHICAL METHOD

In the analytical hierarchical method, the products are also compared in pairs. However, the better product in each pair is also rated in terms of the strength of the attribute it possesses in relation to the strength of the same attribute in the other product in the pair. The strength of the attribute is expressed using a ratio scale. The scale (or the weight) value of 1 is used to denote equal strength of the attribute in both the products in the pair. The scale value of 9 is used to indicate extreme or absolute strength of the attribute in the better product. The product with the weaker strength is assigned the inverse of the scale value of the better product. The following example will illustrate this rating procedure.

Let us assume that there are two products U and R in a pair and the attribute to compare the products is "ease of use." The scale values assigned to the products using the ratio scale would be as follows:

1. If product U is "extremely or absolutely easy" to use as compared with product R, then the weight of U preferred over R will be 9, and the weight of R preferred over U will be 1/9.
2. If product U is "very easy" to use as compared with product R, then the weight of U preferred over R will be 7, and the weight of R preferred over U will be 1/7.

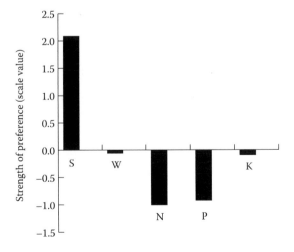

FIGURE 15.2   Scale values of the five products.

3. If product U is "easy" to use as compared with product R, then the weight of U preferred over R will be 5, and the weight of R preferred over U will be 1/5.
4. If product U is "moderately easy" to use as compared with product R, then the weight of U preferred over R will be 3, and the weight of R preferred over U will be 1/3.
5. If product U is "equally easy" to use as compared with product R, then the weight of U preferred over R will be 1, and the weight of R preferred over U will be also 1.

When a decision maker compares two items in a pair for a weight of importance (or preference), Satty (1980) described the 9-point scale by using the following adjectives.

1 = equal importance
2 = weak importance
3 = moderate importance
4 = moderate plus importance
5 = strong importance
6 = strong plus importance
7 = very strong or demonstrated importance
8 = very, very strong importance
9 = extreme or absolute importance

From the viewpoint of making the scales more understandable, usually only the odd-numbered scale values (shown in bold case above) are described and presented to the subjects. To allow the subjects to decide on the weight, the author found that the scale presented in Figure 15.3 works very well. Here the subject will be asked to put a X mark on the scale on the left side if product U is preferable over R. The higher numbers on the scale indicate higher preference. If both products are equally preferred, then the subject will be asked to place the X mark at the midpoint of scale value equal to 1. If product R is preferred over product U, then the subject will use the right side of the scale.

Let us assume that we have to compare six products, namely, U, R, T, M, L, and P, by using the analytical hierarchical technique. A subject will be asked to compare the products in pairs. The fifteen possible pairs of the six products will be presented to the subject in a random order. The subject will be given a preselected attribute (e.g., ease of use) and asked to provide strength of preference ratings for each of the fifteen pairs by using scales such as the one presented in Figure 15.3. The data obtained from the 15 pairs will then be converted into a matrix of paired comparison responses as shown in Table 15.10. Each cell of the matrix indicates the ratio of preference weight of the product in the row over the product in the column. Thus, the ratio 5/1 in the first row and second column indicates that the product in the row (U) was preferred (i.e., considered to be easy: rating weight of 5) over the product in the column (R).

To compute the relative weights of importance of the products, the fractional values in Table 15.10 are first converted into decimal numbers, as shown in the left side matrix in Table 15.11. All the six values in each row are then multiplied together and entered in the column labeled as "Row Product" in Table 15.11. The geometric mean of each row product is computed. It should be noted that the geometric mean of the product of n numbers is the $(1/n)$th root of the product (e.g., 1/6th root of 35.00 is 1.8086). All the six geometric means in the column labeled as "Geometric Mean" are

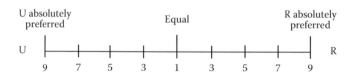

FIGURE 15.3   Scale used to indicate strength of the preference when comparing two products (U and R).

TABLE 15.10
Matrix of Paired Comparison Responses
for One Evaluator

	U	R	T	M	L	P
U	1	5/1	1/1	7/1	1/1	1/1
R	1/5	1	1/2	5/1	1/1	3/1
T	1/1	2/1	1	3/1	5/1	3/1
M	1/7	1/5	1/3	1	1/1	1/3
L	1/1	1/1	1/5	1/1	1	1/3
P	1/1	1/3	1/3	3/1	3/1	1

Note: The value in a cell indicates the preference ratio for comparing the product in a row with the product in a column.

TABLE 15.11
Computation of Normalized Weights of the Product Attribute

	U	R	T	M	L	P	Row Product	Geometric Mean	Normalized Column
U	1.00	5.00	1.00	7.00	1.00	1.00	35.0000	1.8086	0.2580
R	0.20	1.00	0.50	5.00	1.00	3.00	1.5000	1.0699	0.1526
T	1.00	2.00	1.00	3.00	5.00	3.00	90.0000	2.1169	0.3020
M	0.14	0.20	0.33	1.00	1.00	0.33	0.0031	0.3821	0.0545
L	1.00	1.00	0.20	1.00	1.00	0.33	0.0660	0.6357	0.0907
P	1.00	0.33	0.33	3.00	3.00	1.00	0.9801	0.9967	0.1422
							Sum→	7.0099	1.000

then added. The sum, as shown in the table, is 7.0099. Each of the geometric means is then divided by their sum (7.0099) to obtain the normalized weight of the products. It should be noted that due to the normalization, the sum of the normalized weights over all the products is 1.0.

The normalized weights are plotted in Figure 15.4. The figure, thus, shows that the most preferred product (based on the ease of use) was T and the least preferred product was M. The example discussed in this section was based on data obtained from one subject. If more subjects are available, then normalized weights for each subject can be obtained by using the above procedure and then average weights of each product can obtained by averaging over the normalized weights of all the subjects for each product.

SOME APPLICATIONS OF EVALUATION TECHNIQUES IN AUTOMOTIVE DESIGNS

CHECKLISTS

A checklist is used to check if the product being designed meets each applicable ergonomic guideline (or principle or requirement) in the area covered by the checklist. The checklist approach is commonly used during design of many areas such as (1) interior and exterior package design (refer to requirements in Chapters 3, 8, and 9), (2) controls and displays design (refer to requirements in Chapter 5), (3) vehicle lighting design (refer to requirements in Chapter 7), and (4) special

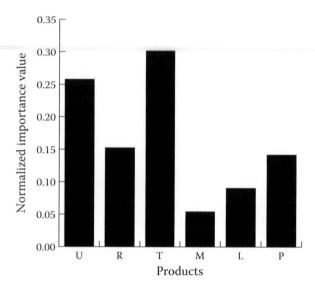

FIGURE 15.4 Normalized weights of the six products.

population issues (refer to Chapter 16). It should be noted that the checklists must be comprehensive and complete and must be completed by trained evaluators. The ergonomic checklists are generally completed by ergonomics experts based on their knowledge or data available from various ergonomic analyses and studies (see Chapters 13 and 14). Pew (1993) has compiled a useful checklist of "poor questions" that should guide the development of any checklist or questionnaire. Some examples of poor questions are the following: (a) they produce a narrow range of answers, (b) they require information the respondent does not know or remember, and (c) their statement is too vague.

OBSERVATIONAL STUDIES

Driver and customer observational studies are conducted to obtain information on issues such as problems encountered while entering and exiting vehicles (see Chapter 8 and Bodenmiller et al., 2002), operating in-vehicle devices (e.g., to study driver understanding of various control functions in audio, climate controls, and navigation systems), and performing vehicle service tasks (e.g., checking fluids, changing fuses and bulbs, refueling, changing a tire).

VEHICLE USER INTERVIEWS

Drivers and other vehicle users are interviewed individually and in groups (e.g., focus group sessions) to understand their concerns, issues, and wants related to various vehicle features. For example, Bhise et al. (2005) asked drivers to develop layouts of center stack and console areas through a structured interview technique (a method of communication).

RATINGS ON INTERVAL SCALES

Rating methods using different interval scales are used for ergonomic evaluations of issues such as (1) interior and exterior package dimensions, (2) characteristics of controls and displays (e.g., acceptability of locations, sizes, grasp areas, feel during movement of controls, compressibility of armrests), and (3) interior materials (e.g., visual and tactile characteristics of materials on instrument panels, door trim, seat areas, and steering wheels; Bhise et al., 2006, 2008, 2009).

Studies Using Programmable Vehicle Bucks

Programmable vehicle bucks are used in early package evaluation studies to assess exterior and interior dimensions such as vehicle width, windshield rake angle, seating reference point location, driver eye location, visibility over the instrument panel, hood and side windows, and height of armrest etc. (Richards and Bhise, 2004).

Driving Simulator Studies

Driving simulators are now routinely used in many automotive companies to evaluate driver workload issues in operating various in-vehicle devices (Bertollini et al., 2010). All the three methods of observation, communication, and experimentation can be used during the simulator tests.

Field Studies and Drive Tests

Various studies under actual driving situations on test tracks and public roads under different road, traffic, lighting, and weather are conducted for evaluation of issues in areas such as seat comfort, field of view, vehicle lighting, controls and displays usage, driver workload (Jack et al., 1995; Owens et al., 2010; Tijerina el al., 1999).

REFERENCES

Bertollini, G., L. Brainer, J. Chestnut, S. Oja, and J. Szczerba. 2010. General Motors driving simulator and applications to human machine interface (HMI) development. SAE Paper 2010-01-1037. Presented at the 2009 SAE World Congress, Detroit, MI.

Besterfield, D. H., C. Besterfield-Michna, G. H. Besterfield, and M. Besterfield-Scare. 2003. *Total Quality Management*. 3rd ed. Upper Saddle River, NJ: Prentice Hall.

Bhise, V., R. Boufelliga, T. Roney, J. Dowd. and M. Hayes. 2006. Development of innovative design concepts for automotive center consoles. SAE Paper 2006-01-1474. Presented at the SAE 2006 World Congress, Detroit, MI.

Bhise, V., R. Hammoudeh, J. Dowd, and M. Hayes. 2005. Understanding customer needs in designing automotive center consoles. *Proceedings of the Annual Meeting of the Human Factors and Ergonomics Society*, Orlando, FL.

Bhise, V., S. Onkar, M. Hayes, J. Dalpizzol, and J. Dowd. 2008. Touch feel and appearance characteristics of automotive door armrest material. *Journal of Passenger Cars—Mechanical Systems*, SAE 2007 Transactions.

Bhise, V., V. Sarma, and P. Mallick. 2009. Determining perceptual characteristics of automotive interior materials. SAE Paper 2009-01-0017. Presented at the 2009 SAE World Congress, Detroit, MI.

Bodenmiller, F., J. Hart, and V. Bhise. 2002. Effect of vehicle body style on vehicle entry/exit performance and preferences of older and younger drivers. SAE Paper 2002-01-00911. Paper presented at the SAE International Congress in Detroit, MI.

Chapanis, A. 1959. *Research Techniques in Human Engineering*. Baltimore, MD: Johns Hopkins Press.

Creveling, C. M., J. L. Slutsky, and D. Antis, Jr. 2003. *Design for Six Sigma—In Technology and Product Development*. Upper Saddle River, NJ: Prentice Hall PTR.

Jack, D. D., S. M. O'Day, and V. D. Bhise. 1995. Headlight beam pattern evaluation: customer to engineer to customer—A continuation. SAE Paper 950592. Presented at the 1995 SAE International Congress, Detroit, MI.

Kolarik, W. J. 1995. *Creating Quality: Concepts, Systems, Strategies, and Tools*. New York: McGraw-Hill.

Owens, J. M., S. B. McLaughlin, and J. Sudweeks. 2010. On-road comparison of driving performance measures when using handheld and voice-control interfaces for cell phones and MP3 players. SAE Paper 2010-01-1036. Presented at the 2010 SAE World Congress, Detroit, MI.

Pew, R.W. 1993. Experimental design methodology assessment. BBN Report 7917, Cambridge: Bolt Beranek & Newman.

Richards, A., and V. Bhise. 2004. Evaluation of the PVM methodology to evaluate vehicle interior packages. SAE Paper 2004-01-0370. Also published in SAE Report SP-1877, SAE International Inc., Warrendale, PA.

Satty, T. L. 1980. *The Analytic Hierarchy Process*. New York: McGraw-Hill.
Thurstone, L. L. 1927. The method of paired comparisons for social values. *Journal of Abnormal and Social Psychology*, 21, 384–400.
Tijerina, L., E. Parmer, and M. J. Goodman. 1999. Driver workload assessment of route guidance system destination entry while driving: A test track study. Transportation Research Center, East Liberty, OH.
Yang, K., and B. El-Haik. 2003. *Design for Six Sigma: A Roadmap for Product Development*. New York: McGraw-Hill.
Zikmund, W. G., and B. J. Babin. 2009. *Exploring Market Research*. 9th ed. Cengage Learning.

16 Special Driver and User Populations

AN OVERVIEW ON USERS AND THEIR NEEDS

The purpose of this chapter is to provide the reader an insight into different populations of vehicle users and the differences in the populations. It must be realized that each population has some unique set of characteristics and needs that must be considered in designing an automotive product for its intended market segment.

The population of vehicle users can be distinguished by considering factors related to (a) vehicle type and body style (e.g., owners of certain types of vehicles have unique needs); (b) geographic locations of the markets (specific countries) in which the vehicle will be sold; (c) type of uses (e.g., personal, family, or work/commercial-related uses); (d) level of luxury based on income, image, and technology expectations of the customers and users; (e) educational and technical/professional background of the users; (f) gender-specific use issues and male-to-female ratio of users; (g) user's-age-related characteristics and needs; and (h) physical abilities and disabilities of the users.

The ergonomics engineers assigned to a vehicle development program need to thoroughly understand the population and subpopulations of the users of the planned vehicle. To understand issues associated in each of the above categories, the ergonomics engineers need to search for information available within the company from sources or databases such as lessons learned from previous vehicle development programs, market research, customer feedback and complaints, and internal and external requirements on the vehicle. In situations where the vehicle type being designed is very different from the existing vehicles, the engineers need to gather information by visiting customers, observing how they use their vehicles, and asking them about their needs and preferences.

For example, when the author was asked to provide ergonomics support to a heavy-truck design program, he not only learned to drive the heavy trucks with their long trailers but also performed many other tasks that the drivers routinely do, such as docking and undocking the trailer, hooking and unhooking air hoses and cables, filling fuel, checking fluids. He also took long trips in the heavy trucks and observed what the drivers did and asked them many questions about their usage situations and problems. The experience helped in designing survey questionnaires and in interviewing the truck drivers who participated in market research clinics and in planning special ergonomics studies.

Some special ergonomic studies conducted to study the heavy-truck driver issues involved (a) asking the truck drivers to create their own full-size instrument panel mock-up using Styrofoam blocks, Velcro strips, and a box full of controls and displays; (b) asking the drivers to enter and exit from the trucks equipped with different step heights and grab handles and video recording their hand and foot movements; and (c) interviewing the drivers on special situations involving field-of-view problems.

UNDERSTANDING USERS: ISSUES AND CONSIDERATIONS

Traditionally, ergonomics engineers will examine the anthropometric characteristics of users to make sure that they can be accommodated in the vehicle space. However, to truly meet the needs of customers and users, ergonomics engineers also need to gather information on many aspects that affect their usage experience. The issues associated with the usage experience are based not just on the demographic and educational background of the user population but also on the users' needs and expectations. The factors to be considered here are grouped according to major influencing variables and are described below.

Vehicle Types and Body Styles

The type of vehicle and body style that a user will select primarily depends on the user's desires (e.g., sportiness, styling, performance, fuel economy, ride comfort, and affordability) and transportation needs (i.e., carrying capacity in terms of number of passengers and load [sizes and weights]). The vehicle type can be classified as passenger car, SUV, crossover, truck (light, medium, and heavy), van, or multipassenger bus. The body style can be discriminated by features such as the seating reference point height from the ground, number of occupants and seating configuration, number of doors and types of doors (e.g., hinged doors, traditional front hinged vs. suicide doors, sliding doors, gull wing doors, hatchback, liftgate, tailgate, dual swing gates), presence or absence of the hood (e.g., a truck with a long hood vs. a cab-over design), and cargo area (e.g., flat bed/open, closed, size of cargo box/bed).

In the United States, the male-to-female ratio of drivers in passenger cars is about 50:50. The smaller and economy passenger cars, typically, have higher percentage of female drivers (about 60%), whereas the larger and more expensive passenger cars have higher percentages of male drivers. The U.S. truck products, overall, have higher proportions of males as drivers. The proportion of males in the populations of the pickup truck drivers, the medium-truck drivers, and heavy-truck drivers are approximately 75%–85%, 85%–90%, and over 95%, respectively. It should be noted that the population of drivers (i.e., actual users) of automotive products could be different from the population of owners or purchasers of the vehicles. This is especially important when considering truck products. The truck owners (or purchasers) are generally different from the passenger car users because higher percentages of trucks are owned by businesses or government agencies.

The pickup trucks in the past were mainly utility vehicles or "work horses" with less concern for driver comfort or ease of use. Over the past 10–15 years, that trend has been changing as many pickup trucks are owned and used by individuals who expect the luxury features and comfort of passenger cars and still want the durability and utility of the trucks. Many models of light trucks strive for the same comfort, efficiency, driveability, and safety as the other passenger vehicle segments. At the same time, light trucks must be designed to meet their primary function of utility and meet the needs that other passenger vehicles are not well suited for, such as hauling, off-road driving, and towing.

There are also vehicles that are truck based and serve specialized functions such as recreational vehicles, vans with different seating arrangements, delivery trucks, ambulances, buses, garbage trucks, fire trucks, cement trucks, etc. They are typically sold as incomplete vehicles (having fully drivable rolling chassis with or without a full cab) by the vehicle manufacturers, and other body builders build bodies with specialized features. The ergonomic considerations of such vehicles are, thus, handled by ergonomics engineers working for the body builders, outside consultants, or the vehicle-acquiring organizations (e.g., transportation companies).

Market Segments

The market segments for passenger cars are typically classified by vehicle size (e.g., subcompact, compact, intermediate, full size), body style, price (e.g., economy, entry luxury, luxury, ultra luxury), and countries where the vehicles will be sold. The needs in the different market segments are driven by a number of considerations.

Some issues underlying the considerations are the following:

1. Anthropometric differences: For example, Asian populations are shorter and require pedals to be located more rearward as compared with the vehicles sold in the United States and Europe.
2. Country-related differences: The narrower roads, traffic congestions, higher fuel prices, lack of available parking spaces, and parking costs have an effect on types and sizes of vehicles used in different countries.

3. Economic differences: For example, manual transmission usages are higher in Europe and Asia (primarily because of the higher fuel prices as compared with the United States).

4. Changing trends and expectations: In recent years, the customers have high expectations of vehicle features across all market segments. Small-car buyers also are loading up on high-tech options, such as the Ford's hands-free Sync multimedia system, which is driving up the transaction prices (Pope, 2009).

FEMALE DRIVERS

Female drivers have different needs and problems during vehicle usage than those of male drivers. On average, females are shorter than are males (see Figure 16.1). They sit more forward and closer to the instrument panel and the steering wheel (if the pedals and the steering wheel are nonadjustable). The shorter females may also find insufficient clearance spaces for their knees (under the steering column and rearward of the knee bolster located below the instrument panel). Some females with shorter lower leg lengths may have difficulties supporting their heels on the floor and reaching the pedals (especially in vehicles with taller seat heights [H30 dimension] such as vans). The seated eye heights of females are lower than those of males. Thus, their view of the road over the steering wheel, instrument panel, and the beltline would be somewhat limited as compared with that of male drivers. Shorter females will also find taller vehicles more difficult to get in and out of as compared with the taller males. They may find it difficult to reach opened liftgates in vans and taller SUVs. Most females are less strong than are males. Thus, lighter forces in operating pedals, hand brakes, liftgates, hoods, etc, must be considered. Many females have longer finger nails (95th percentile finger nail length among U.S. office workers is about 12 mm). Thus, door handles and controls need larger clearances to accommodate longer finger nails. Females carry purses, and they need purse

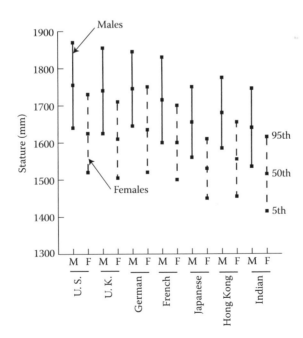

FIGURE 16.1 Comparison of 5th, 50th, and 95th percentile values of stature of male and female adults from seven different countries. (Plotted from data provided in the work of Pheasant, S., and C. M. Haslegrave, *Bodyspace: Anthropometry, Ergonomics and the Design of Work*, 3rd ed., CRC Press, Taylor & Francis Group, London, 2006.)

storage room inside or near the center console. Pregnant drivers would find adjustable (tilt and tele-scopic) steering column useful in obtaining larger stomach clearance.

OLDER DRIVERS

The population of the United States is aging at a rapid pace (see Figure 16.2). By 2030, it is predicted that about 71.4 million or 20% of the U.S. population will be older (aged 65 years and over; U.S. Census Bureau, 2010). The population of older (aged 65 years and over) persons in the United States is estimated to increase from about 13% (40 million) in 2010 to 21% (87 million) in 2050 (refer to Figure 16.2). By the year 2020, drivers aged 65 years and older will account for more than 16.2% of the driving population. The fastest-growing group is the old-old category, which includes drivers aged 75 years and over. Research to improve the quality of life for the elderly is currently quite active, with many studies appearing in journals and books (Charness and Parks, 2000; Fisk et al., 2004).

As humans age, some anthropometric dimensions also change. For example, average stature of males as well as that of females decreases by about 12–15 mm (0.5–0.6 in.) per decade on average after about the age of 30 years. Muscle strength decreases with increase in age (see Figure 2.2). Thus, the older driver will find entering and exiting from vehicles more difficult. Older females, particularly, find entering and exiting from taller vehicles such as the full-size pickup and SUVs more difficult (Bodenmiller et al., 2002). Incorporation of running boards and assist straps and handles should be considered to improve ease during entry/exit from taller vehicles. Tasks requiring higher physical efforts such as lifting rear gates and hoods, folding seats, loading and unloading heavier objects are also more difficult for older drivers. Due to higher incidences of arthritis, the older driver will find the following tasks more difficult: reaching and pulling the seat belt buckle from its B-pillar anchor location, turning the ignition key, operating thumb-activated gear shifters, unlatching seat belts, operating inside and outside door opening handles, operating hand brakes, and turning head while using a side view mirror.

Normal aging is associated with declining visual, attentional, cognitive, and physical abilities. Medical conditions also can accelerate deterioration in driving performance among older drivers. Due to degradation in visual functions with age, the older driver would require larger letter sizes and higher contrast labels and graphics in visual displays (Bhise and Hammoudeh, 2004). The older drivers also have longer detection, search, and decision times, especially in choice reaction time

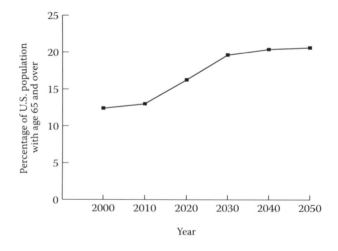

FIGURE 16.2 U.S. population projection of people with the age of 65 years and over. (Data from the U.S. Census Bureau, Population projections, Projected population of the United States, by age and sex: 2000 to 2050 Table 2a. Available at http://www.census.gov/ipc/www/usinterimproj/, 2010.)

situations. Their visibility distances to visual targets at night are considerably shorter than those of younger drivers (see Figure 7.10), and the older drivers are also more affected by the effects of discomfort and disability glare (Bhise et al., 1988). As the drivers age, the hearing abilities of the drivers also decrease, especially in the higher frequencies. Thus, frequencies above 2000 Hz in auditory warning signals should be avoided. The older drivers are also less willing to use new technologies, and they drive less during inclement weather and at night.

In an effort to design vehicles for older buyers, Ford engineers developed a Third Age suit, which helps simulate specific problems associated with age, such as reduced sense of touch, vision, body movements, and muscular strength (Ford Motor Company, 2002). Since wearing the suit adds about 30–40 years to the user's age, after wearing the suit and using the vehicle, the younger vehicle designers and engineers quickly realize the problems experienced by the older drivers. Thus, the Third Age suit can be considered a useful tool for ergonomists in creating awareness of older driver problems within the automotive design community and also to evaluate vehicle designs.

Effect of Geographic Locations of the Markets

As pointed out earlier, the drivers in different geographic locations of different markets in different countries have some unique problems. The drivers' needs differ due to differences in conditions and characteristics related to roads (e.g., road geometry, road surface conditions, traffic speeds), climate (e.g., extreme temperatures, rainfall, snow, sand storms), traffic (e.g., vehicle density, speeds, mix of cars to trucks), languages, economy, culture, user expectations, and government regulations on motor vehicle designs (e.g., requirements provided by ISO, USDOT/NHTSA, European/ECE, Japanese Ministry of Transport).

The differences in anthropometric and strength characteristics between drivers from Western and Asian countries must be considered (see Figure 16.1; Pheasant and Haselgrave, 2006). Natu and Bhise (2005) found differences in the needs of drivers from the United States, India, and China while creating a low-mass vehicle. Some specific issues that need considerations are as follows: (a) driver location in the vehicle (e.g., the left-hand drive [LHD] vs. the right-hand drive [RHD]; note that some countries, namely, the United Kingdom, Australia, India, and Japan, drive on the left side of the road and thus use RHD vehicles), (b) driver positioning with respect to the pedals (note: longitudinal locations of the pedals depend on driver stature, leg lengths, and the presence or absence of the clutch pedal), and (c) distributions of driver characteristics such as age, males-to-females mix, languages, habits, and stereotypes.

The differences in fuel prices, availability of parking spaces, traffic congestion, and roads (geometric designs of the roadways, e.g., road widths and curvatures) between the United States, European, and Asian countries have affected the size of vehicles prevalent in different countries. The Asian and European countries have vehicles with higher percentage of manual-transmission-equipped vehicles, and the percentages are changing with advances in technology and changes in global economy. The effects of recent changes in economic and energy situations are also changing global market trends towards smaller and more-fuel-efficient vehicles. Further, to reduce costs, many manufacturers are sharing components and vehicle platforms and also creating global vehicles. The ergonomics engineers have the challenge to determine the areas where changes in designs are necessary to accommodate drivers in different populations.

Drivers with Disabilities and Functional Limitations

With the increase in the population of older drivers, the demand for increased mobility among the disabled and persons with various functional limitations will continue to increase. To maintain the mobility of drivers who have various disabilities, a number of vehicle modifications are available. The vehicle manufacturers need to consider the requirements to accommodate many common disabilities during vehicle designs so that the vehicles can be more easily modified to accommodate changes such as wheelchair access (modifications to doors/seats, addition of ramps or chair lifts), controls, displays, etc.

With greater use of electronics, wireless, and drive-by-wire technologies, it will be easier in the future to modify vehicles with fewer compromises for space and occupant protection issues. For example, an optimized joystick or lever steering control may offer mobility to persons who have neither the range of movement nor the dexterity to operate a multiple-turn steering wheel with one limb. However, the alternative controls can be difficult to learn and use effectively and impose significantly different workloads in different individuals.

Better fastening systems for wheelchairs and restraint systems for occupants in wheelchairs in vehicles are also major needs. The restraint systems for occupants in wheelchairs are still difficult to use. An ideal system will not require special wheelchair hardware or significant effort to fasten. With further advances in rehabilitation medicine and adaptive technologies, many disabled people who never considered the driving possibility may be able to drive in the future.

ISSUES IN DESIGNING GLOBAL VEHICLES

To reduce vehicle development costs, all manufacturers are considering approaches such as developing vehicles with common platforms, common parts, and common overall designs in creating vehicles that can be introduced in many countries. However, the ergonomics engineers will still face many unique problems in determining what can be commonized and where different designs are needed to meet the different needs in different countries.

Some unique design problems in creating a truly common global vehicle designs are the following:

1. Differences between U.S., European, and Japanese standards: Many design procedures and requirements still are different for different countries (e.g., vehicle lighting and field of view standards) despite of the many actions undertaken to harmonize standards.
2. LHD versus RHD: LHD and the RHD vehicles cannot simply be mirror images of each other. For example, in all vehicles, the accelerator pedal is always operated by the right foot. On the other hand, in many vehicles sold in European countries, the turn signal function is not always placed on the outboard stalk. Conditions of cost and commonality also dictate that certain items, such as instrument clusters and center stack devices, are not made or installed as mirror-imaged components for LHD and RHD vehicles.
3. Vehicle size (passenger cars and truck): The differences in lane widths, road curvatures, traffic congestions, parking spaces, fuel costs, etc, dictate different sized vehicles in different countries.
4. Type of driving, trips, and distance traveled per driver: Availability of alternate-mode travel (e.g., trains and buses) also affects vehicle usages and driving needs in different countries.
5. Differences due to user expectations and stereotypes: For example, "up for on" direction-of-motion stereotype for toggle switches in the United States versus "down for on" in the United Kingdom.
6. Differences due to languages in user interface designs: For example, labels (words used for identification and setting labels, units used in displays [km/h vs. MPH; gallons vs. liters], and voice commands and voice recognition systems must be different to satisfy drivers with different languages.

FUTURING

"Futuring" is a term used by some market researchers in the automotive industry to predict future trends in vehicle designs. Reducing the planning horizon for vehicle development helps to reduce serious errors in determining customer needs for a future vehicle as accuracy in predicting the near-term future is generally greater than in predicting far-term future. The ergonomics engineers need to take into account future changes in user populations, their expectations, and changes due to design trends and technological advances. For example, the introduction of electric vehicles will require additional driver interfaces related to recharging needs, ease during recharging and reducing the "range anxiety," or the uncertainty related to completing the return trips (Bhise et al., 2010).

REFERENCES

Bhise, V., H. Dandekar, A. Gupta, and U. Sharma. 2010. Development of a driver interface concept for efficient electric vehicle usage. SAE Paper 2010-01-1040. Paper presented at the 2010 SAE World Congress, Detroit, MI.

Bhise, V. D. and R. Hammoudeh. 2004. A PC based model for prediction of visibility and legibility for a human factors engineer's tool box. *Proceedings of the Human Factors and Ergonomics Society 48th Annual Meeting*, New Orleans, LA.

Bhise, V. D., C. C. Matle, and E. I. Farber. 1988. Predicting effects of driver age on visual performance in night driving. SAE Paper 890873. Presented at the 1988 SAE Passenger Car Meeting, Dearborn, MI.

Bodenmiller, F., J. Hart, and V. Bhise. 2002. Effect of vehicle body style on vehicle entry/exit performance and preferences of older and younger drivers. SAE Paper 2002-01-00911. Presented at the SAE International Congress, Detroit, MI.

Charness, N., and D. C. Parks. 2000. *Communication, Technology and Aging: Opportunities and Challenges for the Future*. New York: Springer Publishing.

Fisk, A. D., W. A. Rogers, N. Charness, S. Czaja, and J. Sharit. 2004. *Designing for Older Adults: Principles and Creative Human Factors Approaches*. Boca Raton, FL: CRC Press.

Ford Motor Company. 2002. Ford drives a mile in an older person's suit, accessed June 23, 2011, http://www.ncsu.edu/www/ncsu/design/sod5/cud/projserv_ps/projects/case_studies/ford.htm.

Natu, M., and V. Bhise. 2005. Development of specification for UM-D's low mass vehicle for China, India and the United States. SAE Paper 2005-01-1027. Presented at the 2005 SAE World Congress, Detroit, MI.

Pheasant, S., and C. M. Haslegrave. 2006. *Bodyspace: Anthropometry, Ergonomics and the Design of Work*. 3rd ed. London: CRC press, Taylor & Francis Group.

Pope, B. 2009. Ford pulling itself up by its bootstraps. *Ward's AutoWorld*, December 1, 2009, accessed June 23, 2011, http://wardsautoworld.com/ar/auto_ford_pulling_itself/.

U.S. Census Bureau. 2010. Population projections. Projected population of the United States, by age and sex: 2000 to 2050 Table 2a, accessed June 23, 2011, http://www.census.gov/ipc/www/usinterimproj/.

17 Future Research and New Technology Issues

INTRODUCTION

Over the past 25 years, the applications of ergonomics in the vehicle designs have improved user comfort, convenience, and safety substantially. Compare any 1985 model year vehicle with its 2010 model year vehicle. You will find that almost every item that the driver interfaces in these vehicles has changed substantially. For example, the 1985 vehicles had radios with two rotary knobs, five push-button presets, and a sliding bar display. The climate controls were mechanical cable-operated slide or rotary controls. Now, the 2010 model vehicles have radios with many more features such as AM, FM, satellite stations, CD, and a USB interface to connect a number of digital media. In some vehicles, all the radio functions can be operated by voice controls. Further, wireless connectivity is also available in many vehicles to interface cell phones and other devices. The newer climate controls have features such as precise temperature settings for the driver and the passenger, automatic climate control, controls for the rear occupants, and ability to view outside temperature. The climate controls in many 2010 model year vehicles share the center stack display with the audio and navigation system features. In many newer vehicles, the radio and climate controls can be operated by redundant steering-wheel-mounted controls that have lighted labels for night legibility and various electronic displays can be selected and set with these controls. There are at least 50 or more individual controls in front of the driver in newer vehicles (see Figure 5.6) with well-illuminated labels. The change also has led to extending feature content to many economy vehicles.

The above-described situation, thus, illustrates that with the rapid advances in many technologies, the future needs in implementing ergonomic solutions are even greater. The needs are in both areas, namely, in developing improved methods to design and evaluate the products and in developing products with improved functionality, features, and convenience (Parkes and Franzen, 1993; Chaffin, 2001; Reed et al., 2003; Badler et al., 2005).

The purpose of this chapter is to cover future research issues related to improving applications of vehicle ergonomics and implementation of new technologies in designing future vehicles. The chapter addresses the future of automotive ergonomics in terms of what features can be expected in future vehicles and what ergonomic data and tools are needed to design future vehicles.

ERGONOMIC NEEDS IN DESIGNING VEHICLES

To understand the ergonomic needs in designing new vehicles for the future, let us first review the basic driver needs that can be considered to fall in the following broad areas:

1. Providing greater levels of mobility, comfort, and convenience to vehicle users
2. Improving safety (freedom from injuries through crash avoidance and crash protection)
3. Improving productivity and reducing costs (e.g., efficient utilization of time, reduced cost/mile, fuel economy)
4. Providing pleasing environment (e.g., choice in entertainment features for the drivers and passengers, well-crafted and more spacious vehicle interiors)
5. Creating "fun-to-drive" vehicles (i.e., pleasing sensory perceptions during vehicle usages)

Considering the above needs along with the Kano Model of Quality and the Ring Model of Desirability (covered in Chapter 10), the advances in technologies should offer features that would not be just expected by the customers, but they should also increase comfort/convenience, create pleasing perceptions, and delight customers. Now, let us jump directly in the current automotive scene and define the needs and expectations of the users of passenger vehicles that can be introduced in the near future, that is, within the next 5 years. The following section presents a summary of the customer needs in designing the passenger vehicles.

Passenger Vehicles in the Near Future

1. High-tech cars—safer, enjoyable, comfortable, and smart (e.g., can configure functions to individual needs, anticipate unsafe situations, and warn the driver or perform certain functions to assist the driver)
2. Smaller cars—more emphasis on styling and high fuel economy
3. Crossover-type vehicles that can fill the needs met by the SUV and minivan styling but with higher fuel economy
4. Better fitting interiors (greater occupant accommodation, e.g., comfortable driver space that accommodates higher percentage of users by means of adjustable seats, pedals, and tilt/telescoping steering columns)
5. More storage spaces for items to be stored in the vehicles
6. Crash avoidance systems (e.g., improved braking, handling, stability, driver assistance systems [e.g., drowsy driver or alertness monitoring and warning systems], smart lighting and visibility systems)
7. Passive safety systems (e.g., smart airbags, pretensioning belts, side curtains)
8. Comfort systems (e.g., improved climate control systems, heated/cooled seats)
9. Convenience features (e.g., more personalized memory settings of favorite radio stations, seats, mirrors, blind spot sensors, air-conditioning, display options and menus, reconfigurable seating, parking aids, cruise controls)
10. State-of-the-art entertainment systems (e.g., radio with AM/FM, CD, satellite radio, plug-in capabilities for aftermarket entertainment and gaming systems for the rear seat occupants)
11. Information and communication systems (e.g., Bluetooth or other wireless connectivity to cell phones and other handheld devices for a variety of functions, Internet search and e-mailing, navigation systems with real-time traffic, file/data management and storage, vehicle diagnostic system)

The challenges of ergonomics engineers are to work with other team members in meeting the above customer needs by implementing technologically feasible features and simultaneously meet the corporate business needs (primarily keeping the costs and timings under control).

FUTURE RESEARCH NEEDS AND CHALLENGES

Enabling Technologies

Telematics is perhaps the key set of enabling technologies for new features in the driver information interface area. Telematics can be defined as a discipline that has emerged from the coming together of the electronics, communications, and information technologies. The French word "télématique" was coined in the 1970s to denote the combination of "telecommunications" and "informatique 'computing.'" Thus, automotive telematics is an interdisciplinary applied field that deals with the development of devices for communication of information within the vehicle and other external sources through integrated applications of technologies involving wireless data transfer, sensors, microprocessors, databases, displays, lighting, controls, etc., for convenience of vehicle users. The automotive telematics, thus, facilitates creation of a digital car.

Many features or functions in future in-vehicle devices can be created by combining the above capabilities. For example, by combining the GPS capability and databases on locations of gas stations, their service capabilities, and prices, a display screen can provide the driver information on upcoming gas stations on the navigation screen along with their special service capabilities.

Thus, the telematics provides a platform for services and content that could offer value to the vehicle users through combinations of the following:

1. Two-way wireless communication to external and within vehicle sources (e.g., Bluetooth phones)
2. Use of databases (on the Internet and other data centers)
3. Location technologies (GPS, cell phone)
4. Vision systems (with sensors, cameras, and displays with superimposed processed information)
5. Voice technologies (voice recognition, voice controls, and voice displays)
6. Reconfigurable driver interfaces (reconfigurable displays and multifunction controls)
7. Sensors, signal processors, and control units

Other specialized technologies contributing to the advances in-vehicle features can be categorized as follows:

1. Driver interface technologies: reconfigurable displays, touch displays, multifunction controls, voice technologies, steering wheel controls, etc.
2. Driver state measurements and assistance systems: drowsiness detectors, driver workload monitors, intoxicated (under influence of alcohol and/or drugs) driver detection and crash avoidance warning systems, etc.
3. Vehicle status information and diagnostics: hybrid powertrain energy flow display, state of battery charge in an electric vehicle, service alerts, etc.
4. Adaptive control systems: adaptive cruise controls, enhanced stability control, adaptive suspension systems, enhanced braking systems, etc.
5. Lighting technologies: smart headlamps, new technology signal lamps, and interior lighting (LEDs, light piping, electroluminance, etc.)
6. Quality, craftsmanship, and brand sensory perceptions: selection and applications of interior materials with pleasing tactile feel and visual harmony, controls with pleasing tactile feel (e.g., crispness and smoothness felt during operation of switches), and interior lighting for display legibility, locating objects, and pleasing visual effects inside the vehicle

CURRENTLY AVAILABLE NEW TECHNOLOGY HARDWARE AND APPLICATIONS

Many technology applications presently in use or close to near-term introductions are briefly described below.

1. Digital LCD instrument clusters: LCD displays allow for crisper high-contrast graphics and reconfigurability (i.e., depending on the driving situation, different information such as gages, status of different vehicle or trip parameters, or camera views can be presented).
2. Three-dimensional displays: These displays allow presentation of information that can appear to be located at different distances. The apparent depth of objects in the displays can be used for functional grouping of displayed items.
3. Touch screens and multitouch technology: The user can activate different display modes by touching different control areas (e.g., touch buttons) on the screen. Multitouch technology

will allow recognition of multiple touch areas and movements of the touching fingers into preprogrammed control actions (e.g., selecting certain screens or functions).

4. Steering-wheel-mounted controls: Push buttons, rockers, or rotary controls mounted on the steering wheel spokes or hub areas allow the drivers to select and operate the controls without moving their hands away from the steering wheel.

5. Multifunction controls: A combination control that involves multiple controls or a single control with multiple functions assigned to its different activation movements (e.g., pushing, rotating and/or moving in different directions like a joystick with different haptics feedbacks for different switch modes).

6. Projected displays: Projectors mounted behind one or more screens can show (back-projected) displays. The screen can include touch areas as well as programmable (soft) labels for some hard controls.

7. Head-up display: A display projected on a partially reflective glazing surface to form an image focused at a farther distance and located such that it can be viewed by the driver without turning his or her head down (i.e., maintaining the head-up orientation).

8. Dual-view screen: A display screen that when viewed from different directions (e.g., from the driver eye location and a front passenger eye location) provides different images. For example, a center-stack-mounted dual-display screen can provide the driver a navigation display, whereas the front passenger can view a video entertainment channel.

9. Flexible displays: Displays are mounted on a flexible substrate (like a paper). The flexible displays can be formed and applied (or wrapped) on any complex-shaped surface (non-flat). Also, the flexible display can be folded or rolled (like a window shade) for storage purposes.

10. Navigation map display formats: The map can be presented in different formats (e.g., traditional map, map oriented to conform to the forward direction of travel, or a bird's-eye-view perspective with roadside objects or landmarks).

11. Rear passenger entertainment systems: The displays for the rear entertainment system can be roof-mounted flat panel displays, separate flat panel screens for each designated sitting position, or handheld/portable devices that can present TV programs, DVDs, or outputs of game players.

12. Satellite radio: The services offer coast-to-coast, mostly commercial-free radio with a large variety (from rap to opera, sports to children's programming). Currently, XM and Sirius each offer many channels.

13. OnStar System (from General Motors): This system involves a control panel with push buttons mounted inside the vehicle within the driver's reach and vision zones. The OnStar center is contacted with a push button or automatically in case of an accident (through a cellular call). The OnStar center's advisor can contact the driver to provide assistance under a number of situations such as when an air bag is deployed, stolen vehicle tracking (help police with vehicle location), remote door unlock, driving directions, roadside assistance, remote diagnostics, personal calling (hands-free call with voice controls), remote horn and light activation (flashes lights to find the car), accident assist (e.g., it calls police), ride assist (e.g., it calls cab), online concierge (e.g., locates a restaurant), personal concierge (e.g., helps getting tickets), etc. Several other vehicle manufacturers offer systems with similar features.

14. SYNC (from Ford Motor Company): The SYNC is an in-car connectivity system that allows front-seat occupants to operate most popular MP3 players, Bluetooth-enabled phones and USB drives with simple voice commands. The SYNC features include turn-by-turn directions, 911 assist, vehicle health reports, news, sports and weather, real-time traffic, business search, etc.

15. Night vision system: The night vision system helps the driver detect objects before they are visible to the driver under the headlamp illumination. It uses an infrared camera to view

objects beyond the driver's visible areas with headlamps and presents the view to the driver on an instrument-panel-mounted screen or through a head-up display.

16. Lane departure warning system: The system provides warning alerts when the vehicle leaves its lane.

17. Forward collision warning system: The system uses a combination of satellites, radar, and/ or electronic sensors to determine if a driver is approaching a slower or stopped vehicle (or a fixed object in the vehicle's path) too quickly. It alerts the driver with a series of beeps and visual signals on an interior display. The signal warns the driver to brake or make an evasive maneuver.

18. Adaptive cruise control: The cruise control that adjusts the speed of the vehicle to follow a leading vehicle at a preset distance (headway) and maintain speed at or below a preset speed.

19. Rear and side vision aids: These systems provide warning signals if another vehicle or an object (e.g., pedestrian) is within any of the driver's maneuvering or blind areas. Examples of such systems are (a) rear object detection systems, (b) rear and side camera systems, and (c) blind-area (blind-spot) detection systems.

20. Voice controls/recognition systems: The voice systems can (a) recognize spoken words and select settings of control functions, (b) reduce driver's workload by eliminating eye and hand movements to controls, that is, provide for hands-free operation, (c) present long voice messages from the vehicle, and (d) use only selected vocabulary for a faster response. Some problems with such systems are slower response time and poor voice recognition accuracy, especially under noisy moving vehicle environment.

21. Text-to-speech conversion systems: The driver can select different languages as well as voice, for example, male/female. Several text-to-speech software implementations can present voice in different languages and dialects such as U.S. English, Continental French, Latin American Spanish, U.K. English, German, Japanese, Brazilian Portuguese, Americas Spanish, Australian English, and Canadian French.

22. Driver state monitoring systems: The systems can detect the driver's state of alertness by monitoring driver's control actions (e.g., steering wheel movements), eye closures, or physiological state.

23. Memory seats: The seat track location and settings of seat cushion and seatback angles, contouring, and padding can be memorized and readjusted for entry/egress convenience and seating comfort.

24. Heated and cooled seats: These seats can improve the occupant's thermal comfort by adjusting temperatures of seat cushion and seat back.

25. High-intensity discharge headlamps: High-intensity discharge light sources can provide more light flux at reduced wattage and thus can provide beam patterns with improved night visibility as compared with the conventional tungsten filament light sources.

26. Rain sensor wipers: This wiper control system can reduce driver involvement in starting, stopping. or resetting of wiper speeds. The system sensor senses rainfall intensity and adjusts wiper speeds for improved visibility.

27. Remote tire pressure sensing: This system provides warning messages when the air pressure in any of tires is below a preset pressure level.

28. Advanced climate controls: This system provides improved thermal comfort by adjusting temperature, air flow location, direction, flow rates, etc., by measurements from a number of sensors (e.g., temperature and radiant heat sensors).

29. Bluetooth communications: This is a technology that allows users to wirelessly connect electronic devices to their cars and perform tasks like make hands-free calls without having to physically plug the devices into the vehicle.

30. Data transfers: These systems will allow transfer of data wirelessly between the vehicle and other locations such as (a) other vehicles, (b) roadside transmitters and receivers, (c) dealers

and vehicle manufacturers, (d) traffic and weather information centers, (e) other service centers for locations of gas stations, restaurants, banks, stores, etc.

31. Vehicle stability and control systems: These systems improve directional stability of vehicles during maneuvers by selectively controlling wheel speeds and suspension characteristics (if adaptive suspensions are used).

32. Parking aids: These systems will provide the driver warning signals during parking or perform the parallel parking task automatically.

A POSSIBLE TECHNOLOGY IMPLEMENTATION PLAN

A group of six graduate students in the author's Vehicle Ergonomics II class in 2010 were given an assignment to study the technology trends and customer needs and determine feature contents in future vehicles. The students were asked to study an existing late model light vehicle (a passenger car, a light truck, or an SUV) and develop a technology implementation plan for the selected vehicle. Two time frames were considered for the assignment, which included (a) a near-term model that could be introduced in the next 3–5 years and (b) a far-term model that could be introduced in the next 5–10 years. Table 17.1 summarizes the outcomes of the assignment. Since the feature content will depend on whether the selected vehicle is an economy or a luxury version and whether each feature be offered as standard or optional content, the recommendations for the future features were classified as SE (standard economy), SL (standard luxury), OE (optional economy), and OL (optional luxury). Table 17.1 presents 19 features considered in the assignment and their allocation to the near- and far-term vehicle models using the above classification. The table also provides important ergonomic issues, ergonomic advantages, and ergonomic disadvantages considered of each of the 19 features.

The ergonomics engineers are often asked to participate in product planning activities and prepare a feature implementation plan as shown in Table 17.1 for technology planning and feature content planning for future automotive products.

QUESTIONS RELATED TO IMPLEMENTATION OF THE TECHNOLOGIES

The number of possible features that can be introduced in future vehicles will increase rapidly with advances in technologies. The ergonomics engineers will need to consider many issues during designing the systems and deciding whether the benefits that can be claimed would be indeed realized and the costs and disadvantages with their introduction can be minimized.

Many of the issues will require decisions related to details such as the following:

1. Locations of controls and displays
2. Selection of types of controls, displays, and their layouts
3. Driver understanding and operation of the device
4. Compatibility with driver expectations, ease of use, and distracting effects during operation
5. Effect of the new features on operation of other driver interfaces (e.g., it may obstruct an existing control, cause delays or interruptions in operation of other systems, etc.)
6. Priorities in displaying the information related to the new features with respect to information displayed from other features
7. Sharing or reconfiguration of functions within multifunction controls and displays and compatibility between different sensory modalities of displays (e.g., visual, auditory, tactile/touch)
8. Reducing complexity of the features/functions and the resulting driver workload

TABLE 17.1
Technology Implementation and Ergonomic Considerations

No.	Product Feature or System	Implementation Time		Ergonomic Issues	Ergonomic Advantages	Ergonomic Disadvantages
		Near Term (3–5 Years)	Far Term (5–10 Years)			
1	Reconfigurable instrument cluster	SL	OE, SL	Display content, choice of predefined option groupings to customizable displays	Designed for personal preferences, reduced glance durations	Complexity due to more choices
2	Touch screen in the center stack	OE, SL	OE, SL	Legibility, comprehension, visual clutter, feedback on completed action	Simple pointing-finger movement, contact grasp, additional coded feature possibility from multitouch operations	Driver workload, menu structure, pointing errors due to vehicle movements
3	Head-up display	OL	OE, SL	Location, content, luminance, masking of other visual cues and possible distraction	Keep eyes on the roadway, reduced eye time to read displays, easier for older drivers	Masking part of forward field, distracting effects
4	Voice controls	OE, OL	OE, SL	Voice recognition error rates, recognition time	Reduces visual load	Voice recognition accuracy, recognition delays
5	Navigation system with real-time traffic information and alternate route planning	OE, OL	OE, SL	Display content, menus, controls, driver workload	Reduced travel time and reduced driver fatigue	May add more steps and time in route selection; Increased driver workload
6	Data storage (hard disk or other memory)	OL	SL	Type and amount of data to be stored and retrieval procedure, menu structure, use of past/historic data for personalization	Can reduce driver workload and uncertainty by displaying needed information, choice in accessing data according to individual preferences	Added complexity due to data storage and retrieval procedures, options, and controls
7	Rear passenger entertainment system	OL	OL	Additional controls for the driver (decision to delegate control to the rear passenger), placement and size of displays	Reduced conversations and distractions from rear passengers	May increase driver distractions, use of headphones, increased noise
8	Rear climate control system	OE, SL	SL	Additional climate controls for the driver (decision to delegate control to rear passenger)	Increased passenger comfort	Added feature complexity
9	Forward collision warning system	OL	OL	Effectiveness of the warning signal, driver acceptance/trust in the system	Reduced chances of collisions	May increase driver annoyance due to too many warning signals, increased reliance can give false sense of security
10	Lane departure warning system	OL	SL	Effectiveness of the warning signal, driver acceptance/trust in the system	Reduce chances of lane departures and run-off the road collisions	May increase driver annoyance and workload due to too many warning signals

continued

TABLE 17.1 (Continued)
Technology Implementation and Ergonomic Considerations

No.	Product Feature or System	Implementation Time		Ergonomic Issues	Ergonomic Advantages	Ergonomic Disadvantages
		Near Term (3–5 Years)	Far Term (5–10 Years)			
11	Blind area detection system	OE, OL	OE, SL	Warning signal effectiveness based on modality, location, intensity, frequency, and driver acceptance/trust in the system	Detection of targets in blind areas, reduced head movements and accidents related to obscured targets	High frequency of warning signals can lead to drivers disregarding the signals
12	Driver attention monitor	OL	OE, SL	Methods to measure alertness and their accuracy, driver's trust in the system	Alerts the driver before getting into hazardous situations	Increased reliance can give false sense of security
13	360-degree-vision monitoring system		OL	Location and content of the display showing views from multiple cameras	Increased driver awareness of objects around the vehicle (especially useful for commercial truck drivers)	Increased driver workload adds another complex display
14	LEDs in rear signal system	SL	SE, SL	Effectiveness of the signal due to variations in aspect ratios of lighted lamp areas	Improved visibility, quicker rise time, long life, energy savings	Increased variability between rear signals of vehicles in the forward field
15	LED headlamps	OL	OL	Variations in beam patterns between vehicles	Energy saving, can be used with smart head lighting concepts	Uneven luminance distribution on the pavement
16	Smart headlamps	OL	OL	Beam selection and switching criteria and controls, designing beam patterns, prediction of expected and unexpected targets	Improved visibility and reduced glare effects on oncoming drivers	May reduce visibility if smart headlighting does not get correct road and vehicle state information
17	Park assist system	OL	OL	Increase in driver controls and learning to trust the system	Reduced driver workload, parking accuracy	Overreliance on the system and less attentive driver
18	Automatic user profile memory for personalization	OL	SL	High within-subject variability can reduce effectiveness of the feature	Eliminate need to reset personal settings	Increased feature content can increase feature setting difficulty
19	Reclinable rear seats	OL	OL	Passenger accommodation and comfort, control location, range and speed of seat movements	Improve sitting comfort of rear occupants	Need to increase couple distance (longitudinal spacing between front and rear seating reference points)

Driver distraction is one of the key topics of concern to government agencies and the automotive manufacturers (e.g., Regan et al., 2009). Some questions raised during the new technology implementations and driver distraction are the following:

1. How do in-vehicle technologies influence driver distractions? What are the effects of distraction on driving performance and safety? How do distractions from in-vehicle technologies differ from and compare with distractions due to other sources (e.g., talking to other passengers or eating while driving, using cell phones)?
2. What are the methodological challenges in measuring the influence of design features of the devices, their operation, and their impact on driver distraction and driving performance?
3. What methods can be developed to relate measures of driver workload or distraction to the probability of an accident?
4. What actions can be taken by the government, the industry, and the consumers to minimize risks associated with different types of driver distraction?
5. What current and future research issues must be addressed to support actions to minimize driver distractions?

Naturalistic driving research (e.g., Klauer et al., 2006; Sayer et al., 2007) provides a unique means to understand real-world driver behavior and actual risks of crash and near-crash involvement. These data must be considered the means to validate studies conducted in experimental, laboratory, or simulator settings. It is important to keep in mind that a short-duration study in a laboratory or simulator setting can only provide data on driver performance. Driver behavior can only be understood when observed over time in the real world.

Some basic questions regarding the crash avoidance and warning systems include the following:

1. What is the proper time allowed for collision warning? Do multiple-stage warnings make sense for some types of crash hazards? What is an acceptable trade-off between false positive alarms and customer acceptance?
2. How much freedom should drivers have in selecting system functions for their own use?
3. How should multiple warnings for multiple crash threats be coded, combined, or sequenced?
4. How does new technology modify the driver's baseline levels of performance and behavior?

OTHER RESEARCH NEEDS

There are research needs in both the areas of development of improved products and improved methods for evaluating the products to assure that future features will be perceived by the customers to be useful and value-added. The research needs are presented below:

1. Improved anthropometric databases: Larger anthropometric databases involving both additional measurements and measurements of different populations of drivers and users in different market segments are needed. Recent advances in laser scanning equipment such as that used in the CAESAR (Civilian American and European Surface Anthropometry Resource) project administered by the Society of Automotive Engineers (SAE) could be used for collecting whole body scan data (Reed et al., 1999).
2. Tools to accommodate a large percentage of user population with reduced costs and timings:
    a. *Automated occupant packaging and visualization tools.* The tools can perform basic occupant accommodation assessment (e.g., driver positioning, body clearances,

location of major controls entry/egress) and conduct many specialized analyses such as field-of-view analyses for drivers in different populations (note that the present SAE standards are based on the U.S. drivers).

b. *Integrated digital car and digital manikins and visualization tools.* CAD tools with manikin models (digital human models), such as Jack/Jill, SAFEWORK, RAMSIS, SAMMIE, and the UM 3DSSP, are being currently used by different automotive designers to assist in the product development process (Badler et al., 2005; Chaffin, 2001, 2007; Human Solutions, 2010; Reed et al., 2003). Many of these tools are being updated to incorporate additional capabilities. Before using any of the models in the design process, the ergonomics engineer should conduct validation studies to determine if the postures assumed by the selected digital human model and their dimensional outputs indeed match closely with the postures and dimensions of real drivers under different actual usage situations.

c. *Programmable vehicle models (PVMs).* There are many different PVMs that are used in the early stages of vehicle packaging work. The models are physical bucks that can be configured very quickly in about 10–20 s by inputting selected package dimensions in a computer that controls many electric stepper motors used to adjust the buck (e.g., Prefix Corporation, 2010). Such PVMs with improved capabilities to configure different sizes and types of vehicles and interfacing of the bucks with CAD systems would be useful in evaluating a number of vehicle design configurations. The PVMs have the potential of reducing time and costs in the early vehicle development process.

3. Understanding the needs of diverse vehicle users (including special population issues covered in Chapter 16): Systematic studies on different vehicle users' characteristics beyond just the anthropometric and biomechanical characteristics are needed. Future databases with information on characteristics related to driver vision, audition, cognition, touch feel perception, driver familiarity and expectations, preferences on interior materials and operation of controls, etc., for different market segments (and also to create global vehicles) are needed.

4. Driver interface prototyping tools: Computer-aided tools to generate operational prototypes of future in-vehicle devices are needed for evaluation of alternate design concepts in instrumented vehicles and/or driving simulators (Fillyaw et al., 2008).

5. Driver interfaces for electric vehicles: As electric vehicles are expected to gain a larger market share, it is important that their designs be fine-tuned to address customer concerns and needs. Due to limitations of the current battery technologies, recharging needs, availability of electric vehicle servicing needs, and safety concerns, electric vehicles may be perceived and used differently than the traditional vehicles powered by internal combustion engines. Ergonomics engineers, thus, need to address the following four problems: (a) reducing driver uncertainties in using the electric vehicles. Mainly, assuring that the driver will get an accurate estimate of the state of charge of the battery and how far the vehicle can be driven, (b) minimizing energy consumption by providing information that can help the driver in adjusting or adapting his or her driving behavior, (c) minimizing vehicle operating costs, and (d) maintaining the vehicle in its best operating condition. A study by Bhise et al. (2010) provides additional information on the challenges in design and evaluation of driver interfaces for electric vehicles.

6. Methods to measure driver workload: Better methods that can measure driver workload, driver performance, and driver distraction are needed (see Chapters 14 and 15 for more information). Further, criteria levels on a number of driver performance and workload measures are needed to determine acceptable (safe) and unacceptable (unsafe) levels of driver workload during different driving situations.

7. Driver warning systems: New methods to warn and alert drivers are needed to improve effectiveness of future driver aids such as collision warning and driver alertness systems (e.g., drowsiness indicator).

8. Driver perception of quality and craftsmanship: Future research efforts need to be directed in understanding perception of quality and craftsmanship and their relationship to physical characteristics of interior materials, displays, and controls.

CONCLUDING REMARKS

Although a considerable number of research studies, design tools, and methods exist for the evaluation of many ergonomic issues in vehicle design, more research studies and tools are needed to design future vehicles efficiently. Advances in new technologies and changing driver expectations related to capabilities of future vehicle features, perception of quality, and craftsmanship will require ergonomics engineers to expand their knowledge and create new concepts, methods, and tools.

REFERENCES

Badler, N., J. Allbeck, S. -J. Lee, R. Rabbitz, T. Broderick, and K. Mulkern. 2005. New behavioral paradigms for virtual human models. *SAE Transactions: Journal of Passenger Cars—Electronic and Electrical Systems*. Paper 2005-01-2689. Presented at the 2005 SAE Digital Human Modeling Conference, Iowa City, IA.

Bhise, V., H. Dandekar, A. Gupta, and U. Sharma. 2010. Development of a driver interface concept for efficient electric vehicle usage. SAE Paper 2010-01-1040. Presented at the 2010 SAE World Congress, Detroit, MI.

Chaffin, D. B., 2001. *Digital Human Modeling for Vehicle and Workplace Design*. Warrendale, PA: SAE International.

Chaffin, D. B. 2007. Human motion simulation for vehicle and workplace design: Research Articles. *Human Factors in Ergonomics & Manufacturing*, 17(5).

Fillyaw, C., J. Friedman, and S. M. Prabhu. 2008. Testing human machine interface (HMI) rich designs using model-based design. SAE Paper 2008-01-1052. Paper presented at the 2008 SAE World Congress, Detroit, MI.

Human Solutions. 2010. RAMSIS model applications, accessed June 23, 2011, http://www.human-solutions.com/automotive/index_en.php.

Klauer, S., T. Dingus, V. Neale, J. Sudweeks, and D. Ramsey (2006). The impact of driver inattention on near/crash risk: An analysis using the 100-car naturalistic driving study data. DOT HS 810 594. Washington, DC: U.S. Department of Transportation, National Highway Traffic Safety Administration.

Parkes, A. M., and S. Franzen. (eds.). 1993. *Driving Future Vehicles*. London: CRC Press.

Prefix Corporation. 2010. Programmable vehicle model (PV), accessed June 23, 2011, http://www.prefix.com/PVM/.

Reed, M. P., M. B. Parkinson, and D. B. Chaffin. 2003. A new approach to modeling driver reach. Technical Paper 2003-01-0587. *SAE Transactions: Journal of Passenger Cars—Mechanical Systems* (112):709–718.

Reed, M.P., R. W. Roe, and L. W. Schneider. 1999. Design and development of the ASPECT manikin. Technical Paper 990963. *SAE Transactions: Journal of Passenger Cars,* 108.

Regan, M. A., J. D. Lee, and K. L.Young. (eds.). 2009. *Driver Distraction: Theory, Effects, and Mitigation*. Boca Raton, FL: CRC Press.

Sayer, J. R, J. M. Devonshire, and Flanagan, C. A. (2007). Naturalistic driving performance during secondary tasks. *Proceedings of the Fourth International Driving Symposium on Human Factors in Driver Assessment, Training and Vehicle Design,* Stevenson, WA.

Appendix 1

TABLE A1.1
Human Factors Engineering Historic Landmarks

Period	Reference Information	Year	Human Factors Historic Event
Evolutionary process		Early Man	Hand tools and utensils development
		1750	
Age of machines— steam power	James Watt developed condenser for steam engine.	1764	
		1828	Civil engineering was defined to be concerned with "the art of directing the great sources of power in nature to use and convenience of man."—from the Institute of Civil Engineers.
	3010 steam engines in use in the United States— Secretary of State report to the Congress	1838	
	First electric bulb—Edison	1879	
		1880	ASME prepared boiler safety standards.
		1887	Application of scientific method to work in industry—Fredric Taylor
		1890	
Power revolution	Ford's Model T Roadster	1908	
		1912	Frank Gilbreth presented a paper on micromotion study in ASME.
	Model Ts accounted for 50% of the world's motor vehicles.	1924	Harvard Fatigue Laboratory was established in the 1920s.
		1925	Hawthorne studies on relationship between lighting and productivity.
	World War II ended.	1945	Army–Air Force psychology program was established. First Chief of Psychology branch was Dr. Paul Fitts.
Knobs and dials science		1947	Publication of *Applied Experimental Psychology Journal*. Fitts and Jones published the famous paper on Pilot Errors in Operating Controls and Reading Displays.
		1948	Method–time–measurement was developed—human performance prediction model based on hand/body motion analyses
			First Human Factors Laboratory at Bell Labs was began under John Karlin.
		1949	Ergonomics—the word was invented by Murrell, British Navy Labs. Ergonomics Society was formed in Britain.

continued

TABLE A1.1 (Continued)
Human Factors Engineering Historic Landmarks

Period	Reference Information	Year	Human Factors Historic Event
		1950	Formation of Human Factors groups in industry—especially related to the air force/military
		1956	Human Factors Engineering department was established in the Ford Motor Company.
Product		1957	Incorporation of the Human Factors Society in the United States
liability cases begin to accelerate		1960	*Hennigsen vs. Bloomfield Motors* case—each product, by being out on the market for sale, bears an implied warranty that it is reasonably safe to use.
Strict liability		1963	*Greenman vs. Yuba Power Products* case—a manufacturer is strictly liable when the article that he or she places on the market, knowing that it will be sold without inspection for defects, proves to have a defect that causes injury to a human being.
Passage of safety laws		1966	National Traffic Safety and Motor Vehicle Safety Act; Highway Safety Act
		1968	Radiation Control for Health and Safety Act; Fire Research and Safety Act
		1969	Child Protection and Toy Safety Act
		1970	Occupational Safety and Health Act
		1970	*Thomas vs. General Motors* case—the manufacturer is liable for any or all foreseeable unintended uses, misuses, and abuses of the product and even abnormal uses which were foreseeable and could have been designed against or otherwise safe guarded.
		1971	Federal Boat Safety Act
		1972	Consumer Product Safety Act
		1973	*Balido vs. Improved Machinery* case—warnings and directions do not absolve the manufacturer of the liability if there is a defect in design or manufacture.
		1976	*NHTSA vs. GM* case—actual proof of harm is not required for a safety defect to be considered to exist in a vehicle. The government can therefore order defective products to be recalled even if the defect had not caused a collision or injury.
		1979	Three Mile Island Incident
Information age		1980	
	WIMPS (window, icons, menus, and pointers) interface		Macintosh Computers, PCs, Boeing 757
		1984	CRTs in cars
	Ford Taurus/Sable was introduced.	1986	
	User-friendly software issues	1990	The term "usability" was introduced in the literature.
	Internet, cellular phones, and palmtops	1990s	Voice-activated controls; data gloves; prototyping
			Virtual-reality displays
		1995	Chrysler showcases Programmable Vehicle Buck during Super Bowl XXIX commercial

TABLE A1.1 (Continued)
Human Factors Engineering Historic Landmarks

Period	Reference Information	Year	Human Factors Historic Event
	New technologies	2000	Driver Distraction Due to New In-Vehicle Devices Considered as a Major Traffic Safety Problem by NHTSA: Telematics, Bluetooth, Haptics, OLEDs
			Increasing usages of digital cameras, Internet in palm PC, text pagers, iPods, cell phones, and plug-and-play devices (with USB) in cars
		2010	Capacitive touch screens in cars and SUVs

Appendix 2

This Appendix consists of the table on page 296: "Anthropometric Dimensions of Driving Populations from Seven Different Countries."

TABLE A2.1
Anthropometric Dimensions of Driving Populations from Seven Different Countries

Anthropometric Dimension	Percentile Value	Japanese Males	Japanese Females	Hong Kong Males	Hong Kong Females	Indian Males	Indian Females
Stature (mm)	95	1750	1610	1655	1655	1745	1615
	50	1655	1530	1680	1555	1640	1515
	5	1560	1450	1455	1455	1535	1415
Sitting height (mm)	95	950	890	955	900	905	850
	50	900	845	900	840	850	790
	5	850	800	845	780	795	730
Sitting eye height (mm)	95	835	780	840	780	795	745
	50	785	735	780	720	740	690
Sitting shoulder height (mm)	5	735	690	720	660	685	635
	95	635	600	655	610	690	565
	50	590	55605	605	560	640	515
	5	545	510	555	510	590	465
Shoulder breadth (bideltoid) (mm)	95	475	425	470	435	455	390
	50	440	395	425	385	415	350
	5	405	365	380	335	375	310
Hip breadth (mm)	95	330	340	370	365	365	390
	50	305	305	335	330	320	330
	5	280	270	300	295	275	270
Buttock-to-knee length (mm)	95	600	575	595	570	610	590
	50	550	530	550	520	560	540
	5	500	485	505	470	510	490
Knee height (mm)	95	530	480	540	500	565	515
	50	490	450	45	455	515	470
	5	450	420	450	410	465	425
Shoulder-to-elbow length (mm)	95	365	330	370	340	370	340
	50	330	300	340	315	340	310
	5	295	270	310	290	310	280
Elbow-to-fingertip length (mm)	95	475	430	480	440	485	435
	50	440	400	445	400	450	405
	5	405	370	410	360	415	375

Note: Anthropometric dimensions from Bodyspace by Stephen Pheasant (2006).

TABLE A2.1
Anthropometric Dimensions of Driving Populations from Seven Different Countries

Anthropometric Dimension	Percentile Value	US Males	US Females	UK Males	UK Females	German Males	German Females	French Males	French Females	Japanese Males	Japanese Females	Hongkong Males	Hongkong Females	Indian Males	Indian Females
Stature (mm)	95	1870	1730	1855	1710	1845	1750	1830	1700	1750	1610	1775	1655	1745	1615
	50	1755	1625	1740	1610	1745	1635	1715	1600	1655	1530	1680	1555	1640	1515
	5	1640	1520	1625	1505	1645	1520	1600	1500	1560	1450	1585	1455	1535	1415
Sitting height (mm)	95	975	920	965	910	975	930	970	910	950	890	955	900	905	850
	50	915	860	910	850	920	865	910	860	900	845	900	840	850	790
	5	855	800	850	795	865	800	850	810	850	800	845	780	795	730
Sitting eye height (mm)	95	860	810	845	795	850	800	855	800	835	780	840	780	795	745
	50	800	750	790	740	800	740	795	750	785	735	780	720	740	690
	5	740	690	735	685	750	680	735	700	735	690	720	660	685	635
Sitting shoulder height (mm)	95	655	620	645	610	640	570	670	625	635	600	655	610	690	565
	50	600	565	595	555	595	525	620	580	590	555	605	560	640	515
	5	545	510	540	505	550	480	570	535	545	510	555	510	590	465
Shoulder breadth (bideltoid) (mm)	95	515	440	510	435	505	445	515	470	475	425	470	435	455	390
	50	470	400	465	395	465	400	470	425	440	395	425	385	415	350
	5	425	360	420	355	425	355	425	380	405	365	380	335	375	310
Hip breadth (mm)	95	410	440	405	435	385	445	410	430	330	340	370	365	365	390
	50	360	375	360	370	350	375	370	380	305	305	335	330	320	330
	5	310	310	310	310	315	305	330	330	280	270	300	295	275	270
Buttock-to-knee Length (mm)	95	650	625	645	620	640	635	640	610	600	575	595	570	610	590
	50	600	575	595	570	600	580	595	565	550	530	550	520	560	540
	5	550	525	540	520	560	525	550	520	500	485	505	470	510	490
Knee height (mm)	95	605	550	595	540	590	555	575	535	530	480	540	500	565	515
	50	550	505	545	500	545	505	530	495	490	450	495	455	515	470
	5	495	460	490	455	500	455	485	455	450	420	450	410	465	425
Shoulder-to-elbow length (mm)	95	400	365	395	360	395	365	395	360	365	330	370	340	370	340
	50	365	335	365	330	365	335	360	330	330	300	340	315	340	310
	5	330	305	330	300	335	305	325	300	295	270	310	290	310	280
Elbow-to-fingertip length (mm)	95	515	470	510	460	505	470	505	455	475	430	480	440	485	435
	50	480	435	475	430	475	435	470	425	440	400	445	400	450	405
	5	445	400	440	400	445	400	435	395	405	370	410	360	415	375

Note: Anthropometric dimensions from Bodyspace by Stephen Pheasant (2006).

Appendix 3: Verification of Hick's Law of Information Processing

An Excel-based computer program is included in the publisher's website to allow the reader to measure his or her choice reaction times under different equally likely alternatives ranging from 1 to 8.

The procedure to use the program is as follows:

1. Click on the icon "Reaction Time Program" (see Table A3.1 for the screen).
2. Click on "Reaction Time Setup" box (enable macros, if prompted, and set lowest security level through the Trust Center and via "Excel Options").
3. Select number of choices (N) (choose between 1 and 8).
4. Select repetitions (number of trials) (use at least 30 repetitions).
5. Set minimum and maximum delay time (minimum = 1 s; maximum = 3 s)
6. Click on "OK."
7. Click on "Ready."
8. Now, a random number between 1 and the selected number of choices will appear in the white box (to the left of the "Reaction Time Setup" box). Your task is to push the number key corresponding to the displayed number as quickly as possible. Your reaction time (RT; the time between when the number appeared and when you pressed the key) will be displayed.
9. Press "Ready" button as in Step 6 and repeat the procedure until all the repetitions are completed.
10. Note down your data (average and standard deviation).
11. Repeat the above process for a different number of choices (repeat all above steps).
12. Plot the average reaction time versus $\log_2 N$ and fit a straight line through the data points.
13. The y intercept of the line corresponds to constant a, and the slope of the line is equal to the reciprocal of constant b (note: Hick's Law: RT = $a + b \log_2 N$).

Figure A3.1 presents reaction time data obtained by using the above program. The data were obtained by asking a practiced and alerted subject to complete thirty trials in each session where the numbers of choice were 1, 2, 4, and 8. The figure shows mean, mean plus and minus one standard deviation and also a straight line fitted to the data.

TABLE A3.1

Screenshot of the Reaction Time Program

Reaction Time Demonstration		

3

Reaction Time Setup

Ready

Reaction Time =	1.047		Reaction Time Log				
		Trial #		Trial #			
Average =	1.025	1	0.938	9	0.984		
standard deviation =	0.110	2	0.934	10	1.188		
		3	1.027	11	1.109		
number of errors =	0	4	0.859	12	1.047		
		5	1.266	13			
Min	0.859	6	0.938	14			
Max	1.266	7	0.984	15			
		8	1.031				

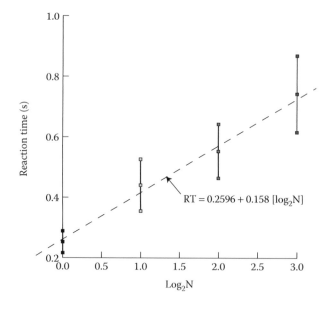

$$RT = 0.2596 + 0.158\,[\log_2 N]$$

FIGURE A3.1   Reaction time data obtained by using the program.

Appendix 4: Verification of Fitts' Law of Hand Motions

Fitts' Law can be verified by collecting measurements on hand movements between circular targets separated by a given distance. Figures A4.1a and A4.1b present 12 targets labeled as T 1 to T 12 that will be used in this experiment. Each target consists of two circular areas placed apart by a distance. The 12 targets are created by a combination of diameters of the circles (W) and distance between the centers of the circles (A). W is varied between 1/8″ and 1″ and A is varied between 2″ and 8″. To conduct this experiment, you will need to first create a magnified copy of Figure A4.1 such that the distance between the centers of the circles of target T#9 is 8″.

In this experiment, a subject is asked to sit down in a chair in front of a desk and asked to place a given target (selected randomly from the 12 targets) on the desk at a comfortable distance and orientation in front and given a pencil to make hit marks in the two circular areas by moving his or her hand back and forth between the circular areas of the targets. The subject's task would be to move his or her hands between the circular areas as fast as possible in a given time interval (5 s) and count the number of hit marks in the interval. The hit mark should not fall outside any of the circular areas. An experimenter should be used to monitor the subject and give the subject "begin" and "end" instructions by using a stopwatch. The subject, thus, is asked to first place his or her pencil inside one of the circles and when told by the experimenter to "begin," make hand movements back and forth between the circles and keep track of number of back-and-forth movements (i.e., by counting "number of times the pencil hits the paper"). The hand movement time is calculated by dividing the time interval (5 s) by the number of hits.

The experimenter should give the subject a few targets to practice the task first and then ask the subject to participate in a number of trials. A different target should be used in each successive trial.

The hand movement time for each target should be plotted against the index of difficulty ($\log_2 [2A/W]$) of the target. If all the data points fall close to a fitted straight line, then Fitts' Law can be considered to be verified.

Figure A4.2 illustrates data obtained by asking one subject to follow the above procedure for four targets, with the index of difficulty ranging from 3 to 6. The figure shows mean and mean plus and minus one standard deviation values of the movement time and a straight line fitting the data.

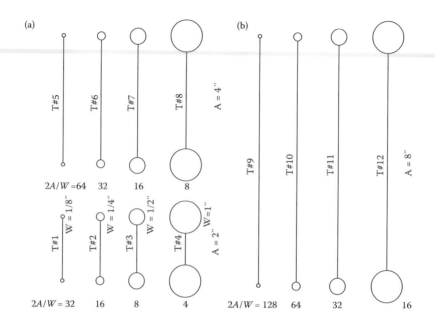

FIGURE A4.1   Targets for verification of Fitts' Law of hand motions.

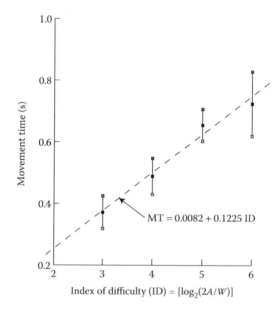

FIGURE A4.2   Illustration of data obtained from the experiment.

Index

Note: Page numbers followed by *f* and *t* refer figures and tables, respectively.

A

Accelerator heel point (AHP), 26, 33
"Accepted" vehicle concept, development of, 31–33
Accident-based safety performance measures, 226
Accommodation, 54–55
Adaptation luminance, 203
Adaptive cruise control, 283
Adjustable seat, 26
Advanced climate controls, 283
Advanced vehicle design stage, 30–31
Air force research, 8
Ambinocular field, 68
Ambinocular vision, 108–111
Analytical hierarchical method, 265–267
Ankle angle, 38
Annoying reflections, 100–101
Anthropometric characteristics, 10
Anthropometric databases, 287
Anthropometric dimensions of driving populations
    from different countries, 296*t*
Anthropometry, in designing vehicles, 13
    percentile values, computation of, 18–19
Armrest height, 27
Armrests, effect of, 22
Aspherical mirrors, 121
Audiometers, 71
Auditory display, 84
Automated occupant packaging and visualization tools,
    287
Automotive exterior interfaces, *see* Exterior interfaces,
    automotive
A-weighting filter, 71

B

Ball of foot (BOF), 34
    and accelerator heel point (AHP)
        horizontal distance between, 43
        horizontal length between, 43
        vertical height between, 43
Behavioral measures, of driver, 226, 228, 246
Belt height, 39, 39*f*
"Between-subject variability," 10
Binocular vision, 108–111, 110*f*
Biomechanical characteristics, 10
Biomechanics applications, in vehicle design, 13, 19
    basic biomechanical considerations, 20–21
    in seat design, 22–25
    seat design considerations related to driver
        accommodation, 25–28
Blackwell contrast threshold curves, 203
Bluetooth communications, 283
BOF-to-SgRP dimension, 43
Bolster height, 27

C

Cameras and display screens, 126
Capless fuel fillers, 171, 172*f*
Cell phones and MP3 players
    hand held versus voice interfaces for, 244
Center speedometer and low radio location, 101
Characteristics of ergonomically designed products,
    systems, and processes, 9
Checklists, 267–268
CHMSL fleet study, 147–150
Choice reaction time, 58–60
Cleanability, 176
Cognitive task (COTA), 248
Comfort and convenience, 178, 182
Comfortable seated posture, 30
Command sitting position, 112
Commission error, 64
Communication method, 174, 176, 254
Comprehensive headlamp environment systems simulation
    (CHESS) model, 142, 143, 143*f*, 144*t*
Cone receptor, 52, 53*f*
Contact points, 162
Continuous versus discrete controls, 79
Contrast multiplier, 205
Contrast threshold, 204
Control, evaluation of
    checklist for, 93–94*t*
Control and display location principles, 88
    frequency of usage principle, 89
    functional grouping principle, 89
    importance principle, 89
    location expectancy principle, 88
    sequence of use principle, 88
    time-pressure principle, 89
Control design considerations, 86
    control size, 86–87
    error-free operation, 87
    identification, 86
    inadvertent operation, 87
    interpretation, 86
    location, 86
    operability, 87
    visibility, 86
Controls and displays interface, 77
    characteristics of good control, 78
    characteristics of good visual display, 78
    design considerations, issues, and location
        principles, 85
        control and display location principles, 88
        control design considerations, 86
        general design considerations, 85
        visual display design considerations, 87
    examples of, 99
        annoying reflections, 100–101

center speedometer and low radio location, 101
door trim panel layout, 101–102
hard-to-read labels and difficult-to-operate radio, 101
power-window location, 99–100, 100*f*
speedometer graphics design, 99, 99*f*
methods to evaluate, 89
checklists for evaluation, 92
ergonomics summary chart, 92, 96*f*
space available for location, 90
types of, 79
in-vehicle controls, 79
in-vehicle displays, 83
Craftsmanship, automotive, 177
attributes of, 177, 180
comfort and convenience, 182
harmony, 181
smell quality, 182
sound quality, 181
touch feel quality, 181
visual quality, 181
importance of, 177
Kano model of quality, 179–180
measurement methods, 182
ring model of product desirability, 178–179
steering wheels, craftsmanship characteristics of, 182–184
Crashworthiness, 19
Critical tracking task (CTT), 247
Curvature, avoiding
in seatback, 25
in seat cushion, 24
Cushion, long seat, 23
Cushioning, 27

D

Dangling feet, avoiding, 23, 24*f*
Data collection methods, 254
communication method, 254, 257
experimentation method, 254, 257*t*, 258
observation method, 254, 257–258
type of measurements, 257*t*
Data collection situations, 253
Data transfers, 283–284
DeBoer discomfort glare index, 209, 211*t*
Dedicated versus programmable display, 84
Definition of ergonomics, 3
Delighters, 179
Descriptive research, 7
Design considerations, issues, and location principles, 85
control and display location principles, 88
control design considerations, 86
general design considerations, 85
visual display design considerations, 87
Design guidelines and requirements, 172–173
Design process, trade-offs in, 194
Design standards, internal, 172
Detection error, 64
Digital LCD instrument clusters, 281
Digital manikins, recent advances in, 28
Disabilities, drivers with, 275–276
Disassociated radio controls with silver background, 101*f*

Discomfort glare prediction, 209–211
Discrete controls, 79
Discrimination error, 64
Dissatisfiers, removal of, 179
Donders' Subtraction Principle, 57, 57*f*
Door and hinge angles, 160–161
Door handles, 156–157
Door trim panel layout, 101–102
Driver accommodation and seat design, 25–28
Driver behavior and performance, comparison of, 245–246
Driver errors, 77, 132, 231–232
Driver information acquisition and processing, 51
applications of information processing for vehicle design, 72–73
driver vision considerations, understanding, 52
accommodation, 54–55
human eye, structure of, 52–54
visual information acquisition in driving, 54
human errors, 64
error, definition of, 64
and SORE model, 65, 65*f*
types of, 64–65
information acquired through other sensory modalities, 71
human audition and sound measurements, 71–72
information processing, 55
generic model of, with three memory systems, 63–64
human information-processing models, 56–62, 63*f*
human memory, 62–63, 62*f*
issues and considerations, 55–56
occlusion studies, 68–70
psychophysics, 65
time, importance of, 51–52
visual capabilities, 66
driver's visual fields, 68
visual acuity, 67–68, 68*f*
visual contrast thresholds, 66–67, 66*f*
Driver interface prototyping tools, 288
Driver interfaces for electric vehicles, 288
Driver package development procedures, 42–49
Driver perception of quality and craftsmanship, 288
Driver performance measurement, 223, 237, 238*t*
applications, 232
characteristics of, 223–224
driving and nondriving tasks, 224–225
errors, 231–232
glance durations, 230–231
in literature, 227
number of glances, 230–231
objectives and driving performance measures, 225–226
range of, 228
standard deviation
of lateral position, 228–229
of steering wheel angle, 229
of velocity, 229
total task time, 230–231
types and categories, 226–227
vehicle speed, 229
Driver state monitoring systems, 283
Drivers with a tall torso, complaints of, 156
Drivers with disabilities and functional limitations, 275–276

Drivers with long legs, complaints of, 156
Drivers with short legs, complaints of, 155
Driver vision considerations, understanding, 52
   accommodation, 54–55
   human eye, structure of, 52–54
   visual information acquisition in driving, 54
Driver vision models, 199
   discomfort glare prediction, 209–211
   legibility, 211
      factors affecting legibility, 211–212
      modeling, 212–214
   light measurements, 200
      light measurement units, 200–201
      photometry and measurement instruments, 202
   systems considerations related to visibility, 199–200
   in vehicle design, 199
   veiling glare effects, 214
      evaluation by design tool, 214–216
      veiling glare prediction model, 216, 217*t*
   visibility computation, steps in, 206–209
   visual contrast thresholds, 202
      Blackwell contrast threshold curves, 203
      computation, 203–204
      effect of glare on, 205–206
      and visibility distance, computation of, 204–205
Driver warning systems, 288
Driver workload measurement, 235, 288
   driving and nondriving tasks, 235–236
   illustrations
      cell phones and MP3 players, hand held versus
         voice interfaces for, 244
      driver behavior and performance, comparison of,
         245–246
      ISO LCT, applications of, 246–249
      navigation systems, destination entry in, 243–244
      text messaging during simulated driving, 245
   mental workload, concepts of, 236–237
   methods, 237–243
   present situation, in industry, 236
Driver's field of view, determining
      through mirrors, 119–121
Driver's side mirror field, 118*f*
Driver's visual fields, 68
Driving simulator studies, 269
Driving tasks, 224–225
   assessment, 235
Dual-view screen, 282
Dynamic display, 83

E

Early design process, steps in, 191–194
Elbow room, 40, 41*f*
Electric vehicles, driver interfaces for, 288
Engine compartment service, 169–171
Engineering anthropometry, in designing vehicles, *see*
      Anthropometry, in designing vehicles
Engineer's responsibilities, in vehicle design, 7
Engine oil, checking, 169
Entrance height, 39, 39*f*
Entry and egress space, 29–30
Entry and exit from automotive vehicles, 155
   effect of vehicle body style on, 166–167
   evaluation methods, 163–164

problems in, 155–156
   task analysis, 164–166
   vehicle features and dimensions, 156
      body opening clearances from SgRP locations,
         160
      door and hinge angles, 160–161
      door handles, 156–157
      heavy-truck cab entry and exit, 162–163
      lateral sections at SgRP and foot movement areas,
         158–159
      running boards, 161–162
      seat bolsters, location, and materials, 161
      seat hardware, 161
      third row and rear seat entry from two-door
         vehicles, 162
      tires and rocker panels, 161
Equivalency-based measures, 226
Ergonomically designed product, 9
Ergonomics, application of, 9
Ergonomics, concept of, 3–4
Ergonomics approach, 4
   "designing for the most," 4
   "fitting equipment to the users," 4
   systems approach, 4–5
Ergonomics research studies, 7
"Ergonomics" coined, 8
Errors, *see* Human errors
Evaluative research, 7
Experimental research, 7
Experimentation method, 174, 176, 254, 258
Exterior interfaces, automotive, 169
   associated tasks and subtasks, 169–172
   methods and issues to study, 172
      checklists, 173
      loading and unloading tasks, biomechanical
         guidelines for, 173–174
      manual lifting models, applications of, 174
      methods of observation, communication, and
         experimentation, 174, 176
      standards, design guidelines, and requirements,
         172–173
      task analysis, 174, 175*t*
Extraneous act error, 64
Eye, human
   comparison with camera, 52*f*
   structure of, 52–54
Eyellipse
   defining44
   location of, 45*f*

F

Female drivers, 273–274
Field of view from automotive vehicles, 105
   definition, 105–106
   forward-field-of-view evaluations, 111
      command sitting position, 112
      obstructions caused by A-pillars, 115–116
      short driver problems, 112–113
      sun visor design issues, 113–114
      tall driver problems, 113
      up- and down-angle evaluations, 111
      visibility of and over the hood, 111–112
      wiper and defroster requirements, 114–115

linking vehicle interior to exterior, 105
methods to measure fields of view, 121
    polar plots, 122–124, 124*f*
mirror design issues, 116
    convex and aspherical mirrors, 121
    determining driver's field of view through mirrors,
       119–121
    inside mirror locations, 118–119, 118*f*
    outside mirror locations, 119
    requirements on mirror fields, 116
origins of data to support required fields of view,
    106
types of, 106
    360-degree visibility, 107–108, 108*f*
    monocular, ambinocular, and binocular vision,
       108–111
visibility issues, 125
    cameras and display screens, 126
    heavy-truck driver issues, 125–126
    light transmissivity, 125
    plane and convex combination mirrors, 125
    shade bands, 125
Field studies and drive tests, 269
Fitts' Law of hand motions, 61–62
    verification of, 299, 300*f*
Flexible displays, 282
FMVSS 111, 116, 119, 121, 128
Forward collision warning system, 283
Forward-field-of-view evaluations, 111
    command sitting position, 112
    obstructions caused by A-pillars, 115–116
    short driver problems, 112–113
    sun visor design issues, 113–114
    tall driver problems, 113
    up- and down-angle evaluations, 111
    visibility of and over the hood, 111–112
    wiper and defroster requirements, 114–115
Four-stage serial information-processing model,
    56, 56*f*
Fovea, 53, 53, 54, 68
Functional product design, 178
Future research and new technology issues, 279
    ergonomic needs in designing vehicles, 279
       passenger vehicles in near future, 280
    future research needs and challenges
       currently available new technology hardware and
          applications, 281
       enabling technologies, 280
       possible technology implementation plan,
          284
       questions related to implementation of
          technologies, 284
Futuring, 276–277

G

Glance durations, 230–231
Global vehicles, issues in designing, 276
Goal of ergonomics, 3
Goal of ergonomics engineers, 189
Good gripping of load, 174
Guessed answer, 5–6

H

H30 dimension, 26
H30 value, 42
Hand controls, types of
    in automotive products, 83
Hand held versus voice interfaces, 244
    for cell phones and MP3 players
Hand reach data, 46–47, 48
Haptics controls, 82
Hardness number, 183
Hard-to-read labels and difficult-to-operate radio, 101
Harmony, 181
Head clearance envelopes, 45–46, 45*f*
Head-down versus head-up visual displays, 84
Headlamps, 128
    design considerations, 129
       new technological advances, 131
       problems with current headlighting systems,
          130–131
       target visibility considerations, 130
    evaluation methods, 139–144
       computer visualization, 141
       dynamic field tests, 141
       photometric measurements and compliance
          evaluations, 139–140
       photometry test points and headlamp beam
          patterns, 134
       static field tests, 140–141
Headroom, 40, 40*f*
Head-up displays (HUDs), 84, 282
    advantages of, 84
    disadvantages of, 84
Heated and cooled seats, 283
Heavy-truck cab entry and exit, 162–163
Heavy-truck driver issues, 125–126
Hick's Law of Information Processing, 59
    verification of, 297–298
High-intensity discharge headlamps, 283
Hip angle, 38
Hip room, 40, 41*f*
History of ergonomics
    in automotive product design, 8–9
    origins and human factors engineering, 8
       air force research, 8
       "ergonomics" coined, 8
       prehistoric times and functional changes in
          products, 8
H-point device (HPD), 36, 36*f*
H-point location curves, 37*f*
H-point location model, 35, 35*f*, 37
H-point machine (HPM), 36
Human audition and sound measurements, 71–72
Human auditory capabilities, 71
Human characteristics and capabilities, 10
    information-processing capabilities, 10
    physical capabilities, 10
Human errors, 64
    error, definition of, 64
    and SORE model, 65, 65*f*
    types of, 64–65
Human eye, structure of, 52–54
Human factors engineering historic landmarks, 291–293*t*
Human hearing thresholds, 71

Human information-processing models, 56–62, 63*f*
    four-stage model, 56, 56*f*
    with three memory systems, 63–64
Human memory, 62–63, 62*f*
Human reaction time, 60–61

I

Illuminance, definition of, 201
Illumination
    computation of, 134
    definition of, 201
    measurement, 133
Impact protection, 19
Implementing ergonomics, 10–11
Importance of ergonomics, 9
Inadequate response error, 65
Incident sunlight and driver's age, effect of, 218
Information processing, 55
    applications of
        for vehicle design, 72–73
    generic model of, with three memory systems, 63–64
    human information-processing models, 56–62, 63*f*
    human memory, 62–63, 62*f*
    issues and considerations, 55–56
Information-processing capabilities, of human, 10
Inside mirror locations, 118–119, 118*f*
Instrument cluster through steering wheel, 87, 88*f*
Integrated digital car and digital manikins and
        visualization tools, 288
Interior and exterior areas, visibility of, 30
Interior dimensions, 38–42
Interior package reference points and dimensions, 34*f*
Interior spaces, evaluating, 13
Interpretation error, 64
Interval scale, 223–224
    examples of, 259, 260*f*
Interval scales, ratings on, 268
In-vehicle controls, 79
In-vehicle displays, 83
ISO Lane Change Test (LCT), 241–242
ISO LCT, applications of, 246–249
IVIS DEMAnD Model, 242, 230–231

K

Kano model of quality, 179–180
Key vehicle dimensions and reference points, definition
        of, 33
    interior dimensions, 38–42
    package dimensions, reference points, and seat-track-
        related dimensions, 33–38
    units, dimensions, and axes, 33
Knee angle, 38
Knee clearance, 42, 42*f*

L

L5/S1 region, load in, 22
Lamp beam pattern, 134
Lamp outputs, photometric measurements of, 133
    headlamp photometry test points and headlamp beam
        patterns, 134
    light measurement units, 133–134

low and high beam patterns, 134–135
    pavement luminance and glare illumination from
        headlamps, 135–139
    photometric requirements for signal lamps, 139
Lane departure warning system, 283
Lateral location of seat, 27
LCT (Lane Change Task) method, 246–249
Legibility, 211
    affecting factors, 211–212
    modeling, 212–214
Legibility error, 65
Leg room, 40, 40*f*
Lighting, automotive, 127
    headlamp evaluation methods, 139–144
        computer visualization, 141
        dynamic field tests, 141
        photometric measurements and compliance
            evaluations, 139–140
        static field tests, 140–141
    headlamps, 128
    headlighting design considerations, 129
        new technological advances, 131
        problems with current headlighting systems,
            130–131
        target visibility considerations, 130
    photometric measurements of lamp outputs, 133
        headlamp photometry test points and beam
            patterns, 134
        light measurement units, 133–134
        low and high beam patterns, 134–135, 137*f*
        pavement luminance and glare illumination from
            headlamps, 135–139
        photometric requirements for signal lamps, 139
    signal lamps, 129
    signal lighting design considerations, 131
        new technology advances and related issues,
            133
        problems with current signal lighting systems,
            132–133
        signal lighting visibility issues, 132
    signal lighting evaluation methods, 144
        analysis of accident data, 147
        CHMSL fleet study, 147–150
        field observations and evaluations, 144–147
        photometric measurements and compliance
            evaluation, 144
        SAE Vehicle Lighting Committee and Motor
            Vehicle Manufacturers Association,
            150–151
Light measurements, 200
    light measurement units, 200–201
    photometry and measurement instruments, 202
Light measurement units, 133–134
Light transmissivity, 125
Linking vehicle interior to exterior, 105
Loading and unloading tasks, biomechanical guidelines
        for, 173–174
Long seat cushion, 23, 24*f*
Long-term memory, 63
Loudness, 71
Low and high beam patterns, 134–135
Lumbar area, 27
Lumbar support, effect of, 22
Luminance, 133, 134, 201

Luminous (light) flux, 133, 201
Luminous energy, 201
Luminous intensity, 133, 201

M

Manual lifting models, applications of, 174
Market segments, 272–273
Maximum reach zone, 90–91
Mean deviation (MDEV), 242
Measurement instruments, 202
Measurement scales, 223–224
Memory seats, 283
Memory system, 62–63, 62*f*
    long-term memory, 63
    sensory memory, 62
    working memory, 62–63
Mental workload, concepts of, 236–237
Minimum reach zones, 91
Mirror design issues, 116
    convex and aspherical mirrors, 121
    determining driver's field of view through mirrors,
        119–121
    mirror locations, 118
        inside mirror locations, 118–119, 118*f*
        outside mirror locations, 119
    requirements on mirror fields, 116
Monitory measures, 26
Monocular vision, 108–111, 109*f*
Multifunction controls, 282
Multifunction switches, 81
Must-have features, 180

N

NASA-Task Load Index (TLX), 239, 240*t*, 246
Naturalistic driving research, 287
Navigation map display formats, 282
Navigation systems, destination entry in, 243–244
Near visual acuity, 67
Night vision system, 282–283
Nomadic task, 48
Nominal scale, 223
Nondriving tasks, assessing, 235
Normal distribution of random variable, 14
NORMDIST function, 18
Number of glances, 230–231

O

Objective measurements, 254, 257*t*
    and data analysis methods, 258
Observational studies, 268
Observation method, 174, 176, 254
Obstructions caused by A-pillars, 115
Occlusion studies, 68–70
Occupant packaging, 29
    driver package development procedures, 42–49
    key vehicle dimensions and reference points, definition
        of, 33
        interior dimensions, 38–42
        package dimensions, reference points, and seat-
            track-related dimensions, 33–38
        units, dimensions, and axes, 33

vehicle package, 29
    layout, 30*f*
    occupant package, developing, 29–30
    occupant package/seating package layout, 29
vehicle package, sequence in development of, 30
    "accepted" vehicle concept, development of, 31–33
    advanced vehicle design stage, 30–31
Older, obese, mobility-challenged drivers, complaints of,
    155–156
Older drivers, 274–275
Olfactory display, 84
Omission error, 64
OnStar System, 282
Operating controls, 30
Ordinal scale, 223
Origins of ergonomics and human factors
        engineering, 8
    air force research, 8
    "ergonomics" coined, 8
    prehistoric times and functional changes in
        products, 8
Outside mirror locations, 119

P

Package dimensions, 33–38
Packaging, *see* Occupant packaging
Padding, 27
Paired-comparison-based methods, 257, 259–262
Parking aids, 284
Passenger vehicle
    in near future, 280
    trunk, designing, 172–173
Pavement luminance and glare illumination from
        headlamps, 135–139
Pedal plane angle (A47), 34, 38, 43
Pedal reference point (PRP), 34
Percentile values, computation of, 18–19
Performance measures, characteristics of, 223–224
Personal interview, 257
Phon, 71
Photometric brightness, 201
Photometric requirements for signal lamps, 139
Photometry, 202
Photosensitive receptors, 52
Physical capabilities, of human, 10
Physical measures, 226
Physiological measurements, 226, 228, 237–238
Pictorial graphics, 83–84
Plane and convex combination mirrors, 125
Pleasing perception enhancements, 178
Polar plots, 122–124, 124*f*
Posture angles, 38
Power seats, 26
Power-window location, 99–100, 100*f*
Prehistoric times and functional changes in products, 8
Problem-solving methodologies, 5–6, 6*f*
Programmable display, 84
Programmable/reconfigurable switches, 81–82
Programmable vehicle models (PVMs), 288
Projected displays, 282
Psychological refractory period, 57–58
Psychophysics, 65, 66
Push button switches, 79

Q

Qualitative display, 83
Quality function deployment (QFD), 32
Quantitative versus qualitative display, 83

R

Radiant energy, 200
Rain sensor wipers, 283
Random variable, normal distribution of, 14
Rating on a scale, 257, 259, 260*f*
Ratio scales, 223, 224
Reaction time, factors affecting, 60
Reaction Time Program, 298*t*
Rear and side vision aids, 283
Rear passenger entertainment systems, 282
Recovered error, 65
Reference points, 33–38
Reflectance, 201
Remote tire pressure sensing, 283
Research
    descriptive, 7
    evaluative, 7
    experimental, 7
Responsibilities, of ergonomics engineer, 190
    design process, trade-offs in, 194
    early design process, steps in, 191–194
    problems and challenges, 194
    support process during vehicle development, 190–191
Retina, structure of, 52
Reversal error, 65
Ring model of product desirability, 178–179
Rocker switches, 81
Rod receptor, 52, 53*f*
Rotary switches, 81, 82*f*
Running boards, 161–162

S

SAE J2364 and J2365 recommended practices, 242–243
SAE Occupant Packaging Committee, 19
SAE Vehicle Lighting Committee and Motor Vehicle
        Manufacturers Association, 150–151
Satellite radio, 282
Satisfiers, 179, 180
Seat, lateral location of, 27
Seatback, avoiding curvature in, 25
Seatback angle, 22, 26–27, 44
Seatback height, 27
Seat bolsters, location, and materials, 161
Seat cushion, avoiding curvature in, 24
Seat cushion angle, 26
Seat cushion length, 26
Seat design, biomechanical considerations in, 22–25
Seat design considerations related to driver
        accommodation, 25–28
Seated posture, anthropometric measurements in, 13, 14*f*
Seat hardware, 161
Seat height, 26
Seating package layout, 29
Seating reference point (SgRP), 26, 34, 35, 36
    and foot movement areas, lateral sections at, 158–159
    locations, body opening clearances from, 160

Seat track length, 27–28, 36–37, 38*f*, 43
Seat-track-related dimensions, 33–38
Seat width, 26
Secondary task performance measurement, 241
Sensory memory, 62
Sequential error, 64
Shade bands, 125
Short driver problems, 112
Shoulder room, 40, 41*f*
Side mirror field, 118*f*
Signal lamps, 129
Signal lighting, 129
    design considerations, 131
        new technology advances and related issues, 133
        problems with current signal lighting systems, 132–133
        signal lighting visibility issues, 132
    evaluation methods, 144
        analysis of accident data, 147
        CHMSL fleet study, 147–150
        field observations and evaluations, 144–147
        photometric measurements and compliance evaluation, 144
        SAE Vehicle Lighting Committee and Motor Vehicle Manufacturers Association, 150–151
Signal lighting requirements, 140*t*
Simple reaction time, 58
Simulated driving, text messaging on, 245
Simulations and prediction of driver's performance, 142
Smell quality, 182
SORE model and human errors, 65, 65*f*
Sound measurements, 71–72
Sound quality, 181
Spare capacity, 237
Special driver and user populations, 271
    futuring, 276–277
    issues in designing global vehicles, 276
    users, understanding, 271
        drivers with disabilities and functional limitations, 275–276
        effect of geographic locations of markets, 275
        female drivers, 273–274
        market segments, 272–273
        older drivers, 274–275
        vehicle types and body styles, 272
    users and their needs, 271
Speedometer graphics design, 99, 99*f*
Standard deviation
    of lateral position, 228–229
    of steering wheel angle, 229
    of velocity, 229
Standing posture, anthropometric measurements in, 13, 14*f*
Static anthropometric dimensions for vehicle design, 19
Static body dimensions of U.S. adults, 13, 15–17*t*
Static versus dynamic display, 83
Steering wheel, 39
    craftsmanship of, 182–184
    instrument cluster through, 87, 88*f*
    location, 48
    location of, 49*f*
    zone defining the visibility through, 91–92
Steering-wheel-mounted controls, 282
Sternberg task, 248

Stone pecking, 161
Storage spaces, 30
Studies using programmable vehicle bucks, 269
Subjective measurements, 226, 254, 257*t*
  and data analysis, 258
    analytical hierarchical method, 265–267
    paired-comparison-based methods, 259–262
    rating on a scale, 259
    Thurstone's method of paired comparisons,
      262–265
Subjective Workload Assessment Technique (SWAT), 239,
  240*t*
Subjective workload measurement techniques, 239–241
Substitution error, 65
Subtasks, 164, 165, 166
Summary chart, of ergonomics, 92, 96–98*f*
Sun visor design issues, 113
Surrogate reference task (SURT), 247
Switches, pleasing operational feel of, 182
Symbolic/pictorial graphics, 83–84
SYNC, 282
Systems engineering "V" model of vehicle design process,
  187–189

T

Tactile display, 84
Tall driver problems, 113
Task analysis, 164–166, 174
  for checking engine oil level, 175*t*
Text messaging during simulated driving, 245
Text-to-speech conversion systems, 283
Thigh room, 42, 42*f*
Thirty-five-degree down-angle cone, 92
Three-dimensional displays, 281
360-degree visibility, systems consideration of, 107–108, 108*f*
Thurstone's method of paired comparisons, 261, 262–265
Time, importance of, 51–52
Time error, 64
Tire changing, steps in, 171
Tires and rocker panels, 161
Torso angle, 38, 44
Total task time, 230–231
Touch feel quality, 181
Touch screen displays, 80–81, 80*f*
Touch screens and multitouch technology, 281–282
Transmittance, 201
Trunk-compartment-related tasks, 171
Two door trim panels, layout of, 102*f*
Two-door vehicles, third row and rear seat entry from, 162

U

Unrecovered error, 65
Unspoken wants, 180
U.S. Air Force, 8
User population, anthropometric data of, 13
Users, understanding, 271
  drivers with disabilities and functional limitations,
    275–276
  effect of geographic locations of markets, 275
  female drivers, 273–274
  market segments, 272–273
  needs of users, 271

older drivers, 274–275
vehicle types and body styles, 272

V

Vehicle, refueling, 171
Vehicle design, ergonomics in, 3
  ergonomics approach, 4
    designing for the most, 4
    fitting equipment to the users, 4
    systems approach, 4–5
  ergonomics engineer's responsibilities in vehicle
    design, 7
  ergonomics research studies, 7
  problem-solving methodologies, 5–6, 6*f*
  systems engineering "V" model of, 187–189
Vehicle development process
  ergonomic evaluations in, 254, 255–256*t*
  ergonomics support process in, 190–191
  support process during, 190–191
Vehicle evaluation, 189
  applications
    checklists, 267–268
    driving simulator studies, 269
    field studies and drive tests, 269
    interval scales, ratings on, 268
    observational studies, 268
    studies using programmable vehicle bucks, 269
    vehicle user interviews, 268
  communication methods, 257–258
  evaluation issues, overview on, 253
  evaluation measures, 189
  evaluation methods, 254, 257*t*
  experimental methods, 258
  goal of ergonomics engineers, 189
  objective measures and data analysis methods, 258
  observational methods, 254, 257
  subjective methods and data analysis, 258
    analytical hierarchical method, 265–267
    paired-comparison-based methods, 259–262
    rating on a scale, 259
    Thurstone's method of paired comparisons,
      262–265
  tools, methods, and techniques, 189–190
  vehicle development, ergonomic evaluations during,
    254, 255–256*t*
Vehicle package, 29
  evaluation, by direction magnitude and acceptance
    rating scales, 261*t*
  layout, 30*f*
  occupant package, developing, 29–30
  occupant package/seating package layout, 29
  sequence in development of, 30
    "accepted" vehicle concept, development of, 31–33
    advanced vehicle design stage, 30–31
Vehicle packaging, parametric model developed for, 193*f*
Vehicle service, 30
Vehicle speed, measuring, 229
Vehicle stability and control systems, 284
Vehicle state, measuring, 226
Vehicle types and body styles, 272
Vehicle user interviews, 268
Veiling glare, 214
Veiling glare coefficient (VGC), 215, 216

Veiling glare prediction model, 216, 217*t*
Visibility, systems considerations related to, 199–200
Visibility issues, 125
    cameras and display screens, 126
    heavy-truck driver issues, 125–126
    light transmissivity, 125
    other visibility-degradation causes, 125
    plane and convex combination mirrors, 125
    shade bands, 125
Visibility model, application of, 208*t*, 210*t*
Visibility prediction model, 208, 214, 216
Visual capabilities, 66
    driver's visual fields, 68
    visual acuity, 67–68, 68*f*
    visual contrast thresholds, 66–67, 66*f*
Visual contrast thresholds, 202
    Blackwell contrast threshold curves, 203
    computation, 203–204
        steps in, 206–209
    effect of glare on, 205–206
    and visibility distance, computation of, 204–205
Visual display, 84
    evaluation, checklist for, 94*t*

Visual display design considerations, 87
    findability and location, 87
    identification, 87–88
    interpretability, 88
    legibility, 88
    reading performance, 88
    visibility, 87
Visual field, center of, 53
Visual information acquisition in driving, 54
Visual quality, 181
Visual search task, 247
Visual spare capacity, 237
Voice controls, 82–83
Voice controls/recognition systems, 283

W

Windshield rake angle, 214
Wiper and defroster requirements, 114
"Within-subject variability," 10
Working memory, 62–63
Workload Profile (WP), 239, 241
"Wow" features, 180

An environmentally friendly book printed and bound in England by www.printondemand-worldwide.com

#0113 - 260213 - C0 - 254/178/17 [19] - CB